INTRODUCTORY ASTROCHEMISTRY

INTRODUCTORY ASTROCHEMISTRY

FROM INORGANIC TO LIFE-RELATED MATERIALS

AKIO MAKISHIMA

*Institute for Planetary Materials, Okayama University,
Misasa, Japan*

ELSEVIER

Elsevier
Radarweg 29, PO Box 211, 1000 AE Amsterdam, Netherlands
125 London Wall, London EC2Y 5AS, United Kingdom
50 Hampshire Street, 5th Floor, Cambridge, MA 02139, United States

ISBN: 978-0-443-23938-0

For Information on all Elsevier publications
visit our website at https://www.elsevier.com/books-and-journals

Publisher: Candice Janco
Acquisitions Editor: Peter Llewellyn
Editorial Project Manager: Aleksandra Packowska
Production Project Manager: Rashmi Manoharan
Cover Designer: Matthew Limbert

Typeset by MPS Limited, Chennai, India

Contents

Preface   **vii**

1. Basic knowledge of astroscience   **1**

1.1 Introduction to astrophysics   1
1.2 Introduction to astronomy   12
1.3 Introduction to inorganic astrochemistry   29
1.4 Introduction to mineralogy and petrology   68
1.5 Introduction to organic astrochemistry   88
1.6 Statistics   114
References   118

2. Origin of elements   **121**

2.1 The origin of the universe   121
2.2 Origin of elements   126
2.3 Formation of the proto-Earth in the solar nebular   137
References   149

3. Materials on the Moon   **151**

3.1 Characteristics of the materials of the Moon   151
3.2 The age of the Moon   161
3.3 The giant impact model for formation of the Earth-Moon system   166
3.4 Constraints from the stable isotope astrochemistry for the giant impact   170
3.5 Summary   203
3.6 The "synestia" model   204
3.7 C, N, and S were delivered by a giant impact   205
References   207

4. Materials on the Earth   **215**

4.1 The late veneer   215
4.2 The oldest geological records on the Earth   224
4.3 The oldest evidence of Life on the Earth   234
References   245

5. Origins of life-related molecules on Earth   **251**

5.1 Transportation of life-related molecules of extraterrestrial origin on Earth 251

5.2 Synthetic experiments of life-related molecules on the Earth   253
References   261

6. Life-related molecules on Venus   263

6.1 Life-related molecules on Venus atmosphere   263
References   264

7. Materials on Mars   265

7.1 Exploration of Mars   265
7.2 Life-related molecules (LRMs) in the atmosphere on Mars   267
7.3 Organic compounds in the Martian meteorites   269
7.4 Fossils of microorganisms were found in the Martian meteorite,
ALH 84001?!   270
7.5 The least contaminated Martian meteorite, Tissint fell   282
References   284

8. Comet and asteroid materials   289

8.1 Asteroid and comet explorations   289
8.2 Sample-return missions from the comets and asteroids   298
8.3 Organic materials and life-related materials in asteroids and comets   305
8.4 Mineralogy and inorganic chemistry of samples recovered from the
asteroid Ryugu   306
8.5 Organic materials and life-related materials in the recovered samples
from the asteroid Ryugu   309
8.6 Origin of life-related materials in asteroids and comets by laboratory
experiments   318
References   328

9. Liquid water on the moons of Jupiter and Saturn   335

9.1 Exploration of Jupiter and Saturn   335
9.2 Four contrasting moons of Jupiter   339
9.3 Moons of Saturn   347
References   354

10. Exosolar materials and planets   357

10.1 Life-related molecules observed in space and interstellar medium   357
10.2 Exosolar planets in the habitable zone   361
10.3 Atmospheric molecules in exoplanets   369
References   372

Index   373

Preface

This book project started as the second edition of *Origin of the Earth, Moon, and Life* published by Elsevier in 2017. Since then, there have been lots of findings and progress in astro-sciences, such as astronomy, astrobiology, astrogeology, astrophysics, and astrochemistry. Especially, there are needs for introductory textbooks for astrochemistry, and to understand the up-to-date discussion deeply. Unfortunately, such an interdisciplinary textbook did not exist, and still does not exist.

In the previous edition, the book contained too much content even though the target was only the Origin of the Earth, Moon, and Life. To explain basic knowledge, the previous book used the box, which was not well organized. Thus, in this book, which corresponds to the new edition, all basic knowledge is gathered in Chapter 1, which explains the basic knowledge of the five astro-sciences. The reader can read them one by one, or refer them when the new knowledge necessary to understand appears.

The next point is that the book is based on the materials science. Each chapter and section are (or maybe) based on materials that we can touch or infer from data of materials. Thus the discussion on the origin of Life which contained many speculations disappeared, and the discussion used in this book should have become more concrete and practical.

As the author is basically a chemist, and is mainly interested in astrochemistry, therefore, the book chased my curiosity in astrochemistry, and the author wrote the book following the policies:

1. The author wants to explain and present the most accepted ideas in "easier" words. Understanding this book only requires the literacy of undergraduate to graduate students, or, even high school students who have large curiosity are worth tackling this book.
2. Many problems related to not only astrochemistry but also other astro-sciences, such as astronomy, astrophysics, astrobiology, and astrogeology are treated, although the title of this book is *Introductory Astrochemistry*. However, each astro-science uses specific words to describe its knowledge, which is often utterly incomprehensible for people with other disciplines. Therefore, in this book, such knowledge is explained in Chapter 1, and the author tries to break or

destroy such boundaries or discrimination. The author hopes that this book could work as pipelines or synapses connecting each professional discipline.

3. Nowadays, the speed of research is very fast, and new results in each discipline are issued day by day. Therefore it becomes very difficult to follow the latest ideas. The author wants to present the current, up-to-date but generally accepted information on the universe, stars, planets, including the Earth, Moon and satellite moons, life-related molecules (LRMs), in 2023.

4. The author assumes that readers are undergraduate to graduate students in astro-sciences who want to increase literacy in astrochemistry. Therefore this book is suitable for undergraduate to graduate students. Another target of this book is middle-aged scientists or business people who want to understand up-to-date information in astro-sciences. Especially, those who have questions why so much money is spent in astro-sciences. Of course, pure chemists, astronomers, biologists, geologists, and physicists are welcome, because astro-literacy can be obtained by this book without using advanced mathematics employed in astronomy and astrophysics, and only qualitative results are presented.

Recent progresses in astrochemistry in these five years are dramatic: (1) James Webb Space Telescope (JWST) which has six times larger primary mirror than that of Hubble Space Telescope (HST) was launched in 2021 and successfully started observation (the final cost was US\$10 billion!); (2) stable isotope chemistry for various transition metals have advanced very much by multicollector ICP-MS, resulting in new constraints on the origin of the Moon; (3) the Hayabusa2 spacecraft fetched the sample of carbonaceous asteroids, which contained various LRMs in 2022; and the OSIRIS-REx spacecraft will fetch another carbonaceous asteroid samples in 2023; (4) liquid water ocean under ice crust in some moons of Jupiter and Saturn has become more likely, and the Jupiter Icy Moons Explorer (JUICE) was launched by ESA in 2023 and will arrive at Jupiter in 2031; (5) the many exoplanets with the Earth size in the habitable zone are found and confirmed; (6) merging of two neutron stars plays very important roles in the elemental synthesis, especially, gold, and platinum group elements (PGEs).

Thus almost all sections are rewritten based on these new findings and this book is composed of 10 chapters as follows:

Chapter 1 is the basic knowledge to read this book smoothly. If the technical words with which you have no acquaintance appear, try to find them in Chapter 1. The author simply hopes that the readers will enjoy learning new knowledge of space-sciences, especially astrochemistry. For graduate students with interdisciplinary knowledge, Chapter 1 would be a handy encyclopedia for learning basic knowledge of other disciplines. You can skip Chapter 1 when you have confidence in knowledge levels of astro-sciences.

Chapter 2 describes the elemental synthesis. All the materials are made of the elements. The topic starts with the Big Bang. Then we learn about the Life of stars. The stars produce energy by burning hydrogen and helium, and heavier elements up to nickel are synthesized. The stars die quietly as white or brown dwarfs or flamboyantly as supernovae. For a long time, heavier elements than nickel have been synthesized by the "r-process" in the supernovae. However, recent observation of supernovae has not detected the r-process. Instead, the merging of two neutron stars (so-called the "neutron star merger") is the main process of the element synthesis heavier than strontium (Sr), especially gold and PGEs.

In Chapter 3, we learn how the solar nebular, where planets from Mercury to Neptune, including the proto-Earth, were formed, and how the Sun started to shine. We describe not "the Earth" but "the proto-Earth," because the "proto-Earth" was destroyed by a Martian-sized planetesimal named "Theia" via a giant impact, forming the present Earth—Moon system. In Chapter 3, the materials and history of the Moon are discussed, especially the recent results on stable isotopic composition of various elements are described, resulting in the new formation model of the Earth—Moon system.

Chapter 4 deals with interesting materials of the Earth. To explain the high abundance of the highly siderophile elements in the Earth's mantle, the idea of the late veneer is proposed. In addition, the oldest rock on the Earth and the oldest record of Life on the Earth are discussed.

Chapter 5 describes how the origins of LRMs appeared on the Earth. The transportation from extraterrestrial LRMs is discussed. This topic also appears again in Chapter 8.

Chapter 6 describes the LRMs on Venus. Venus is considered to be a dead planet by the greenhouse effect with high pressure (~ 90 bar), temperature (~ 750 K), and sulfuric acid rain. However, high up in the atmosphere (~ 55 km), the pressure and temperature are about 1 bar and 15°C, resulting in the habitable condition. As a very reductive material, phosphine (PH_3) has been observed recently in the clouds of Venus, there should be some unknown synthetic processes. Several researchers think that this is possibly caused by some kinds of Life.

Chapter 7 is about the materials of Mars, which is the most probable planet to have Life. The LRMs in the Martian meteorites are found and the Life on Mars is discussed in this chapter. Mars is barren today, but a few Gyrs ago, lots of water and atmosphere existed on Mars, and Life could have flourished. Many spacecraft landed and investigated Mars, and the sample return project is ongoing, which can finalize the discussion on the Life on Mars.

In Chapter 8, comet and asteroid materials are related. The cometary and asteroidal materials are considered to have fallen on the Earth as the meteorites, which have transferred many LRMs and water on the Earth. In 2021 the Hayabusa2 mission successfully collected and fetched the samples of the C-type asteroid (a carbonaceous asteroid), Ryugu. LRM analyses of Ryugu samples were done by many laboratories, which are shown in this chapter. In 2023 the OSIRIS-REx sample-return mission will also return to the Earth with large amounts of samples of the B-type asteroid (also a carbonaceous asteroid), Bennu.

Chapter 9 deals with the liquid water in the moons of Jupiter and Saturn. These two huge gas planets affect their moons by tidal force. Three moons of Jupiter (Ganymede, Callisto, and Europa), and two moons of Saturn (Enceladus and Mimas) have the possibility to have an inner sea, in which the cores are floating. If such an inner sea exists, the energy can be transferred more easily from the mother planet by the tidal force. The floating core can generate magnetic fields, which can protect the solar wind. If the liquid ocean exists under the ice of the moon, there is a possibility to nurture the Life in it. To scrutinize the Jovian moons, the spacecraft, JUICE has been launched, and it will arrive in Jupiter in 2031.

Chapter 10 is about the exosolar planets and materials. Exosolar planets are hot topics in astronomers since planets other than our Solar system have been found. Furthermore, as the observational technology improved, exoplanets with the Earth's size in habitable zones were also found. Astrobiologists were also excited about the findings of such exoplanets because not only Life but also civilization could be found! In this chapter, molecules found in clouds of exoplanets are also shown, and the habitable exoplanets are introduced.

The author hopes that the readers will understand the enthusiasm of astro-scientists, and why so much money is spent in astro-sciences. Furthermore, some of the readers would find the carrier as scientists or science-related professions, pursuing their own curiosity.

Akio Makishima
Institute for Planetary Materials, Okayama University, Misasa, Japan

Notice: In this book, abbreviations of NASA (the National Aeronautics and Space Administration), JPL (NASA's Jet Propulsion Laboratory), ESA (the European Space Agency), and JAXA (the Japan Aerospace Exploration Agency) are used without comment. The author owes a great deal to Wikipedia for various knowledge, but of course, the responsibility for all contents in this book is on the author.

1

Basic knowledge of astroscience

1.1 Introduction to astrophysics

1.1.1 The SI units, the coherent derived units, the defining constants, and the SI prefix

In science, the international system of units (SI units) is the standard unit defined by the General Conference on Weights and Measures (GCWM). The SI units are composed of six base units (see Table 1.1): kg (kilogram; weight), m (meter; length), s (second; time), A (ampere; electric current), K (kelvin; thermodynamic temperature), mol (mole; amount of substance), and cd (candela; luminou9s intensity).

The derived units are composed of powers, products, or quotients of the base SI units, which are infinite. Twenty-two derived units are called as "coherent derived units," which have special names and symbols. For example, Hz (Hertz, s^{-1}; frequency), N (Newton, $kg\,m/s^2$; force, weight), Pa (Pascal, $kg/m/s^2$; pressure, stress), J (Joule, $kg\,m^2/s^2$; energy, work, heat), W (Watt, $kg\,m^2/s^3$), C (Coulomb, s A; electric charge), Wb (Weber, $kg\,m^2/s/A$; magnetic flux), T (Tesla, $kg/s^2/A$; magnetic flux density), °C (degree, Celsius, K; temperature), and Bq (Becquerel, s^{-1}; activity referred to a radionuclide) are the coherent derived units.

The SI units are defined by seven defining constants: (1) the speed of lights in vacuum, c; (2) the hyperfine transition frequency of cesium, Dn_{Cs}; (3) the Planck constant, h; (4) the elementary charge, e; (5) the Boltzmann constant, k; (6) the Avogadro constant, N_A; and (7) the luminous efficacy of 540 THz radiation, K_{cd}. These values defined in 2019, are listed (see Table 1.2).

When the SI unit is too large or small, the SI prefixes are used together with the SI unit. The SI prefixes are summarized in Table 1.3. In this chapter, Gy (billion years), My (million years), Ga (billion years ago),

Introductory Astrochemistry
DOI: https://doi.org/10.1016/B978-0-443-23938-0.00001-7

TABLE 1.1   The SI base units.

Name	Measure	Symbol
Weight	Kilogram	kg
Length	Meter	m
Time	Second	s
Electric current	Ampere	A
Thermodynamic temperature	Kelvin	K
Amount of substances	Mole	mole
Luminous intensity	Candela	cd

TABLE 1.2   The SI defining constants.

Defining constant	Symbol	Exact value	Unit
The speed of light in a vacuum	c	299,792,458	$m\,s^{-1}$
The hyperfine transition frequency of cesium	Dn_{Cs}	9,192,631,770	Hz
The Planck constant	h	$6.62607015 \times 10^{-34}$	Js
The elementary charge	e	$1.602176634 \times 10^{-19}$	C
The Boltzmann constant	k	1.380649×10^{-23}	JK^{-1}
The Avogadro constant	N_A	$6.02214676 \times 10^{23}$	mol^{-1}
The luminous efficacy of 540 THz radiation	K_{cd}	683	$lm\,W^{-1}$

TABLE 1.3   The SI prefixes.

10^{21}	zetta	Z	10^{-1}	deci	d		
Base 10	Name	Symbol	Base 10	Name	Symbol		
10^{18}	exa	E	10^{-2}	centi	c		
10^{15}	peta	P	10^{-3}	milli	m	Empirical expression	
10^{12}	tera	T	10^{-6}	micro	μ	ppm	$\mu g\,g^{-1}$
10^{9}	giga	G	10^{-9}	nano	n	ppb	$ng\,g^{-1}$
10^{6}	mega	M	10^{-12}	pico	p	ppt	$pg\,g^{-1}$
10^{3}	kilo	k	10^{-15}	femto	f	ppq	$fg\,g^{-1}$
10^{2}	hecto	h	10^{-18}	atto	a		
10^{1}	deca	da	10^{-21}	zepto	z		

Ma (million years ago), and km (kilometer) are often used. Empirical expressions such as ppm, ppb, etc. are often used especially when the concentrations of elements or chemical compounds are indicated. However, it is not recommended to use them.

In astronomy, au (or AU; astronomical unit; the distance between the Earth and the Sun; 150 Gm), ly (or l.y.; light-year; 9.46 Pm or 63.2 kilo au), and pc (parsec; 30.9 Pm or 3.26 ly) are sometimes used.

1.1.2 Classical mechanics

The classical mechanics is very simple. The classical mechanics assumes Euclidean geometry for the structure of the space, and the time is absolute. The velocity vector or the rate of change ($\mathbf{v}$; all bold and italic characters indicate vectors in this section) of a particle P is defined as:

$$\mathbf{v} = \frac{d\mathbf{r}}{dt} \tag{1.1}$$

where $\mathbf{r}$ is a vector of the position of P. The acceleration ($\mathbf{a}$) is:

$$\mathbf{a} = \frac{d\mathbf{v}}{dt} = \frac{d^2\mathbf{r}}{dt^2} \tag{1.2}$$

The relation between force ($\mathbf{F}$) and momentum ($m\mathbf{v}$) is known as Newton's second law:

$$\mathbf{F} = \frac{d(m\mathbf{v})}{dt} = m\mathbf{a} \tag{1.3}$$

If a constant force $\mathbf{F}$ is applied to a particle P and makes a displacement of $\Delta\mathbf{r}$, the work done by the force is defined as the scalar product:

$$W = \mathbf{F}\Delta\mathbf{r} \tag{1.4}$$

When the position of P moved from $\mathbf{r}_1$ to $\mathbf{r}_2$ along a path C, the work done on the particle is:

$$W = \int_C \mathbf{F}(\mathbf{r})d\mathbf{r} \tag{1.5}$$

Kinetic energy (E_k) of a particle of mass m at speed $\mathbf{v}$ is given as:

$$E_k = \frac{1}{2}m\mathbf{v}^2 \tag{1.6}$$

When the total work W moved the particle P from $\mathbf{r}_1$ to $\mathbf{r}_2$ with the change of velocity $\mathbf{v}_1$ to $\mathbf{v}_2$,

$$W = \int_C \mathbf{F(r)} d\mathbf{r} + \frac{1}{2} m \left(\mathbf{v}_2^2 - \mathbf{v}_1^2 \right) \qquad (1.7)$$

In 1687, Newton described gravity in the book "Mathematical Principles of Natural Philosophy":

$$\mathbf{F} = G \frac{m_1 m_2}{r^2} \qquad (1.8)$$

where $\mathbf{F}$ is the force, m_1 and m_2 are the masses of the interacting objects, r is the distance between the centers of the masses, and G is the gravitational constant.

1.1.3 The three-body problem and Lagrange points

In classical mechanics, the three-body problem appeared to study the movement of the Sun, the Earth, and the Moon. The problem must satisfy Newton's law of motion Eq. (1.3), Newton's law of gravitation Eq. (1.8), and the law of conservation of the total energy. This problem is a special case of the N-body problem, and no general solutions exist.

As a restricted case of the three-body problem, Lagrange points were found. Lagrange points are the stable orbits of a small body (a spacecraft) where one massive body (the Earth) goes around another massive body (the Sun) by the gravitational forces (see Fig. 1.1). These points are

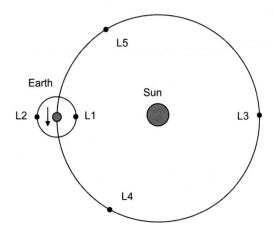

FIGURE 1.1 **Lagrange points (not to scale).** The Earth is rotating around the Sun. L1 and L2 are ~1.5 million km or ~0.01 au (au is the distance between the Sun and the Earth) from the Earth. L3 resides at the opposite side of the Earth on the Earth's orbit. L4 and L5 are the apexes of two equilateral triangles formed by the Sun, the Earth, and the L4/L5 points on the Earth orbit.

ideal positions for spacecraft where the minimum energy is required to keep staying around the Earth. The solutions are five points (L1−L5) shown in the figure, which are called the Lagrange points. L1, L2, and L3 are on the line connecting the Earth and Sun. L1 and L2 are on the rotating orbits around the Earth. L3, L4, and L5 are on the Earth orbit, and L4 and L5 are 60 degrees ahead and behind of the Earth orbit. When the first body is Jupiter, there are many small asteroids at the L4 and L5 positions called Greek and Trojan camps or Greeks and Trojans, respectively.

1.1.4 Beyond the classical mechanics

Since the quantum theory of light was found, the limitation of classical mechanics has been noticed. In Fig. 1.2, the present status of classical mechanics is schematically shown. In the special relativity theory, the momentum of a particle (p) is given as:

$$p = mv/\sqrt{1 - (v/c)^2} \tag{1.9}$$

where m is the mass of the particle, v is the velocity of the particle, and c is the speed of light (3.0×10^8 m/s). As the velocity of the particle approaches that of light, the effect of Eq. (1.9) cannot be neglected, and relativistic mechanics is required (see Fig. 1.2). The size of the particle also affects the classic mechanics. When the size of the particle becomes 10^{-9} m levels, quantum mechanics must be used. In the special

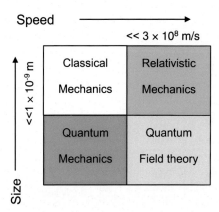

FIGURE 1.2 **Validity of the classical mechanics.** Limitation of the classical mechanics is shown, based on the size and speed indicated in the figure. The classical mechanics is only valid when the size is far larger than nm, and the speed is far less than that of the light (3×10^8 m s^{-1}). See text for details.

relativity theory, space and time are treated as a unified structure, and become a spacetime.

In the 20th century, QFT was recognized as the standard model including, the weak and strong interactions, the gauge theory, etc. The four fundamental forces are determined: (1) gravity, (2) electromagnetism, (3) the weak [nuclear] force, and (4) the strong [nuclear] force.

In the standard model, the elementary particles are shown in Fig. 1.3. The elementary particles are six types of quarks (up, down, charm, strange, top, and bottom), six types of leptons (electron, muon, tau, electron neutrino, muon neutrino, and tau neutrino), and four types of gauge bosons (gluon, photon, Z boson, and W bosons), which govern fundamental interactions, and Higgs boson, which gives the mass to the elementary particles.

The elementary particles are divided into two groups: bosons (integer spin) and fermions (with odd half-integer spin). The elementary particles that form ordinary matter are (leptons and quarks) fermions, while bosons play special roles in particle physics, such as force carriers or mass carriers. There are antiparticles (antiquarks, antileptons) for all fermions (quarks and leptons) and bosons. They have opposite charges and spin.

There are five standard bosons. One is a scalar boson (spin = 0) named Higgs boson, which gives the phenomenon of mass. The others

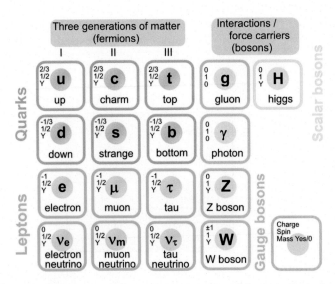

FIGURE 1.3 **Elementary particles in the Standard models.** In modern physics, there are six types of quarks and leptons, four types of gauge bosons, and a Higgs boson. Each particle has charge, spin, and mass (no mass particles exist), which are shown on the left side shoulder of each particle.

are four vector bosons (spin = 1) that act as force carriers: (1) photon is the force carrier of the electromagnetic field; (2) gluons (eight types) are the strong force carriers; (3) Z boson (neutral weak boson) is the weak force carrier; and (4) $W^{\pm}$ bosons (charged weak bosons, two types) are also the weak force carriers. A tensor boson (spin = 2) named graviton (G) is hypothesized, but not found.

The element is made of the atom, which is composed of the nucleus and surrounding electrons. In Fig. 1.4, a ^{4}He atom is depicted as an example. The electrons exist as possibilities around the space of the nucleus, thus, we call them as electron clouds. The size of the atom is defined as the size of the electron clouds with the size of 100 pm (10^{-10} m). The sphere-shaped nucleus with the size of fm (10^{-15} m) resides in the center of the atom. As a result, the atom is occupied by almost vacant space. The nucleus is made of protons and neutrons. Many models exist as to why nuclear can keep its shape. The proton and neutron are composed of three quarks: two up and down quarks, and one up and two down quarks, respectively. As the total charges of quarks in the proton and neutron are +1 and 0, the proton and neutron have +1 and no charge, respectively. The quarks attract each other by the strong [nuclear] force which is mediated by the gluons.

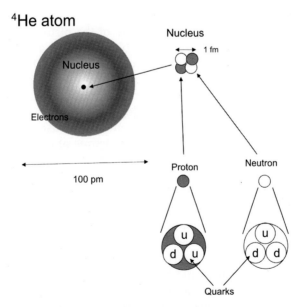

FIGURE 1.4 **The fine structure of a ^{4}He atom.** The atom is composed of a nucleus and electron clouds. Electrons exist as a possibility, therefore, they are called as electron clouds. The diameter of the nucleus is 10^{-5} times smaller than that of the electron clouds. Both the protons and neutrons are composed of up and down quarks.

The weak [nuclear] forces are carried by Z, W^+, and W^- bosons, which play important roles in the radioactive decay of a nucleus such as β^--decay, and the fusion of hydrogen into helium. The radioactive decay is explained in Section 1.3.2.

Recently, the group of (Krasznahorkay et al., 2023) reported new evidence for the existence of the new boson named as "X17" which has 17 MeV and ~33 times of mass of an electron with a life of 10^{-14} s was found. This new particle was first reported in 2016 by Krasznahorkay et al. This particle cannot be explained by the standard model based on the four forces, suggesting the existence of "the fifth force," but waiting for a lot of scrutiny.

1.1.5 High-temperature and high-pressure experiments

To know the mineral composition, phase relation, and partitioning coefficient of the target element to silicates or melts at high-temperature and high-pressure, experiments with the same chemistry at the same conditions are the best methods. There are three types of high-pressure and high-temperature experiment apparatuses in the high-pressure research group in our institute, IPM (Institute for Planetary Materials, previously named as ISEI, Institute for Study of the Earth's Interior) at Okayama University.

1.1.5.1 The piston cylinder apparatus

The piston cylinder apparatus is made of the piston and cylinder. In the center of the water-cooled cylinder block, there is a hole through the cylinder, in which the sample is placed (see Fig. 1.5). The sample is sandwiched by pistons from the top and bottom. The pistons are pressed by oil pressure and the sample is compressed. For simulation experiments for relatively shallow geological phenomena, the piston cylinder apparatus is very effective equipment because the operation is easy and the homogeneous pressure and temperature volume is large. The piston cylinder can be used up to 1.5 GPa and 1500°C.

1.1.5.2 The Kawai-type high-pressure apparatus

The Kawai-type high-pressure apparatus, which is also called as the multi-anvil apparatus is shown in Fig. 1.6. In the one-axial press, there are six-split-sphere guide blocks (see Fig. 1.7). The cubic assembly composed of the eight tungsten carbide (WC) cubic anvils is set in the guide-block. Each WC anvil is truncated, and in the center of the eight cubic anvils, the octahedral sample assembly is placed as shown in Fig. 1.8A. From one face to another face, as shown in Fig. 1.8B, the octahedral sample assembly is made of $MgO + Cr_2O_3$. In this case, the sample shown as blue in the figure is stuffed in a Re foil, covered with MgO, and then $LaCrO_3$ in a ZrO_2 tube. The thermo-couple for measuring temperature is placed in the

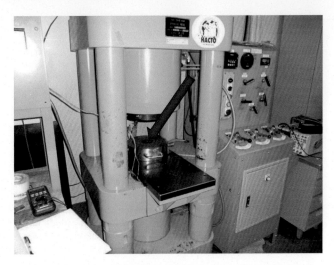

FIGURE 1.5 **A piston cylinder high-pressure apparatus.** The red arrow in the photo shows the cylinder block, which is water-cooled. In the center of the block, there is a hole in the cylinder through which the sample is placed. The sample is sandwiched by pistons from the top and bottom. The pistons are pressed by oil pressure and the sample is compressed. *Courtesy: Dr. D. Yamazaki, the Hacto group, IPM, Okayama University.*

FIGURE 1.6 **A photograph of the Kawai-type high-pressure apparatus in IPM.** The red arrow indicates the place of the guide-block, where the cubic anvils with the sample assembly are placed and pressed. *Courtesy: Dr. D. Yamazaki, the Hacto group, IPM, Okayama University.*

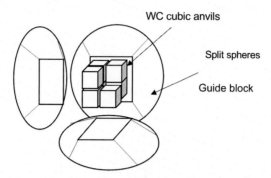

WC cubic anvils

Split spheres

Guide block

FIGURE 1.7 A conceptual image of the guide-block. The guide block is based on the split spheres in the figure, to press the assembly of the tungsten carbide (WC) cubic anvils. Only three split spheres are drawn. Actually, three faces pressing the cubic assembly are connected on each side of the guide-block, and one-axial press is used to press the cubic anvils.

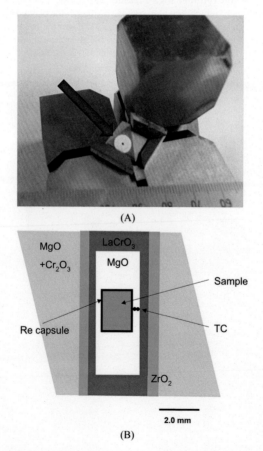

(A)

MgO
+Cr$_2$O$_3$

LaCrO$_3$

MgO

Sample

Re capsule

TC

ZrO$_2$

2.0 mm

(B)

FIGURE 1.8 The sample assembly. (A) A photograph of the disassembled WC cubic anvils and the sample assembly indicated by the red arrow. The WC anvils are truncated to put the octahedral sample assembly. (B) An example of the sample assembly. TC indicates the thermo-couple to measure temperature. *Courtesy: Dr. D. Yamazaki, the Hacto group, IPM, Okayama University.*

Re foil bag. Lanthanum chromite ($LaCrO_3$) is used as a heater by supplying the current. This assemblage is placed in the guide block of the press as shown in Fig. 1.5. The sample is compressed and heated alternately to the target pressure and temperature. The apparatus can achieve ~ 27 GPa and $\sim 2600°C$ (the upper mantle and lower mantle boundary). To achieve higher pressure, sintered-diamond anvils are used.

1.1.5.3 Diamond anvil cell

To increase the pressure, the harder material is more advantageous. Therefore, the diamond is the best material, and the diamond anvil cell was invented.

In Fig. 1.9A, the photograph of the diamond anvil cell is shown. There is a sample at the place indicated by the red box. A right-side cartoon is the enlarged cross-section of the sample assembly. The sample is held by the Re gasket. The initial pressure is given by the cell assembly with four small screws (6 mm Φ) as shown in the Fig. 1.8B, and the red arrow indicates the place of the diamonds. The sample is generally heated by the laser light from the top of the sample through the diamonds to >100 GPa and 3000°C.

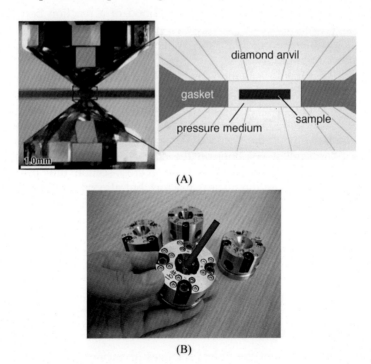

(A)

(B)

FIGURE 1.9 **The diamond anvil cell.** (A) Two diamonds are facing with sandwiching a sample (red arrow). The right-side figure is a cross-section of the sample, gasket, and diamond anvil. (B) The diamond anvil cell assembly. The red allow shows the place of the diamond anvils. *Courtesy: Dr. S. Tateno, the Hacto group, IPM, Okayama University.*

The largest merit of the diamond anvil cell is that we can observe directly the sample at high temperature and pressure conditions. The demerit is that the high-pressure area is too small to take out highly-pressurized samples. Therefore, the sample is studied in situ using lights (infrared (IR) lights, visible lights, and X-rays) through diamonds. 400 GPa and 6000K can be available.

1.1.6 The Roche limit

Assume that one celestial body (B), which forms its shape by gravitational self-attraction, is approaching another larger celestial body (A). The Roche limit of A is a distance from A, at which B will disintegrate by tidal forces of A. For example, consider that A and B are the Sun and a comet, respectively. If a comet approaches the Sun, and the comet goes into the Roche limit of the Sun, the comet breaks up and forms the ring along the Roche limit around the Sun.

1.1.7 Neutron cross-section

The neutron cross-section is the likelihood of a reaction between an incident neutron and a target nucleus. It is expressed as a "barn" unit, which is a dimension of 10^{-28} m^2. When the cross-section is larger, the reaction occurs more.

We can assume a situation that an incident neutron hits a target nucleus. On the Earth, this takes place in an atomic reactor. On the Moon or the asteroid, neutrons from the Sun or cosmic neutrons in space hit the Moon and the asteroid surfaces, and neutron addition of the atom occurs.

The probability of the reaction between the neutron and the target nucleus is dependent not only on the characteristics of the nucleus (for example, zirconium has a very low neutron cross-section, thus used in the tube to put uranium in the nuclear reactor), but also on the energy (velocity) of the neutron. There is a threshold of the velocity for the nuclear reaction. Low energy neutron (called thermal neutron corresponds to a temperature of ~ 290K) is effectively absorbed by the nucleus, while the faster neutron causes not the reaction but the neutron scattering.

1.2 Introduction to astronomy

1.2.1 Overview of astronomy

Astronomy is one of the oldest sciences in ancient civilizations, such as in Mesopotamia, Greece, Persia, India, China, Egypt, and Central America, etc. The stars were observed by the naked eye. In 1609, Galileo Galilei built and used his telescope, of which the observation drastically improved. In 1668, Isaac Newton built reflecting telescopes using parabolic mirrors without spherical aberration. In 1733, the achromatic lens was invented to correct color aberrations. The largest reflecting telescopes have objectives of >10 m. The telescopes have been evolving to see smaller and further objects. Today, the telescopes are placed in the space to eliminate the effects of the air.

The modern astronomy is split into two branches: observational and theoretical blanches. Observational astronomy is further branched by the wavelength of light (electromagnetic spectrum; see Fig. 1.10), because the observation equipment (telescopes and detectors) are different from each other according to the wavelength. Here, the observational astronomy is explained from shorter to longer wavelength lights which corresponds higher energy to lower energy photons.

1.2.1.1 Gamma-ray astronomy

Gamma-ray astronomy observes the shortest wavelengths of the electromagnetic spectrum. Gamma-rays are observed by satellites such as the Compton Gamma Ray Observatory (CGRO; from 20 keV to 30 Gev; 1991–2000; see Fig. 1.11A) and Imaging Atmospheric Cherenkov

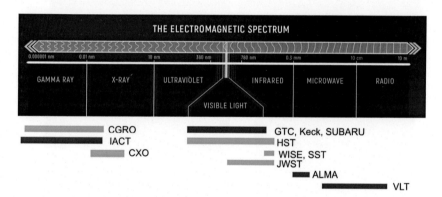

FIGURE 1.10 **The electromagnetic spectrum and observatories.** The wavelength of the electromagnetic spectrum is shown in the horizontal top bar. The wavelength increases from left to right (note that they are not in the linear scale!). As the wavelength increases, the energy of the photon decreases. Each wavelength (or energy) range has the specific name, such as "gamma ray," etc. The gamma-ray is included in the X-ray. The underbar indicates the range of detection of each satellite (blue) or terrestrial (red) observatory. *CGRO*, Compton Gamma Ray Observatory; *IACT*, Imaging Atmospheric Cherenkov Telescopes; *CXO*, Chandra X-ray observatory; *GTC*, the Gran Telescopio Canarias; *Keck*, W.M. Keck Observatory; *SUBARU*, the Subaru Telescope; *HST*, the Hubble Space Telescope; *WISE*, the Wide-field Infrared Survey Explorer; *SST*, the Spitzer Space Telescope; *JWST*, the James Webb Space Telescope; *ALMA*, the Atacama Large Millimeter/submillimeter Array; *VLT*, Karl G. Jansky Very Large Array. *Source: Modified from CSA, ESA, NASA, and Leah Hustak (STScI). https://webbtelescope.org/contents/media/images/01F8GF4P4S9TQPQHE15MAG23ZG.*

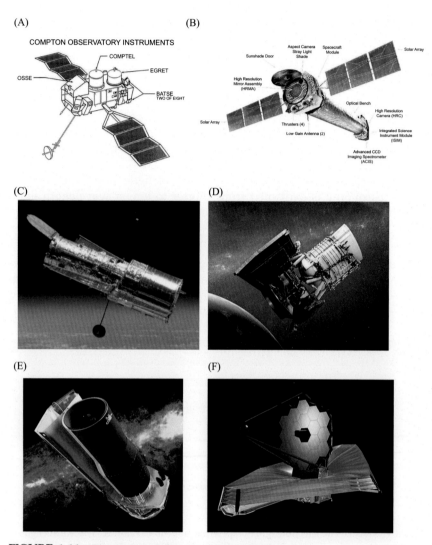

FIGURE 1.11 **The spacecraft for astronomy and astrophysics.** (A) Compton Gamma Ray Observatory (CGRO). It was launched in 1991 by NASA's Space shuttle. NaI(Tl) and CsI(Na) scintillation crystals were equipped for gamma-ray detection. (B) Chandra X-ray observatory (CXO). The CXO was launched in 1999. The CXO was 100 times sensitive than any X-ray telescope on the Earth's surface, because the Earth's atmosphere absorbs most X-rays. The CXO was named after S. Chandrasekhar. (C) The Hubble Space Telescope (HST). The HST was launched into the Earth orbit in 1990. The mirror was 2.4 m, and the main instruments can observe the near-ultraviolet, visible, and near infrared spectra. The telescope was named after the astronomer E. Hubble. The HST can take high-resolution images without atmospheric distortion and background light. The HST was built by NASA, with contributions from ESA. (D) The Wide-field Infrared Survey Explorer (WISE) and NEOWISE. The Wide-field Infrared Survey Explorer (WISE, 2009–11; 3–23 μm) scans the entire sky in IR light, picks up the glow of 100 million of objects, and produces

Telescopes (IACT; from 20 keV to 50 TeV; see Fig. 1.12A) placed in the Canary Islands, Spain. The Cherenkov telescopes do not detect the gamma rays directly but detect the flashes of visible light produced when the gamma rays are absorbed by the Earth's atmosphere. Main gamma-ray emitting events are gamma-ray bursts, in which very strong gamma rays are emitted only a few msec before fading away. Only ∼10% of gamma-ray sources are nontransient sources, which are pulsars, neutron stars, and black holes in the active galactic centers.

1.2.1.2 X-ray astronomy

X-ray astronomy uses X-ray, which is mainly produced by synchrotron radiation or superheating of gases above 10^7 K. Since X-rays are absorbed by the Earth's atmosphere, all X-rays must be observed from high-altitude balloons, rockets, or X-ray astronomy satellites (e.g., Chandra X-ray observatory; 0.2–10 keV; 1999–2022; see Fig. 1.11B). X-ray sources are X-ray binaries, pulsars, supernova remnants, black holes, elliptical galaxies, clusters of galaxies, and active galactic centers.

1.2.1.3 Ultraviolet astronomy

Ultraviolet astronomy employs ultraviolet wavelengths between 10 and 320 nm, which are completely absorbed by the atmosphere, requiring observations from the upper atmosphere or space. Ultraviolet astronomy is best suited to the study of hot blue stars (OB stars) in other galaxies.

---

millions of images. The mission found the coolest stars, the most luminous galaxies, and the darkest near-Earth asteroids and comets. By measuring the infrared light of the object, the size distribution of the asteroid population was obtained to know the hazardous asteroids. The WISE detected brown dwarfs, which are gas ball like Jupiter. The NEOWISE project (2011–14) succeeded the WISE project. (E) The Spitzer Space Telescope (SST; 2003–20). The SST was launched in 2003 for the detection of infrared by Infrared Array Camera (IRAC), Infrared Spectrograph (IRS), and Multiband Imaging Photometer for Spitzer (MIPS). As the on-board liquid helium was exhausted to cool the infrared detector in 2009, the planned mission at 3.6∼37 μm was finished. However, the two shortest-wavelength equipment of IRAC cameras are still available without liquid helium, so the SST could work as the Spitzer Warm Mission. (F) The James Webb Space Telescope (JWST), which is a space telescope designed primarily for infrared astronomy, was launched by NASA in 2021 and placed at L2. It just started operating in 2022. *Source: (A) NASA. https://heasarc.gsfc.nasa.gov/docs/cgro/images/epo/ gallery/cgro/cglro_line_labels.jpg. (B) NASA/CXC. https://www.nasa.gov/sites/default/files/thumb-nails/image/craft_lable_0.jpg. (C) NASA. http://www.nasa.gov/sites/default/files/thumbnails/image/ 345535main_hubble1997_hi.jpg. (D) NASA/JPL-Caltech. https://commons.wikimedia.org/wiki/File: WISE_artist_concept_(PIA17254,_crop).jpg. (E) NASA/JPL-Caltech. https://commons.wikimedia. org/wiki/File:Spitzer_space_telescope.jpg. (F) NASA. https://commons.wikimedia.org/wiki/File: James_Webb_Space_Telescope_2009_top.jpg.*

FIGURE 1.12   Telescopes on the Earth. (A) The MAGIC gamma-ray telescope in IACT in the Canary Islands, Spain. (B) The Gran Telescopio Canarias (GTC; visible & $0.365 \sim 1 \, \mu m$ near-IR). 12 segmented mirrors form a 10.4 m primary mirror. (C) W.M. Keck Observatory, on Mauna Kea, Hawaii (visible & $0.97–2.41 \, \mu m$ near-IR). Both Keck I and Keck II have 10 m primary mirrors. (D) The Subaru Telescope, on Mauna Kea, Hawaii (visible & $0.3–5.4 \, \mu m$ near-IR). The primary mirror is 8.2 m, which is supported by 261 computer-controlled actuators to correct the distortion of the mirror. (E) The future Atacama Large Millimeter/submillimeter Array (ALMA; $0.32–3.6 \, \mu m$), (the artist rendering). Antenna sizes are 12 and 8 m in diameter. (F) Karl G. Jansky Very Large Array (VLA) in New Mexico, US. The VLA comprises 28 25-m radio telescopes placed in a Y-shaped array. The frequency coverage is 0.7–400 cm. *Source: From (A) Raedts, https://commons.wikimedia.org/wiki/File:MAGIC_Telescope_-_La_Palma.JPG. (B) Pachango, https://commons.wikimedia.org/wiki/File:Grantelescopio.jpg. (C) NASA/JPL. https:// commons.wikimedia.org/wiki/File:Kecknasa.jpg. (D) Denys. https://commons.wikimedia.org/wiki/ File:MaunaKea_Subaru.jpg. (E) ESO/ NAOJ/ NRAO /L. Calçada (ESO). https://commons.wikime-dia.org/wiki/File:The_future_ALMA_array_on_Chajnantor.jpg. (F) Hajor. https://en.wikipedia.org/ wiki/File:USA.NM.VeryLargeArray.02.jpg.*

1.2.1.4 Optical (visible light) astronomy

Visible light astronomy is the oldest form of astronomy. The basic function is to take visible light images by CCDs, and to record and distribute the images by worldwide networks. In addition, dispersed light spectrums can be obtained. Not only visible lights from 400 nm to 700 nm but also near-IR and near-ultraviolet (UV) lights can be detected both by the Hubble space telescope (HST) and by modern optical telescopes on the Earth (see Fig. 1.12B−D).

1.2.1.5 Infrared astronomy

Infrared astronomy is based on the detection and analysis of IR, which corresponds to the wavelength of rotation or vibration signals of molecules. IR is useful for studying cold objects or objects in dust particles such as young stars, early planets in solar nebulars, and the cores of galaxies, because the planets are too cold to emit visible lights and IR is not blocked by dust clouds. IR radiation is heavily absorbed by the atmosphere or masked by emission from the atmosphere itself. Consequently, IR observatories have to be located in high, dry places on the Earth or in space. Some molecules radiate strongly in the IR. This allows the study of the chemistry of space or it can detect water in comets. Furthermore, the visible light is transformed into IR by the redshift, therefore, very far stars and galaxies (i.e. very old stars and galaxies in the history of the universe) can be observed.

Observations of IR from space by the Wide-field Infrared Survey Explorer (WISE; Fig. 1.11D) and the Spitzer Space Telescope (SST; Fig. 1.11E) were successful to unveil numerous galactic protostars and their host star clusters, and to find numerous dark asteroids around the Sun.

To see earlier and more distant objects, James Webb Space Telescope (JWST) was designed (Fig. 1.11F). The primary mirror is 6.5 m, compared with a 2.4 m mirror of HST. JWST covers from orange to mid-IR (0.6−28.3 μm), while HST covers near-ultraviolet, visible, and near-IR (0.1−1.7 μm). For this purpose, JWST is kept 50K placed at Lagrange point L2 which is in the shadow of the Earth and on the kite-shaped five-layer sunshields. As a result, JWST can detect 100 times fainter objects than HST and back to redshift (z; see Section 1.2.3) $z = 30 \sim 20$ (180 Myrs after the Big Bang) while HST could $z = 11.1$ (400 Myrs).

IR can be observed at dry and high places on the Earth, such as the Atacama Desert in Chile, where the Atacama Large Millimeter/submillimeter Array (ALMA; see Fig. 1.12E) composed of an astronomical interferometer of 66 radio telescopes to observe electromagnetic radiation at 0.32−3.6 μm wavelengths.

1.2.1.6 Radio astronomy

It uses radio waves, which is longer than 1 mm. Such waves are detected by radio telescopes. The Karl G. Jansky Very Large Array (VLA; see Fig. 1.12F) is the largest cm-wavelength radio astronomy observatory located in New Mexico, US. Radio astronomy targets radio galaxies, quasars, pulsars, supernova remnants, gamma-ray bursts, radio-emitting stars, the Sun and planets, black holes, etc. Interstellar gas molecules also produce various spectral lines. In particular, a hydrogen 21 cm line (H I line; H I line is a spectral line from hydrogen atom) is used because it penetrates cosmic dust. Broadening of this line can be used for measuring velocity by the Doppler effect (see Section 1.2.3).

1.2.2 Stellar classification

1.2.2.1 Hertzsprung–Russell diagram (H–R diagram)

H–R diagram is a classic plot of the effective surface temperature of stars versus the luminosities of stars, which is shown in Fig. 1.13. The horizontal axis is the effective surface temperature, the spectral classes, or color index B-V. The vertical axis is luminosity, which is $L/L_\odot$ where $\odot$ means the Sun. Most stars are plotted along the main sequence. The asymptotic giant branch stars (AGB stars) also appear in this diagram.

1.2.2.2 Harvard spectral classification

The Harvard system is a one-dimensional classification by A.J. Cannon. Stars are ordered and simplified by the alphabet (see Table 1.4), optionally subdivided by Arabic numbers (0–9), where 0 is the hottest star and 9 is the coldest one in a given class. For example, our Sun is classified as G2.

1.2.3 Doppler effect and the redshift (z)

The siren of an ambulance car becomes higher than the original tone when the car is approaching, and lower when the car is getting farther. This is called the Doppler effect. The Doppler effect is given by the following equation:

$$\lambda' = \lambda_0 (V - v_s)/(V - v_r) \tag{1.10}$$

where λ_0 and λ' are wavelengths at the source and the receiver, respectively; V is the velocity of the sound in medium (air); v_s and v_r are the velocities of the source and the receiver, respectively (the direction from the source to the receiver is positive; see Fig. 1.14A). When the source is coming to the receiver moving with the velocity

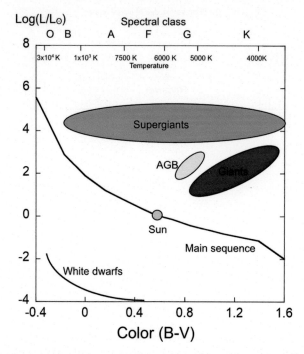

FIGURE 1.13 Hertzsprung–Russell diagram (H–R diagram). The horizontal axis is the effective surface temperature of stars. The vertical axis is the luminosities of stars. The effective surface temperature is a color index from B to V or spectral classes. The luminosity is the absolute bolometric magnitude, or $L/L_\odot$ where the luminosity of the star (L) is divided by that of the Sun ($L_\odot$). Most stars are plotted along the main sequence. The asymptotic giant branch stars (AGB stars) appear between Giants and Supergiants.

TABLE 1.4 Harvard spectral classification.

Class	Effective temperature (K)	Vega-relative chromaticity	Chromaticity (D65)	Solar mass ($M_\odot$)	Solar-radius ($R_\odot$)	Luminosity (bolometric; $L_\odot$)
O	>30,000	Blue	Blue	>16	>6.6	>30,000
B	10,000–30,000	Blue white	Deep blue white	2.1–16	1.8–6.6	25–30,000
A	7500–10,000	White	Blue white	1.4–2.1	1.4–1.8	5–25
F	6000–7500	Yellow white	White	1.04–1.4	1.15–1.4	1.5–5
G	5200–6000	Yellow	Yellowish white	0.8–1.04	0.96–1.15	0.6–1.5
K	3700–5200	Light orange	Pale yellow-orange	0.45–0.8	0.7–0.96	0.08–0.6
M	2400–3700	Light orange red	Light orange red	0.08–0.45	<0.7	<0.08

"Chromaticity" is the quality of color regardless of its luminance. The chromaticity is defined by two independent parameters, "hue (h)" and colorfulness (s). "D65" is "CIE standard illuminant D65," which is a commonly used standard illuminant defined by the International Commission on Illumination (CIE). D65 corresponds to the average midday light in Western Europe.

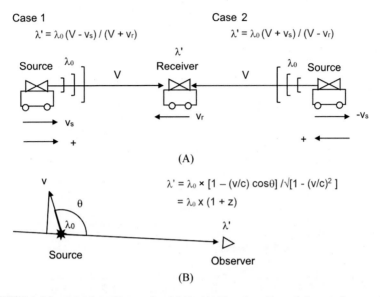

Case 1
$$\lambda' = \lambda_0 (V - v_s) / (V + v_r)$$

Case 2
$$\lambda' = \lambda_0 (V + v_s) / (V - v_r)$$

(A)

$$\lambda' = \lambda_0 \times [1 - (v/c)\cos\theta] / \sqrt{[1 - (v/c)^2]}$$
$$= \lambda_0 \times (1 + z)$$

(B)

FIGURE 1.14   **Doppler effect and redshift.** (A) Doppler effect of the usual wave. Case 1. The source is approaching the receiver. Case 2. The source is going away from the receiver. The direction of the source to the receiver is positive. (B) Doppler effect of the light. When the light source directly comes to the observer, the angle $\theta = 0$. The redshift, z is defined as $(\lambda' / \lambda_0) - 1$.

of v_r ($v_r > 0$) with the velocity of v_s ($v_s > 0$ and $v_s > v_r$; Case 1 in Fig. 1.14A), λ' is $\lambda_0 (V - v_s)/(V + v_r) < \lambda_0$. After the overtake of the source or the source is leaving from the receiver (Case 2), λ' is $\lambda_0 (V + v_s)/(V - v_r) > \lambda_0$, resulting in stretching the wavelength or lowering the frequency, because $V = \lambda \times \nu$ (λ and ν are the wavelength and frequency, respectively).

The Doppler effect also occurs for light. However, the Doppler effect of the light is different from that of the sound, because the velocity of the light is constant and the time dilation occurs in the special relativity theory. The Doppler effect of light is expressed as:

$$\lambda' = \lambda_0 \times [1 - (v/c)\cos\theta] / \sqrt{1 - (v/c)^2} \tag{1.11}$$

where λ_0 and λ' are wavelengths of the source and that by an observer; c is the velocity of the light in vacuum; v is the absolute velocity of the light source; θ is the direction of the light source (when the light is coming directly to the observer, $\theta = 0$) (see Fig. 1.14B).

It is interesting that even if the light source has no component of its velocity to the observer ($\theta = 90$ degrees), the wave length/frequency is affected by the Doppler effect. As is clear from Eq. (1.11), when the light source is approaching the observer, the wavelength of the light becomes shorter (blue shift), and when the light source is getting farther, the wavelength becomes longer (redshift).

Generally, the wavelength of the absorption lines by hydrogen in the emitted light by a star becomes longer as the light-emitting star is moving farther from us. Based on this fact, we could know that the universe is swelling. When the star exists farther from us (it also means that the star is older), the star is going away from us faster, and the wave length of light becomes longer (more reddish).

The redshift is defined by z, which is:

$$z = \left(\lambda'/\lambda_0 - 1\right) \tag{1.12}$$

where λ' and λ_0 are wavelengths of the observed and emitted lights, respectively.

1.2.4 Metallicity of stars

Baade (1944) proposed two groups of stars exist in the Milky Way: one population (Pop I) is composed of blue stars in spiral arms, and the other (Pop II) of yellow stars in the center of the galaxy (bulge) or in the elliptical galaxies. He referred that this was already noticed by Oort in 1926. The Pop III was added in 1978 as the very low metallicity stars, however, not yet found.

This classification was later combined with the difference in chemical composition or metallicity of stars. It should be noted that the metallicity of stars means the abundance of elements heavier than hydrogen and helium, and not those of metallic elements such as iron. However, metallicity is often defined as the following relation:

$$[\text{Fe/H}] = \log_{10}(N_{Fe}/N_H)_{star} - \log_{10}(N_{Fe}/N_H)_{sun} \tag{1.13}$$

where N_{Fe} and N_H are the numbers of iron and hydrogen atoms per unit volume, respectively.

Pop III, Pop II, and Pop I are associated with the metallicity as well as the age of stars, and defined as the first (very old) stars with very low metallicity, old stars with low metallicity, and recent stars with high metallicity, respectively.

In 2022, the Earendel (Old English for "morning star") star with $z = 6.2 \pm 0.1$, which meant the star was just 900 Myrs after the Big Bang, was found by the HST. The star was magnified by a factor of thousands by the gravitational lensing of foreground galaxy cluster lens WHL0137-08 ($z = 0.566$). The mass of the star was estimated to be 50 times heavier than that of the Sun (Welch et al., 2022). The star could be a Pop III star (Schauer et al., 2022).

1.2.5 Definition of asteroids and comets

Both asteroids and comets are defined as "small solar system bodies (SSSB)." The asteroids and the comets are separated by the existence of a tail and envelope. The comet has the tail and envelope that are made of gases and dust which are sublimated or emitted from the nucleus of the comet (coma) by the solar wind. When the comet is observed by the telescope, it looks dim. However, when the comet is far away from the Sun, it cannot be separated from the asteroid. As the asteroid is the source of meteorites, to study meteorites means to study asteroids.

1.2.6 Classification of asteroids

There are three major classification methods of asteroids, which are the source of the meteorite. The first classification is the orbital classification. The second one is the classification of Tholen, which uses albedo (degree of reflection of solar radiation) and IR spectra. The third one is the SMASS (Small Main-belt Asteroid Spectroscopic Survey) classification which is only based on IR spectra, keeping the classification of Tholen.

1.2.6.1 The orbital classification

The first classification is families or groups, which are based on the elements of the asteroidal motion. The families or groups were first found by the Japanese astronomer, Kiyotsugu Hirayama in 1918. In Fig. 1.15, the semi-major axis (a) versus the inclination (i) of the asteroids are shown. The semi-major axis (a) and the inclination (i) to express the motion of the asteroid are explained in Fig. 1.16.

The main asteroid groups are shown in Table 1.5, together with the types (related later in this section) and the famous asteroid in each group. The asteroids in 2.1−3.3 au form the major asteroid belt, which is separated into inner, middle, and outer asteroid belts. Each group is separated by the Kirkwood gaps after the name who found them. Especially, gaps at 2.06, 2.5, 2.8, and 3.27 au are clearly observed in Fig. 1.15, which is made by resonance between the asteroid and Jupiter (or mean motion resonance—MMR) of 4:1, 3:1, 5:2, and 2:1, respectively.

The Apollo group asteroids cross the Earth's orbit and, therefore, potentially collide with the Earth. Such asteroids are called near-Earth asteroids (NEAs). In 2021, 27,323 asteroids were known as NEAs, and 2224 are sufficiently large and potentially hazardous.

1.2.6.2 The classification of Tholen

The second classification is the classification of Tholen (1989), which is shown in Table 1.6. Tholen established the classification of asteroids by an albedo (degree of reflection of solar radiation) and an IR spectrum

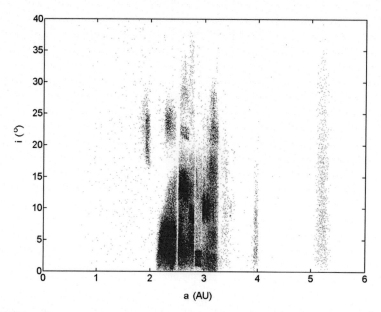

FIGURE 1.15 **The distribution of the asteroid.** The horizontal and the vertical axes indicate the distance of the semi-major axis (a) and the inclination (i), respectively, which are explained in Fig. 1.16. Only a < 6 au is shown. Asteroids with a semi-major axis of 2.1–3.3 au and the inclination of 0–20 degrees are called as the main belt. The red asteroids are the main belt asteroids, containing 94% of all the asteroids. The black asteroids are the Trojan asteroids. The Mars orbit is 1.67 au, and the Jupiter orbit is 4.95–5.46 au. *Source: From Piotr Deuar. https://upload.wikimedia.org/wikipedia/commons/e/ed/Main_belt_i_vs_a.png.*

between 0.31 and 1.06 μm obtained by the Eight-Color Asteroid Survey. He classified the asteroids into three major types; C-type (carbonaceous meteoritic type), S-type (silicates or stony type), and X-type (including M-type which is metallic iron type).

1.2.6.3 The Small Main-belt Asteroid Spectroscopic Survey classification

The third one is the SMASS classification by Bus and Binzel (2002). This is shown in Table 1.7. They established the classification of asteroids based on 1447 asteroids of SMASS of 0.44–0.92 μm following the classification of Tholen, but without taking albedo into account. The SMASS classification of Bus and Binzel (2002) classified asteroids into 27 subtypes.

1.2.6.4 Asteroid number and naming of asteroids

Those who have found the asteroid have naming rights. When you report the minor-planet position for more than two nights to the

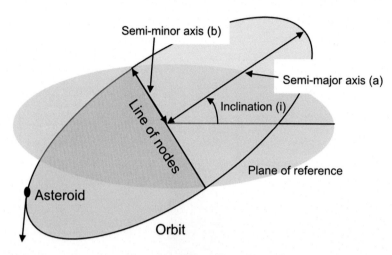

FIGURE 1.16 The motion of the asteroid. The orbit of an asteroid is shown in the figure with the plane of reference (usually the movement plane of the Earth). The orbit of the asteroid (shown in pale green) is an ellipse that intersects the reference plane at the two points (nodes). The line connecting the two nodes is called as the line of nodes. The angle between the orbit plane of the asteroid and the reference plane is called as the inclination (i), and the largest and the shortest distances from the center of the line of nodes to the orbit is called as the semi-major axis (a) and the semi-minor axis (b), respectively. The eccentricity (e) is defined as $(1 - b^2/a^2)^{0.5}$.

TABLE 1.5 Asteroid belts from 1.8 to 5.3 au.

Groups	Distance (au)	The Kirkwood gap	Type	Famous asteroid
Hungaria	1.8–2.0		E, C, S	
	2.06	--- 4:1 MMR ---		
Major asteroid belt				
Inner	2.1–2.5		V, S, M, D	Vesta
	2.5	--- 3:1 MMR ---		
Middle	2.5–2.8		C, B, P	Ceres
	2.8	--- 5:2 MMR ---		
Outer	2.8–3.3		C, B, P, M	Hygeia
		--- 2:1 MMR ---		
Cybele	3.3–3.5		P, C, D	
		--- 5:1 MMR ---		
Hilda	3.9–4.0		P, C, D	
Trojan	5.1–5.3		D, P, C	

TABLE 1.6 The classification of asteroids of Tholen.

Major classification	Minor classification	Characteristics
C-type asteroids		[C] comes from Carbonaceous; dark and like carbonaceous chondrites, >2.7 au.
	B-type	No 0.5 μm absorption. Clay minerals with water, magnetite, sulfide, organic matter, etc.
	F-type	Like B-type, but no 3 μm absorption of hydrated minerals.
	G-type	Strong <0.5 μm absorption. Some absorb 0.7 μm of clay minerals.
	C-type	Not B, D, F, G-types.
	D-type	Low albedo ([D] comes from dark), featureless spectrum, steep red slope, ice/water.
S-type asteroids		[S] comes from Stony or Silicates; albedo is 0.10∼0.22, relatively bright; inside the main asteroid belt.
	A-type	Moderate albedo, steep red slope 0.75 μm, strong 1 μm of olivine.
	Q-type	1 and 2 μm absorption of olivine and pyroxenes, between V- and S-types, ordinary chondrites?
	R-type	Middle bright, 1 and 2 μm absorption of olivine and pyroxenes, between V- and A-types.
X-type asteroids		
	M-type	[M] comes from metallic; albedo is 0.10∼0.18; relatively bright; the metallic core of planetesimals?
	E-type	Albedo is >0.3, very bright; the core of the planetesimal? source of enstatite chondrites?
	P-type	Very dark, albedo <0.1; organic matters, carbon, silicates, ice.
	T-type	Reddish, absorption of ∼0.85 μm, no water, no meteorites.
	V-type	0.75 and 1 μm absorption of olivine, more pyroxene rich than S-type, [V] comes from Vesta, like HED meteorites.

International Astronomical Union's (IAU) Minor Planet Center (MPC), they will issue the provisional number, which is composed of the finding year, space, two alphabets, and number (e.g., 1999 JU_3). There are rules to determine the alphabets and number, but they are skipped here. When the orbit of the asteroids is determined as the new one, the

TABLE 1.7   The SMASS classification.

Major classification	Subtypes and characteristics
C-type asteroids	Dark carbonaceous asteroids.
	B (Tholen's B and F), C (Tholen's C)
	Cg, Ch, Cgh (Tholen's G)
	Cb (between C and B)
S-type asteroids	Stony or siliceous asteroids.
	A (Tholen's A), Q (Tholen's Q), R (Tholen's R)
	K ($0.75\ \mu m$ absorption, like CV and CO chondrites)
	L, Ld ($0.75\ \mu m$ absorption)
	S (the most popular S-type)
	Sa, Sq, Sr, Sk, and Sl (between S and A, Q, R, K, and L, respectively)
X-type asteroids	Mainly metallic asteroids.
	X (Tholen's M, E, P), T (Tholen's T), D (Tholen's D), V (Tholen's V)
	Ld (spectra are more extreme than D)
	O (deep $0.75\ \mu m$ absorption, like L6 and LL6 meteorites)
	Xe, Xc, Xk (between X and E, C and K, respectively)

asteroid number is given (e.g., 162173 Ryugu), and it is named by the finder. The name must be inoffensive and not connected with recent political or military activity.

1.2.7 The Sloan Digital Sky Survey

The Sloan Digital Sky Survey (SDSS) is making the most detailed three-dimensional maps of the Universe, using deep multi-color images of one-third of the sky, and spectra for more than three million astronomical objects. The A.P. Sloan Foundation has been supporting the SDSS. The Sloan Foundation has a 2.5 m Telescope at Apache Point Observatory (APO) in New Mexico, USA, and the Irenee du Pont Telescope at Las Campagnas Observatory (LCO) in Chile.

In its first 5 years of operations (SDSS-I/II; 2000−08), the SDSS carried out deep multi-color imaging over 8000 square degrees and measured spectra of more than 700,000 celestial objects. There are three measurements; Legacy, Supernova, and SEGUE (Sloan Extension for Galactic Understanding and Exploration)-1 surveys.

SDSS-III (2008−14) undertook a major upgrade of the venerable SDSS spectrographs and added new powerful instruments to execute inter-weaved sets of four surveys; APOGEE (Apache Point Observatory Galactic Evolution Experiment), BOSS (Baryon Oscillation Spectroscopic Survey), MARVELS (Multi-object APO Radial Velocity Exoplanet Large-area Survey) and SEGUE-2.

SDSS-IV (2014−20): SDSS is extending three main measurements: APOGEE-2, eBOSS (extended BOSS), and MaNGA (Mapping Nearby Galaxies at APO).

SDSS-V (2020): APO and LCO began gathering data by the Milky Way Mapper, the Black Hole Mapper, and the Local Volume Mapper which targets interstellar gas.

Both spectra and images are available on the internet. The map taken by SDSS covering 35% of the sky is summarized in Fig. 1.17. Photographs of about 1 billion objects have been taken, and spectra of over 4 million objects have been obtained.

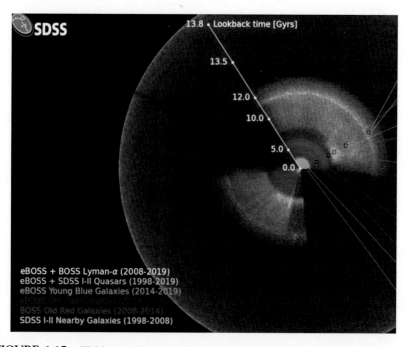

FIGURE 1.17 **SDSS map showing the observable Universe in rainbow colors.** The scale shows lookback time (Gy). Each color corresponds to the SDSS project. The outer sphere shows the fluctuation in cosmic microwave background radiation (CMBR, see 2.7.1). *Credit: Anand Raichoor (EPFL), Ashley Ross (Ohio State University) and SDSS. https://www.mpe.mpg.de/7474540/news20200720.*

FIGURE 1.18   **Illustration of Kepler space observatory.** Kepler found over 5000 exosolar planets. See Chapter 10 for its success. *Source: From NASA. https://commons.wikimedia.
org/wiki/File:Kepler_Space_Telescope_spacecraft_model_2.png.*

1.2.8 Kepler space observatory

Kepler space observatory (2009–18; see Fig. 1.18) launched by NASA
was designed to search for Earth-sized extrasolar planets in Milky Way
stars. The Kepler instrument is only a photometer that continuously
monitors the brightness of >145,000 main sequence stars in a fixed
view. Then data are analyzed to find periodic dimming of stars by
crossing extrasolar planets in front of the host stars.

1.2.9 The Herschel space observatory

The Herschel space observatory was built by ESA, which was operated from 2009 to 2013 to observe far IR and submillimeter wavelengths
(55–672 µm). This space observatory was placed on the second
Lagrangian point (L2; see Fig. 1.1) of the Sun and the Earth.

The existence of molecular oxygen was first confirmed by the
Herschel space observatory. It was also suggested that the water on the
Earth could have initially come from cometary impacts. The water
vapor on the dwarf planet, Ceres was detected and published in 2014
(Küppers et al., 2014). The observatory ended in 2013.

1.2.10 The planetary system

The silicate planetesimals or planets with the metallic core were formed in the early stage of the solar nebular. The inner part of the solar nebula is hot, where water and methane exist as gas and rocky planets (Mercury, Venus, Earth, Mars, and S-type asteroids) formed. In contrast, the outer part of the nebula (outside the snow line where water exists as ice) is cold, so that ices of water and methane condensed, forming giant planets such as Jupiter, Saturn, Uranus, and Neptune. In Table 1.8, planets, satellites, orbital characteristics, physical characteristics, and special characteristics in our solar system treated in this chapter are summarized.

Outside of Neptune at 30−50 AU from the Sun, there is a circumstellar disk like the asteroid belt. It is called as the "Kuiper belt" or "Edgeworth-Kuiper belt," which is far massive than the asteroid belt (20−200 times). They are made of small bodies which are remnants when the solar system formed. The Kuiper belt objects (KBOs) are mainly composed of frozen volatiles such as methane, ammonia, and ice. There are large dwarf planets such as Pluto, Haumea Quaoar, and Makemake, and more than 100,000 KBOs over 100 km in diameter exist.

1.2.11 Lyman-α lines

The Lyman-α lines are the strongest UV spectral lines of hydrogen. The Lyman-α lines are the doublet (121.5668 and 121.5674 nm) which are emitted by the electron transitions of hydrogen of $2p \rightarrow 1s$ and $2s \rightarrow 1s$, respectively. The redshift can be calculated from the position and the distances of two Lyman-α lines. The very distant star light can be detected as visible light or IR by the redshift. However, they are easily attenuated by the absorption of interstellar dust and hydrogen gas. In such cases, they are detected by the satellite observatory.

1.3 Introduction to inorganic astrochemistry

1.3.1 Overview

Astrochemistry is one of the sections of chemistry, and similar to geochemistry. The method in astrochemistry and geochemistry is the same, but if the target sample is terrestrial, it is called geochemistry. If the target is extraterrestrial, it is called astrochemistry. The author named this section as "inorganic astrochemistry," because there is a large difference between "inorganic" and "organic" chemistries. Basically, the astrochemist as well as the geochemist are analytical chemists, whose science is based on the data

TABLE 1.8 The selected planets, moons, and dwarf planets in our solar system.

Target	Category	Orbital characteristics		Physical characteristics		Characteristics
		Semi-major axis (AU)	Orbital period (× Earth)	Radius (R_E)	Mass (M_E)	
		(1.49×10^8 km)	(365.25 d)	(6.357×10^3 km)	(5.97×10^{24} kg)	
Earth	Planet					Life
Moon	Moon, Earth	0.00257	27.3	0.273	0.0123	The largest relative to its parent
Venus	Planet	0.723	0.613	0.952	0.815	Dense CO_2 air
Mars	Planet	1.52	1.88	0.532	0.107	Life?, SNC meteorites
Ceres	Dwarf-planet	2.76	4.6	0.0744	0.00015	The largest asteroid, the bright spot
Vesta	Dwarf-planet	2.36	3.63	0.0413	0.00004	HED meteorites
Jupiter	Planet	5.203	11.86 y	11	318	Largest planet
Io	Moon, Jupiter	0.00283	1.77 d	0.286	0.015	Sulfur-volcanoes
Europa	Moon, Jupiter	0.0045	3.55 d	0.246	0.008	Ice-covered
Ganymede	Moon, Jupiter	0.0995	7.15 d	0.414	0.0248	Salty sea
Callisto	Moon, Jupiter	0.0722	16.7 d	0.379	0.018	Ice-covered
Saturn	Planet	9.58	29.46 y	9.16	95.15	Ring
Mimas	Moon, Saturn	0.0012	0.942 d	0.03	6.3×10^{-6}	Liquid water?
Enceladus	Moon, Saturn	0.0016	1.37 d	0.04	1.8×10^{-5}	Salty sea, life?
Titan	Moon, Saturn	0.0082	15.9	0.04	3.4×10^{-5}	Atmosphere

analysis of the samples. Thus, there is also "organic astrochemistry," which is based on the data analysis of the organic materials. Organic astrochemistry includes astrobiology, but in this chapter, conventional carbon−hydrogen−oxygen−nitrogen−sulfur isotopic astrochemistry is included in inorganic astrochemistry. However, when carbon behaves as an organic material, it is obviously included in organic astrochemistry.

Inorganic astrochemistry has evolved according to the evolution of analytical chemistry or analytical methods. Such evolution includes new strategies to determine as many elements as possible to measure isotopic ratios and to analyze as smaller amounts of samples as possible.

Four strategies exist in inorganic astrochemistry: (1) the major element astrochemistry; (2) the trace element astrochemistry; (3) the stable isotope astrochemistry; and (4) the radiogenic isotope astrochemistry. For each strategy, suitable analytical methods exist. In this section, the basic knowledge of the astrochemistry is briefly explained.

1.3.2 Atomic number, element symbol, and element

As shown in Fig. 1.4, atoms are composed of protons and neutrons. The number of protons is called the atomic number, which characterizes the chemical and physical properties of the atom called as an element. The atomic number, element symbol, and elements are shown in Table 1.9. Presently the element with up to 118 atomic number have been synthesized. The synthesized elements by nuclear reactions are shown with "asterisk (*)" at the atomic number in the table. Technetium (Tc), promethium (Pm), and heavier elements than uranium (U) that are called as transuranium elements are the synthesized elements.

The periodic table (see Fig. 1.19) is a table in which each element is placed according to the atomic number. The row and array of the table are named the period and group, respectively. The chemical property of each element has periodicity according to the atomic number, which is shown as the vertical array (group). The periodicity of the table is based on the electron configuration of each element. Groups 1−2 and 13−18 elements are called as the main group element, and groups 4−11 elements are called as transition metals, which have partially d-subshell (see the next paragraph) filled elements (the IUPAC definition). But sometimes groups 3 and 12 elements are also included in the transition metals.

A neutral element (atom) has the number of electrons which is the same as that of the proton, namely, the atomic number. According to quantum mechanics, the electrons are placed in the shell, in which the number of $2n^2$ ($n = 1, 2, 3, 4, 5, \ldots$ which are named as K, L, M, N, O... shells) electrons can be placed, thus, the total electrons become 2, 8, 18, 32, 50, ..., respectively.

TABLE 1.9   Atomic number, element symbol, and element.

At.N.	E.S.	Element	At.N.	E.S.	Element	At.N.	E.S.	Element
1	H	Hydrogen	41	Nb	Niobium	81	Tl	Thallium
2	He	Helium	42	Mo	Molybdenum	82	Pb	Lead
3	Li	Lithium	43*	Tc	Technetium	83	Bi	Bismuth
4	Be	Beryllium	44	Ru	Ruthenium	84	Po	Polonium
5	B	Boron	45	Rh	Rhodium	85	At	Astatine
6	C	Carbon	46	Pd	Palladium	86	Rn	Radon
7	N	Nitrogen	47	Ag	Silver	87	Fr	Francium
8	O	Oxygen	48	Cd	Cadmium	88	Ra	Radium
9	F	Fluorine	49	In	Indium	89	Ac	Actinium
10	Ne	Neon	50	Sn	Tin	90	Th	Thorium
11	Na	Sodium	51	Sb	Antimony	91	Pa	Protoactinium
12	Mg	Magnesium	52	Te	Tellurium	92	U	Uranium
13	Al	Aluminum	53	I	Iodine	93*	Np	Neptunium
14	Si	Silicon	54	Xe	Xenon	94*	Pu	Plutonium
15	P	Phosphorus	55	Cs	Cesium	95*	Am	Americium
16	S	Sulfur	56	Ba	Barium	96*	Cm	Curium
17	Cl	Chlorine	57	La	Lanthanum	97*	Bk	Berkelium
18	Ar	Argon	58	Ce	Cerium	98*	Cf	Californium
19	K	Potassium	59	Pr	Praseodymium	99*	Es	Einsteinium
20	Ca	Calcium	60	Nd	Neodymium	100*	Fm	Fermium
21	Sc	Scandium	61*	Pm	Promethium	101*	Md	Mendelevium
22	Ti	Titanium	62	Sm	Samarium	102*	No	Nobelium
23	V	Vanadium	63	Eu	Europium	103*	Lr	Lawrencium
24	Cr	Chromium	64	Gd	Gadolinium	104*	Rf	Rutherfordium
25	Mn	Manganese	65	Tb	Terbium	105*	Db	Dubnium
26	Fe	Iron	66	Dy	Dysprosium	106*	Sg	Seaborgium
27	Co	Cobalt	67	Ho	Holmium	107*	Bh	Bohrium
28	Ni	Nickel	68	Er	Erbium	108*	Hs	Hassium
29	Cu	Copper	69	Tm	Thulium	109*	Mt	Meitnerium
30	Zn	Zinc	70	Yb	Ytterbium	110*	Ds	Darmstadtium

(Continued)

TABLE 1.9 (Continued)

At.N.	E.S.	Element	At.N.	E.S.	Element	At.N.	E.S.	Element
31	Ga	Gallium	71	Lu	Ruthenium	111*	Rg	Roentgenium
32	Ge	Germanium	72	Hf	Hafnium	112*	Cn	Copernicium
33	As	Arsenic	73	Ta	Tantalum	113*	Nh	Nihonium
34	Se	Selenium	74	W	Tungsten	114*	Fl	Flerovium
35	Br	Bromine	75	Re	Rhenium	115*	Mc	Moscovium
36	Kr	Krypton	76	Os	Osmium	116*	Lv	Livermorium
37	Rb	Rubidium	77	Ir	Iridium	117*	Ts	Tennessine
38	Sr	Strontium	78	Pt	Platinum	118*	Og	Oganesson
39	Y	Yttrium	79	Au	Gold			
40	Zr	Zirconium	80	Hg	Mercury			

At.N., Atomic number; *E.S.*, element symbol; *, synthesized elements.

Group	1	2	3	4	5	6	7	8	9	10	11	12	13	14	15	16	17	18

Period																		
1	H																	He
2	Li	Be											B	C	N	O	F	Ne
3	Na	Mg											Al	Si	P	S	Cl	Ar
4	K	Ca	Sc	Ti	V	Cr	Mn	Fe	Co	Ni	Cu	Zn	Ga	Ge	As	Se	Br	Kr
5	Rb	Sr	Y	Zr	Nb	Mo	Tc	Ru	Rh	Pd	Ag	Cd	In	Sn	Sb	Te	I	Xe
6	Cs	Ba	L	Hf	Ta	W	Re	Os	Ir	Pt	Au	Hg	Tl	Pb	Bi	Po	At	Rn
7	Fr	Ra	A	Rf	Db	Sg	Bh	Hs	Mt	Ds	Rg	Cn	Nh	Fl	Mc	Lv	Ts	Og

Lanthanides	L	La	Ce	Pr	Nd	Pm	Sm	Eu	Gd	Tb	Dy	Ho	Er	Tm	Yb	Lu
Actinides	A	Ac	Th	Pa	U	Np	Pu	Am	Cm	Bk	Cf	Es	Fm	Md	No	Lr

FIGURE 1.19 **The periodic table of the elements.** The row and array of the table are named the period and group, respectively.

Each shell has a subshell, in which $2(2l + 1)$ electrons can accommodate. The first "2" comes from the electron spin, which has two types, $+1/2$ (upper spin) and $-1/2$ (lower spin). The subshells are named as s, p, d, f, and g orbitals according to $l = 0, 1, 2, 3,$ and 4 with total electron numbers of 2, 6, 10, 14, and 18, respectively. Thus, electrons are filled from lower electron energy levels of shells and subshells, namely, 1s, 2s, 2p, 3s, 3p.

For example, $_1H$, $_3Li$, $_8O$, $_{15}P$, and $_{18}Ar$ (the atomic number is often written in the lower case of the elemental symbol) have the electron configuration of $1s^1$, $1s^2 2s^1$, $1s^2 2s^2 2p^4$, $1s^2 2s^2 2p^6 3s^2 3p^3$, and $1s^2 2s^2 2p^6 3s^2 3p^6$, respectively. The number of electrons in each electron orbital is written in the upper case after the orbital. $_{19}K$, $_{20}Ca$, and $_{21}Sc$ can be expressed as [Ar] $4s^1$, [Ar]$4s^2$, and [Ar]$4s^2 3d^1$. The reason why $_{21}Sc$ is not [Ar] $4s^2 4p^1$ is that the configuration of the electron is determined by the order of the energy levels of electrons, which is basically in the order of Fig. 1.20. This is called the Aufbau principle or the Madelung rule.

Many elements have different neutron numbers with the same proton number. It is called an isotope. A sum of proton and neutron numbers is called a mass number. Isotopes are shown as the superscript of the element symbol, e.g., ^3He, ^4He, ^{235}U, ^{238}U, etc. Hydrogen has three isotopes, ^1H, ^2H, and ^3H, but ^2H and ^3H have special symbols of ^2D (deuterium) and ^3T (tritium). There are stable isotopes and unstable isotopes that are called radioisotopes (RI). The isotopes of each element generally observable in nature are shown in Fig. 1.21 as atomic abundances (%).

1.3.3 Decays of radioactive elements

There are unstable isotopes in nature. They are called as radioactive isotopes, and they break into more stable isotopes. This is called as radioactive decay (see Table 1.10 for radioactive isotopes used in astrochemistry). There are six main radioactive decay types of the radioactive isotope; α-decay, β^--decay, double β^--decay, β^+-decay, electron capture

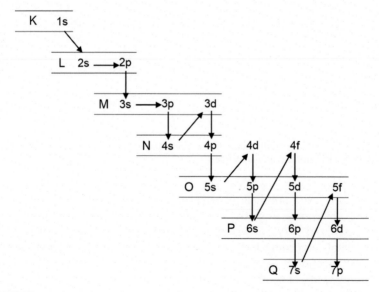

FIGURE 1.20   The Madelung rule or the Aufbau principle. Electrons are filled according to this order.

(EC), and spontaneous fission (SF). The isotope made by the radioactive isotope (sometimes called as the "mother" isotope) is called as "radiogenic" or "daughter" isotopes.

The α-decay is the decay where α-particle (^{4}He) is emitted. Two proton and two neutron numbers decrease, and the four mass numbers decrease. The β^--decay is electron emission. Therefore, one proton

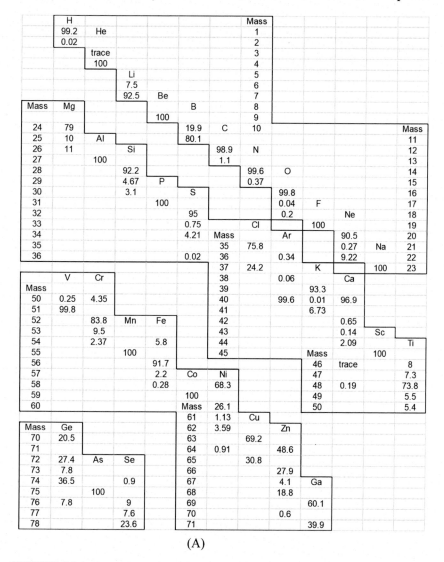

(A)

FIGURE 1.21 **Stable isotopic abundances of each element.** (A) Mass number from 1 to 78. (B) Mass number from 78 to 142. (C) Mass number from 143 to 238. The atomic abundance is shown in %.

Isotopic abundance chart (values in %):

Se, Br, Kr

Mass	Se	Br	Kr
78			0.35
79		50.7	
80	49.7		2.25
81		49.3	
82	9.2		11.6
83			11.5
84			57
86			17.3

Rb, Sr, Y, Zr

Mass	Rb	Sr	Y	Zr
84		0.56		
85	72.2			
86		9.86		
87	27.8	7		
88		82.6		
89			100	
90				51.5
91				11.2
92				17.2
94				17.4
96				2.8

Nb, Mo, Ru

Mass	Nb	Mo	Ru
92		14.8	
93	100		
94		9.25	
95		15.9	
96		16.7	5.52
97		9.55	
98		24.1	1.88
99			12.7
100		9.63	12.6
101			17
102			31.6
104			18.7

Rh, Pd, Ag, Cd

Mass	Rh	Pd	Ag	Cd
102		1.02		
103	100			
104		11.1		
105		22.3		
106		27.3		1.25
107			51.8	
108		26.5		0.89
109			48.2	
110		11.7		12.5
111				12.8
112				24.1
113				12.2
114				28.7
116				7.49

In, Sn, Sb, Te, I

Mass	In	Sn	Sb	Te	I
112		0.97			
113	4.3				
114		0.65			
115	95.7	0.36			
116		14.5			
117		7.68			
118		24.2			
119		8.58			
120		32.6		0.1	
121			57.3		
122		4.63		2.6	
123			42.7	0.91	
124		5.79		4.82	
125				7.14	
126				19	
127					100
128				31.7	
130				33.8	

Xe, Cs, Ba, La, Ce, Pr, Nd

Mass	Xe	Cs	Ba	La	Ce	Pr	Nd
124	0.1						
126	0.09						
128	1.91						
129	26.4						
130	4.1		0.11				
131	21.2						
132	26.9		0.1				
133		100					
134	10.4		2.42				
135			6.59				
136	8.9		7.85		0.19		
137			11.2				
138			71.7	0.09	0.25		
139				99.9			
140					88.8		
141						100	
142					11.1		27.1

(B)

FIGURE 1.21   Continued.

Gd	Tb	Dy	Mass	Er	Ho
14.8			155		
20.5		0.06	156		
15.7	Tb		157		
24.8		0.1	158		
	100		159		
21.9		2.34	160	Er	
		18.9	161		
		25.5	162	0.14	
		24.9	163		Ho
		28.2	164	1.61	
			165		100

Nd	Sm	Eu	Gd	Mass
12.2				143
23.8	3.1			144
8.3				145
17.2				146
	15			147
5.76	11.3			148
	13.8			149
5.64	7.4	Eu	Gd	150
		47.8		151
	26.7		0.2	152
		52.2		153
	22.7		2.18	154

Mass	Re	Os	Ir	Pt
184		0.02		
185	37.4			
186		1.58		
187	62.6	1.6		
188		13.3		
189		16.1	Ir	Pt
190		26.4		0.01
191			37.3	
192		41		0.79
193			62.7	
194				32.9
195				33.8
196				25.3
197				
198				7.2
199				
200				

Mass	Er	Tm	Yb	Lu	Hf	Ta	W
166	33.6						
167	23	Tm					
168	26.8		0.13				
169		100					
170	14.9		3.05				
171			14.3				
172			21.9				
173			16.1	Lu	Hf		
174			31.8		0.16		
175				97.4			
176			12.7	2.59	5.21		
177					18.6		
178					27.3		
179					13.6	Ta	W
180					35.1	0.01	0.13
181						99.9	
182							26.3
183							14.3
184							30.7
185							
186							28.6

Mass	Au	Hg	Tl	Pb	Bi
196		0.14			
197	100				
198		10			
199		16.8			
200		23.1			
201		13.2			
202		29.8	Tl	Pb	
203			29.5		
204		6.85		1.4	
205			70.5		
206				24.1	
207				22.1	Bi
208				52.4	
209				100	

Mass	Th
232	100

Mass	U
234	0.01
235	0.72
236	
237	
238	99.3

(C)

FIGURE 1.21 Continued.

number increases with the same mass number. The double β^--decay occurs when a single β^--decay nucleus has higher energy or spin-forbidden. The β^+-decay is a positron (anti-electron) emission. Therefore, one proton number decreases with the constant mass number. The electron capture (EC) is when one electron is absorbed in the nucleus. Therefore, one proton number decreases with the constant mass number. The SF is when that nucleus decays into two nuclei. This occurs only in heavy nuclei such as ^{235}U, ^{238}U, or ^{240}Pu. The changes in the N (neutron number)$-$Z (proton number) plot are shown in Fig. 1.22.

TABLE 1.10 Radioactive isotopes and half-lives used in astrochemistry.

Radioactive isotope	Decay scheme	Daughter isotope	Half-life (y)
^{40}K	EC	^{40}Ar	1.19×10^{10}
	β^-	^{40}Ca	1.40×10^{9}
^{87}Rb	β^-	^{87}Sr	4.88×10^{10}
^{138}La	EC	^{138}Ba	1.56×10^{11}
	β^-	^{138}Ce	3.03×10^{11}
^{147}Sm	α	^{143}Nd	1.06×10^{11}
^{176}Lu	β^-	^{176}Hf	3.71×10^{10}
^{187}Re	β^-	^{187}Os	4.16×10^{10}
^{190}Pt	β^-	^{190}Os	4.69×10^{11}
^{232}Th	Decay chain	^{208}Pb	1.40×10^{10}
^{235}U	Decay chain	^{207}Pb	7.04×10^{8}
^{238}U	Decay chain	^{206}Pb	4.47×10^{9}

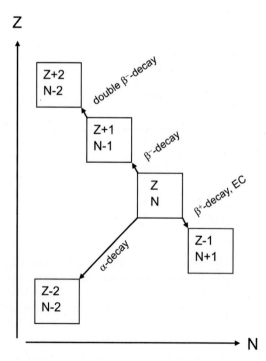

FIGURE 1.22 Change of neutron (N) and atomic (Z) numbers by radioactive decay.
The horizontal and vertical axes are the neutron and the atomic numbers, respectively.
The mother isotope is at the position of (N, Z), and moves to positions of (N − 1, Z + 1),
(N − 2, Z + 2), (N + 1, Z − 1), and (N − 2, N − 2) by β^--decay, double β^--decay, β^+-decay/
EC, and α-decay, respectively.

When we hit a proton to a ^{2}H atom, it becomes a ^{3}He atom with the emission of gamma-ray. We can describe the reaction as:

$$^2\text{H}(p, \ \gamma)^3\text{He} \tag{1.14}$$

This description is very useful, when reactions occur successively. For example, $^2\text{H} \rightarrow {}^3\text{He} + \text{p} + \gamma \rightarrow {}^3\text{He} + {}^4\text{He} \rightarrow {}^7\text{Be} + \gamma$ and $^7\text{Be} + \text{e}^- \rightarrow {}^7\text{Li} + \nu$ occur (ν is neutrino), we can describe as:

$$^2\text{H}(p, \ \gamma)^3\text{He}(^4\text{He}, \ \gamma)^7\text{Be}(\text{e}^-, \ \nu)^7\text{Li} \tag{1.15}$$

1.3.4 Absorption and emission of lights

We assume that there are two energy levels of the ground and excited states of the electron. When there is an electron in the ground state (see Fig. 3.2A, left), and the light with the energy of $h\nu$ comes to the electron, the electron absorbs the energy and goes into the excited state. This is the "absorption" of light. In contrast, when there is an electron in the excited state, the electron goes down to the ground state by emitting light with the energy of $h\nu$ (see Fig. 1.23A, right). This is the "emission" of light. In Fig. 1.23B, energy (wavelength) spectra at absorption and emission are shown. The photo-spectroscopy is the method in which the abundance of the target element is determined by the intensity of the specific lights emitted or absorbed from the target elements from the photo-spectra.

When the electron transition is between K, L, M, N, ... shells (e.g., L→K, M→L, etc.), the energy between each shell is very high and becomes X-ray. This energy is specific to each element; thus, the element determination can be done. This X-ray is called as the characteristic X-ray and is used in the determination of elements in SEM-EDX, EPMA (Section 1.3.5.2), and XRF (Section 1.4.7).

Unfortunately, there are many energy levels in the photo-spectra because there are similar energy levels in many electrons. Therefore, the assignment and overlaps of the spectra become more complicated than mass spectrometry. The mass spectra are only a few sheets (Fig. 1.21), but the light spectra become one book or more. This is one of disadvantages of the photo-spectrometry.

In astronomy, a similar spectrometry method is applied. The collected lights of a star or galaxy by the telescope are dispersed by their wavelength or color. The sharp absorption or emission lines in the spectra are observed, and used as the existent evidence of elements. Interestingly, photo-spectrometry is applied for analysis from hand specimen to the furthest galaxy!

(A) Absorption and emission of lights

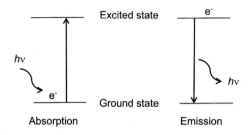

(B) Energy (wavelength) spectra of lights at absorption and emission

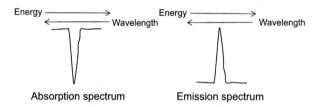

FIGURE 1.23 **Absorption and emission of lights and energy (wavelength) spectra.** (A) Absorption and emission of lights. For simplicity, only two energy states (ground and excited ones) are shown. The vertical arrow shows electron transfer from one energy state to the other energy state. The left-hand figure is absorption, in which the electron absorbs energy and goes from the ground energy state to the excited one. The right-hand figure is emission, in which the electron goes from the excited energy state to the ground one. The curved line shows the absorption or emission of the lights (energy) corresponding to the difference between the two energy states. (B) Energy (wavelength) spectra of lights at absorption and emission. The left-hand figure shows the energy spectra when the absorption of light occurred. The right-hand figure shows the spectra when the emission occurred. As the wavelength (λ) is c/ν (c is the speed of light), the direction of the wavelength spectrum is the opposite direction of the energy spectrum.

1.3.5 Mass spectrometry

1.3.5.1 Introduction

Mass spectrometry is the method, which uses the specific mass of each element as the evidence of the element. For example, when the mass number of 238 appears in the mass spectrum, uranium is judged to exist. The mass spectrometry uses the mass spectrometer (MS), which is basically composed of three units: (1) the ion source; (2) the mass separator, and (3) the detector. Each unit is differently designed according to the target element or the purpose of analysis.

The ion source ionizes the target material into ions and accelerates the ions to the velocity (v [m s^{-1}]) by the accelerating voltage (E [V]). Various methods are used to ionize according to the target element/materials. Using the energy conservation law, the velocity (v) of the ion becomes:

$$v = \sqrt{2E \times z/m} \qquad (1.16)$$

where m and z are the mass (kg) and charge (usually $z = +1$ or $1 \times 1.6 \times 10^{-19}$ [C]), respectively. The mass separator unit separates the ions according to mass over charge (m/z). The simplest mass separator is a magnet which makes a magnetic field (B [T or Wb/m^2]). By Fleming's left-hand rule, the accelerated ion receives the force, zvB [N]. When the direction of the magnetic field is downwards, and the charge z is positive, the ion curves to the right because the force (zvB) becomes the centrifugal force $mr\omega^2$ (ω is an angular velocity):

$$zvB = mr\omega^2 = mv^2/r \qquad (1.17)$$

Using Eqs. (1.16) and (1.17):

$$r = mv/zB = (\sqrt{2Em/z})/B \qquad (1.18)$$

Therefore, the radius of the ion path is proportional to the root of m/z when both the accelerating voltage and the magnetic field are constant. Or, when the detector position is fixed, the position of m/z can be chosen by changing the magnetic field.

The mass spectrometry is roughly classified into two groups: one is for the inorganic analysis and the other is for the organic analysis. The details are described in Sections 1.3.5.2 and 1.3.5.3, respectively.

1.3.5.2 Inorganic mass spectrometry

The first purpose of inorganic mass spectrometry is inorganic (or refractory) elemental determination in the silicate samples. Combined with chemical digestion and ion exchange chromatography, the detection limits in silicates are from μg g^{-1} to fg g^{-1} level. For this purpose, inductively coupled plasma mass spectrometer (ICP-MS) is the most popular machine.

As related in Section 1.3.3, some isotopes of the inorganic elements are radioactive, and decay into other isotopes (daughter isotopes). To detect the change of daughter isotope ratios caused by the radioactive decay, inorganic mass spectrometry is employed. For this purpose, a thermal ionization mass spectrometer (TIMS) is used.

Another purpose of mass spectrometry is to determine the fractionation of the isotope ratios of volatile elements precisely. The isotope ratios change by mass fractionation, which is the natural process to change isotope ratios. For example, when water (H_2O) evaporates, lighter isotopes (^{1}H and ^{16}O) easily evaporate than the heavier isotopes (^{2}D, ^{17}O, and ^{18}O). Such mass fractionation occurs in most physical processes, such as evaporation, melting, dissolution, and precipitation. In addition, biogenic mass fractionation also occurs, which is one of the evidence for the element to be related to biogenic processes. For this purpose, especially for the volatile elements, such as H, C, O, N, and S (CHONS), the isotope ratio mass spectrometer (IRMS) with a gas separation line is used.

For the determination of the isotopic fractionation of non-volatile or refractory inorganic elements, inorganic mass spectrometry is applied. As the inorganic elements are relatively heavy compared to CHONS, the fractionation is very small, thus, the highly precise determination ($<0.05‰$) methods are required. For this purpose, a multicollector ICP-MS (MC-ICP-MS) is widely used.

The noble gases, such as helium (He), neon (Ne), argon (Ar), and xenon (Xe) behave differently from other elements, and special techniques have been developed for noble gas mass spectrometry. The noble gas mass spectrometer with the noble gas purification line is used for quantitative and isotope ratio analyses of the noble gases.

Mass spectrometers are generally very heavy, because the sector magnet is a chunk of metals. However, the mass spectrometer for determination of the D/H ratio is sometimes equipped on a spacecraft because the magnet for hydrogen is very light. For detection of the life on planets such as Mars, the mass spectrometer for the determination of isotopic ratios of hydrogen, carbon, nitrogen, and oxygen was loaded on the spacecraft. Instead of the sector electro-magnet, the Q-pole mass filter (Q-pole) is often used for the mass separator.

1.3.5.3 Organic mass spectrometry

As the mass spectrometry, organic mass spectrometers are explained here, although, it is included in organic astrochemistry. For organic materials, liquid chromatography, high-performance liquid chromatography, and gas chromatography are combined with organic mass spectrometers. The organic molecules are partially decomposed when they are ionized, and specific signals are observed according to the functional groups of the target materials. When multiple functional groups exist, repetitive signals appear. Using the computer, the database of ionization patterns of many organic materials helps the complex signal analysis and identification of materials. The organic mass spectrometers need high mass resolution (~ 1 million) with high mass scan speed, which is achieved by the combination of two mass spectrometers (MS/MS), such as a Q-pole type mass spectrometer and sector field mass spectrometer.

1.3.6 Elemental analytical methods

1.3.6.1 *Bulk analysis versus spot analysis*

In astrochemistry, the elemental and isotopic analyses of samples are divided into bulk elemental/isotopic analyses and spot elemental/isotopic analyses (see Fig. 1.24; abbreviations of analytical methods are shown in the figure caption). In the bulk analysis, significant amounts of the sample are digested. "The significant amounts" means the recognizable or visible amounts that can be measured accurately by a balance. Usually, 1 mg to 1 g are used in the bulk analysis. It is difficult to digest >1 g perfectly by chemical methods. Less than 1 mg sample is also difficult to handle or to measure weight precisely. After the sample is perfectly decomposed and dissolved into the sample solution, the target element concentrations are determined. When precise isotope ratios are required, the target element is purified by column chromatography, and isotope ratios of the target element are measured by mass spectrometry. Fig. 1.25 shows the summary of the bulk elemental/isotopic analyses.

In the spot elemental analysis, electron beam, ion beam, or laser beam are bombarded onto a spot area, of which the diameter is from 10 to 100 μm (10^{-6} to 10^{-9} g). The summary of spot elemental analyses is

	Analyzed amount (Analyzed area)	Exciting method (Analytical method)	Detection method
Bulk analysis	$1 \sim 10^{-3}$ g	Flame (FAAS)	Light (Visible)
		Heat (FLAAS/TIMS)	Light (Visible)/Ions
		Plasma	
		(ICP-AES/ICP-MS)	Light (Visible)/Ions
		X-ray (XRF)	
Spot analysis	$10^{-6} \sim 10^{-9}$ g	Electron (EPMA)	Light (X-ray)
		Ion (SIMS)	Mass
	$(10 \sim 100~\mu m~\phi)$	Light (IR~UV Laser)	
		(LA-ICP-MS)	Mass

FIGURE 1.24 **Comparison of bulk analysis and spot analysis.** *FAAS*, Flame atomic absorption spectrometry; *FLAAS*, FLameless atomic absorption spectrometry; *TIMS*, thermal ionization mass spectrometry; *ICP-AES*, inductively coupled plasma atomic emission spectrometry; *ICP-MS*, inductively coupled plasma mass spectrometry; *XRF*, X-ray fluorescence spectrometer; *EPMA*, electron probe micro analyzer; *SIMS*, secondary ion mass spectrometry; *LA-ICP-MS*, laser ablation inductively coupled plasma mass spectrometry.

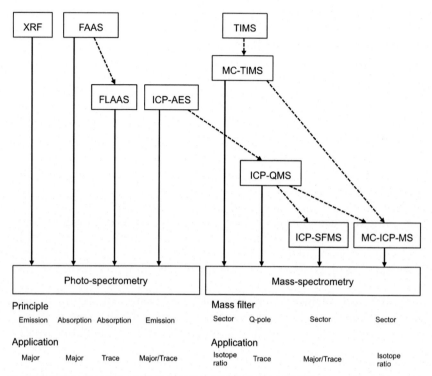

FIGURE 1.25 **Bulk analytical methods used in astrochemistry together with principles for photo-spectrometry, mass filters for mass spectrometry, and application.** The solid arrows show the evolution (or no evolution) of each analytical method. The broken arrows show the transfer of technology. Each photo-spectrometry is based on a principle; emission [Emission] or absorption [Absorption] of lights. Each mass spectrometry is based on a mass filter made of a Q-pole mass filter [Q-pole] or a sector magnet [Sector]. These methods are applied to major element determination in rocks [Major], trace element determination in rocks [Trace], or precise isotope ratio determination [Isotope ratio]. The square bracket means the abbreviation in the figure. *MC-TIMS*, multicollector-thermal ionization mass spectrometry; *ICP-QMS*, Q-pole type inductively coupled plasma mass spectrometry; *ICP-SFMS*, sector field type inductively coupled plasma mass spectrometry; *MC-ICP-MS*, multicollector-inductively coupled plasma mass spectrometry.

shown in Fig. 1.26. The elements bombarded by electrons emit the characteristic X-rays, and element compositions are determined (electron probe micro analyzer—EPMA or SEM-EDX/WDX). In secondary ion mass spectrometry (SIMS), the ion beam is bombarded, and secondary ions are collected and detected. In LA-ICP-MS, the laser beam is ablated on the sample surface, and small sample particles are produced and detected. Infrared, visible, and ultraviolet laser beams are used. ICP-QMS, ICP-SFMS, and MC-ICP-MS are used depending on analytical purposes.

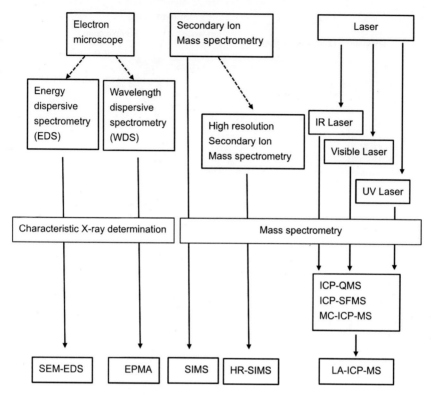

FIGURE 1.26 Detailed classification of spot analyses. *SEM-EDS*, scanning electron microscope; *EPMA*, electron probe microanalyzer; *SIMS*, secondary ion mass spectrometry; *HR-SIMS*, high-resolution secondary ion mass spectrometry; *LA-ICP-MS*, laser ablation inductively coupled plasma mass spectrometry; *IR*, infraRed; *UV*, ultra violet.

1.3.6.2 Bulk analysis methods

In astrochemistry, popularly used bulk analytical methods are XRF, TIMS, ICP-QMS, and MC-ICP-MS. In this section, ICP-QMS, TIMS, and MC-ICP-MS except XRF are briefly explained. XRF is explained in Section 1.4.6.

ICP-QMS: A schematic diagram and explanation of ICP-QMS is shown in Fig. 1.27. ICP-QMS is suitable for the determination of most inorganic trace elements (from $\mu g\ g^{-1}$ to $ng\ g^{-1}$ in silicate samples) after acid digestion of the sample.

TIMS: A schematic diagram of TIMS is shown in Fig. 1.28 As the multicollectors became the standard option of TIMS, TIMS means MC-TIMS. The chemically separated element (Sr, Nd, and Pb) by chromatography from the sample is loaded on the filament. This machine is

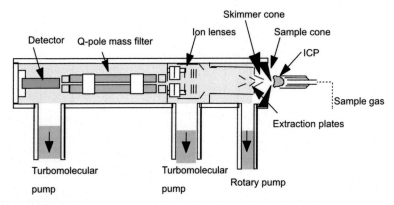

FIGURE 1.27   A schematic diagram of Q-pole type ICP-MS (ICP-QMS). Mist of the sample solution is put into the Ar inductive coupled plasma of ~7000K, and most elements are ionized. The air pressure pushes the plasma including the ions into a small hole in the center of a sample cone. The tip of the sample cone is made of Cu, Ni, or Pt. The back space of the sample cone is evacuated by a large rotary pump to be ~1 Pa. The flow of the plasma gas further hits the skimmer cone, which also has a hole in the center to lead the ions into the second chamber. The skimmer cone is also made of Cu, Ni, or Pt. Then the ions pass through ion lenses in which the ions are curved and separated from strong UV lights from the Ar plasma, because lights go straight. After the separation of lights, the low background (<10 counts s^{-1}) can be achieved. Then the ions go into the Q-pole mass filter. By the three-stage pumping system, the vacuum around the Q-pole and the detector is less than ~10^{-6} Pa. Although various m/z ions are introduced into the Q-pole, only one m/z can pass through the Q-pole filter. Q-pole can usually scan 10 times from Li to U in 1 s. The disadvantage of the Q-pole mass filter is that the peak shape has no flat peak top, therefore, the precision of isotopic ratios becomes >0.3%. Finally, selected mass ions are detected by secondary electron multiplier (SEM) and counted by the ion counting system.

mainly used for the precise isotope ratio analysis of Sr, Nd, and Pb. In the negative mode (N-TIMS), the Os isotope ratio is measured. The precision is 0.002% (2 SD) for Nd and Sr. Ions are generated on the surface ionization or thermal ionization process on high-temperature filaments made of Re, Ta, and W, and at about 1500°C. Although high repeatability and precision can be obtained, only a few elements that thermally ionize on the filament are available.

MC-ICP-MS: A schematic diagram of MC-ICP-MS is shown in Fig. 1.29 The fluctuation of each isotope signal by the plasma was perfectly canceled out by simultaneous measurement by multicollectors, and the precision drastically improved to the level of TIMS. Furthermore, MC-ICP-MS has an advantage over TIMS that fractionation of two-isotope elements such as Li, V, Cu, and Tl, etc. which are impossible to be measured by TIMS.

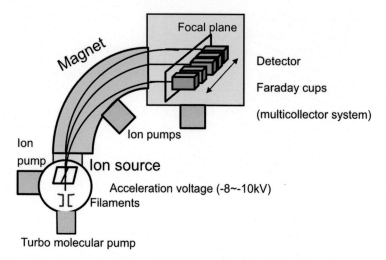

FIGURE 1.28 **A schematic diagram of TIMS (MC-TIMS).** The ion source is composed of filaments, acceleration voltage supply (-8 to -10 kV for positive ion; $8-10$ kV for negative ion), and focusing plates. A turbo molecular pump, a rotary pump, and a cryopump are equipped at the ion source. The purified target elements are loaded on the W, Ta, or Re filament and heated. Thermally produced ions are accelerated and go into the magnetic field and are separated according to m/z. Along the focal plane, where ions of each m/z focus, multiple Faraday cups are placed. Each Faraday cup can be moved on the focal plane from outside of the vacuum. The flight tube and the detector units are evacuated by the ion pumps.

1.3.6.3 Spot analysis methods

As the spot analytical methods, SEM-EDS, EPMA, SIMS, HR-SIMS, and LA-ICP-MS are popularly used. Therefore, in this section, these techniques are explained briefly.

SEM-EDS: A schematic diagram of the scanning electron microscope (SEM) is shown in Fig. 1.30. The characteristic X-rays are detected by energy dispersive spectrometry (EDS), which is enabled by the semiconductor detector that can separate energy and intensity at once. The merits of SEM-EDS are as follows: (1) the spectrum for all elements is taken at once, therefore, the measurement time is very short; (2) secondary and backscattering electron images are better than that of EPMA, because the detector positions are designed to collect these electrons; and (3) the price of SEM-EDS is $1/3-1/4$ of EPMA.

EPMA: A schematic diagram of EPMA is shown in Fig. 1.31. The spot analysis of EPMA is simple. The polished and carbon-coated sample is set on the sample stage. The electron beam is bombarded on the sample and the characteristic X-rays are emitted. The position and material of the diffracting crystals are changed according to the wavelengths

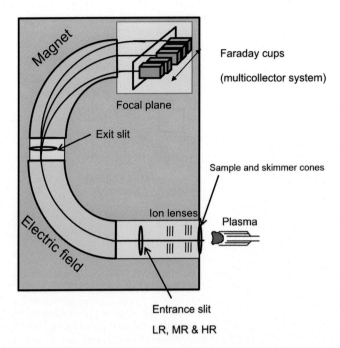

FIGURE 1.29　A schematic diagram of MC-ICP-MS. The ions produced by ICP passed through the cones and accelerated by −8 kV. For the acceleration of the ions, the plasma is the ground voltage, and the whole machine is kept at −8 kV (the orange area). An entrance slit is placed in front of the electric field to change the resolution. Then the ions go into the electric field to make the ion energy to be the same, and then into the magnet. The ions are separated according to m/z and detected by a multicollector system. For the mass discrimination correction of two-isotope elements, the standard-sample-standard bracketing method (SSB method) and the fractionation correction method by neighboring elements were developed. Nowadays MC-ICP-MS is used in space sciences for fractionation determination of two-isotope elements such as V, Cu, and Tl, etc., most of which cannot be determined by TIMS as well as multiple (>2) isotope elements.

of characteristic X-rays of the target elements. This is called the wavelength dispersive spectrometry (WDS). The concentration of major elements is calculated by a ZAF correction method using the standard materials. Z, A, and F mean influences from atomic number, X-ray absorption, and secondary fluorescence, respectively. The precision of elemental analysis is highly dependent on the standard materials. The scanning (mapping) analysis is a feature of SEM-EDS and EPMA. The scanning measurement by WDS takes a very long time, therefore, only a few spot analyses should be done by WDS. As the electron beam diffuses in the sample, the size of the analytical area is $\sim 2.5\ \mu m^2$, therefore, setting the stage steps of <2.5 μm is meaningless.

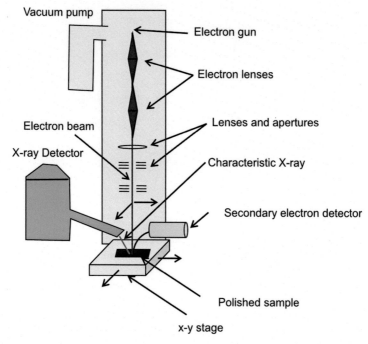

FIGURE 1.30 **A schematic diagram of SEM-EDS.** The electron beam is made at the top of the column, and the shape of the electron beam is reformed using electron lenses and apertures. Finally, the beam is projected to the sample. The electron beam is scanned in the small area, and the secondary electrons, scattered electrons, and the characteristic X-rays are emitted. The electrons are detected by electron detectors. The characteristic X-ray is detected by the semiconductor detector. This configuration is called the energy dispersive spectrometry (EDS). In case the electron beam does not cover the whole sample, an x−y stage is equipped to move the whole sample.

SIMS: When the negative ion is accelerated and bombarded into the target (sample), the structure of the target is destroyed and neutral target particles scatter around into space. This phenomenon is called sputtering (see Fig. 1.32). A few of them ($\sim 0.01\%$) are positively charged (secondary ions). If there is an electric field to extract the charged particles, such ions are accelerated. Then the sputtering phenomena can be the ion source. This is the principle of SIMS. Details of the SIMS machine are shown in Fig. 1.33.

As the primary ion beam, both positive (Cs^+) and negative (O^-) ions are available. The primary ions can be concentrated to $50-10\ \mu m\ \Phi$. As the secondary ion generation efficiency can be different depending on matrix composition (matrix effects), the preparation of standards with similar matrix composition is required. In addition, various molecular

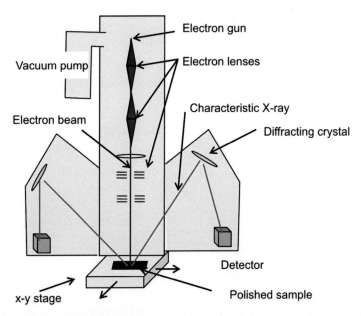

FIGURE 1.31   A schematic diagram of EPMA. The electron gun, lenses, and apertures are almost the same as those of SEM. The difference is the method how to collect and determine characteristic X-rays. The characteristic X-rays emitted from the sample are diffracted by the crystals and determined by X-ray detectors. Five sets of diffracting crystals are maximally placed, and ten elements can be determined in two scans. The merits of WDS are (i) the resolution of the characteristic X-ray is higher; therefore, (ii) the background is lower, and (iii) the X-ray diffraction is independent of the detector. Therefore, the diffracting crystal and the detector are chosen separately. Thus (iv) detection and diffraction correction are independent of the X-ray wavelength. If more than five elements are needed to be measured, the diffraction crystals and their positions are changed, and the characteristic X-rays are measured again. The stage is moved in x and y directions on the x—y stage.

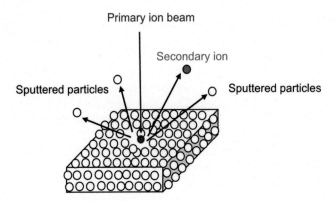

FIGURE 1.32   A schematic diagram of the sputtering phenomenon. The sample is bombarded by the primary ions. Most sputtered atoms are neutral, but ~0.01% of sputtered atoms are ionized (the secondary ions).

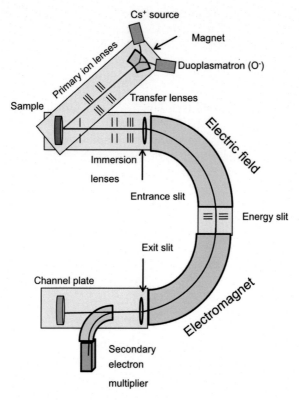

FIGURE 1.33 **A schematic diagram of SIMS, Cameca ims-5f.** From Cs^+ source or duoplasmatron, Cs^+ or O^- ions are made, accelerated, and focused on the sample. Then the secondary ions are generated and extracted by immersion lenses. The accelerated ions are passed through the electric field and magnetic field, and finally go into a channel plate to obtain a secondary ion image. Or the ion can go into the secondary electron multiplier for ion counting of the target elements.

ions are generated, and peak assignment is not easy. Thus, in addition to the high price of the machine, the application of SIMS is limited.

HR-SIMS: In the U−Pb dating of a single zircon (see Section 1.3.14) by SIMS, not only the determination of precise Pb isotope ratios but also precise Pb/U concentration ratios are required. The largest problem is molecular interferences of $^{90}Zr_2{}^{28}Si^+$, $^{174-176}Yb^{16}O_2{}^+$, $^{178-180}Hf^{28}Si^+$, etc. To resolve such as $^{90}Zr_2{}^{28}Si^+$, $^{174}Yb^{16}O_2{}^+$, and $^{178}Hf^{28}Si^+$ from $^{208}Pb^+$, $^{206}Pb^+$, and $^{206}Pb^+$, mass resolutions of 1092, 4502, and 3827 are required, respectively. By using conventional SIMS, operation at such high resolution (HR) is possible, but the count rate is not enough to date zircon precisely.

To enable accurate dating, an HR SIMS was designed and constructed by the Australian National University group. It was named

"Sensitive High Mass Resolution Ion Probe (SHRIMP)," but the size was huge (~ 3 m $\times \sim 7$ m). The Cameca Company also built similar size ion probes, ims-1280 (see Fig. 1.34). Now, the largest problem is not the mass spectrometer but that you need to prepare homogeneous standard zircons with precisely known amounts of ppm levels of U and Pb.

LA-ICP-MS: LA-ICP-MS (see Fig. 1.35) is a hybrid analytical system of ICP-MS (ICP-QMS or MC-ICP-MS) with laser ablation (LA) system. The key to LA-ICP-MS is the stable and small elemental fractionation in LA. Therefore, shorter wavelength laser, such as 193 nm ArF excimer laser and femtosecond laser are popularly used.

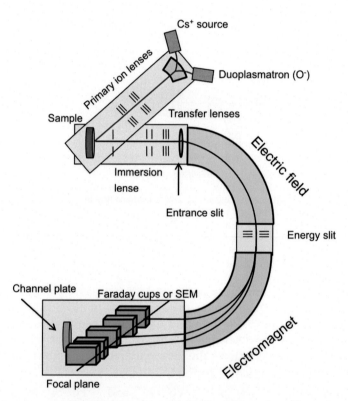

FIGURE 1.34   A schematic diagram of HR-SIMS, Cameca ims-1280. The primary ion gun is the same as the small SIMS, ims-5f (Fig. 1.33), but a larger electro analyzer and a magnet are designed to obtain higher transmission at high resolution. As the magnet became bigger (r = 1280 mm; 12 tons), multiple collectors (Faradays cups or SEMs) can be placed on the focal plane. As the SEMs can be movable, not only ^{204}Pb, ^{206}Pb, ^{207}Pb, and ^{208}Pb, but also other elements such as ^{16}O, ^{17}O and ^{18}O, ^{10}B and ^{11}B, ^{6}Li and ^{7}Li, etc. can be measured by the same machine.

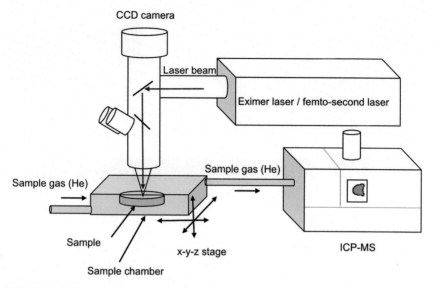

FIGURE 1.35   **A schematic diagram of LA-ICP-MS.** In a commercial ArF excimer laser system, the bottom of the laser pit becomes flat from ∼10 to ∼100 μm in diameter. As oxygen gas absorbs the UV light, and to cool the evaporated materials and pits effectively, the laser beam line is filled with He gas (shown in yellow-green in the figure). Not the melting but the sublimation of samples is required to reduce the elemental fractionation, the short-time ablation laser such as the femtosecond laser is preferred. The sample chamber can be controlled to x−y−z direction by the computer.

1.3.7 Classic classification of elements

Modern astrochemistry was started by V.M. Goldschmidt (1888−1947), who is called as Father of geochemistry. He categorized elements into lithophile, chalcophile, siderophile, and atmophile (see Fig. 1.36). The lithophile elements reside in silicate parts of Earth such as the mantle and the continental crust. The chalcophile elements are preferentially combined with sulfur, mainly found in sulfide ores or minerals. The siderophile elements prefer metallic iron and, thus, reside in the core. The atmophile elements are volatile elements or noble gases, which reside in the atmosphere. This classification is empirical; however, this classic classification of elements is still used in modern astrochemistry.

1.3.8 The rare earth element pattern

The rare earth element (REE; La, Ce, Pr, Nd, Sm, Eu, Gd, Tb, Dy, Ho, Er, Tm, Yb, and Lu) pattern (REE pattern) is the plot that the logarithms

1. Basic knowledge of astroscience

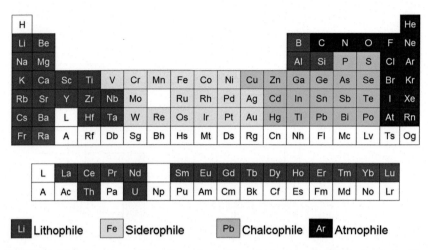

FIGURE 1.36 Four geochemical categorization of elements by V.M. Goldschmidt.
Goldschmidt separated elements into four groups: lithophile, siderophile, chalcophile, and
atmophile.

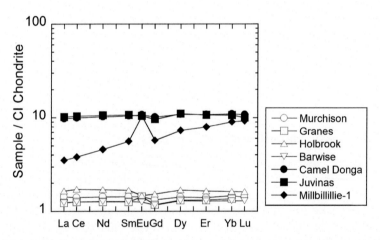

FIGURE 1.37 The rare earth element patterns (REE patterns) of meteorites. The horizontal axis is the REE elements. The vertical axis is the value of concentrations of samples
divided by those of the CI chondrite (average or the Leedey chondrite). Murchison,
Granes, Holbrook, and Barwise are chondrites of CM2, L6, L6, and H5, respectively.
Camel Donga, Juvinas and Millbillillie are eucrites. For the classification of meteorites. The
data were obtained by thermal ionization mass spectrometry (TIMS) (**Makishima and
Masuda, 1993**). *Source: Data are from Makishima, A., Masuda, A. 1993. Primordial Ce isotopic
composition of the solar system. Chem. Geol. 106, 197–205.*

of samples' REE concentrations normalized to the chondritic values (the
vertical axis) are plotted in the order of the atomic number (the horizontal axis) (see Fig. 1.37; it is also called as a Masuda–Coryell diagram).

The REE pattern utilizes the smooth decrease of ionic radii of each REE according to the increase of the atomic number.

Usually, REEs have generally ionic charge of +3. However, Ce (cerium) and Eu (europium) show anomalous values of +4 and +2, respectively. When Eu becomes +2 (in both extraterrestrial and terrestrial conditions), it can substitute for the site of Ca^{2+}. As the host of Ca in partially molten magma is plagioclase, when plagioclase crystallizes in magma, plagioclase shows a positive Eu anomaly in the REE pattern, and, in contrast, a negative Eu anomaly appears in the residual melt. For example, the positive Eu anomaly is observed in the REE pattern of Millbillillie-1 (eucrite) in Fig. 1.37.

Ce is usually +3, but in oxic conditions such as in surface weathering or shallow seawater, Ce can be +4, and Ce can show positive or negative anomalies. Most Ce^{4+} is easily hydrolyzed and removed from the surface seawater, which shows a negative Ce anomaly. In contrast, the hydrolyzed elements are finally deposited on manganese nodules, they show a large positive Ce anomaly. However, most cosmochemical processes occur in vacuum or non-oxic conditions, extraterrestrial samples show no Ce anomaly. If a Ce anomaly appeared in a meteorite, we can conclude that it should be caused by weathering on the terrestrial surface, because there is no place in the space to be oxic condition except the Earth's atmosphere with the free oxygen gas.

1.3.9 The trace element geochemistry

Nowadays, all REEs, as well as other lithophile trace elements such as Li, B, Rb, Sr, Y, Zr, Nb, Mo, Cs, Ba, Hf, Ta, Tl, Pb, Bi, Th, and U in ten samples, can be determined within 2 weeks, of which most time is sample digestion and drying-up (Lu et al., 2007; Makishima and Nakamura, 2006).

As other lithophile elements' data have become available with ease, the REE pattern was expanded and evolved into the trace element pattern (see Fig. 1.38). These orders of elements were determined partly empirically, partly by the partition coefficients (K_d; see Section 1.4.8) into major rock-forming minerals. This horizontal order is called as the incompatibility, and the left-side element has higher incompatibility than the right-side element. The vertical axis is the log plot of the trace element concentrations normalized by the primitive mantle values (PM-normalized trace element patterns; McDonough and Sun, 1995). The smoothness of the trace element pattern comes from the smoothness of K_d (partition coefficients) of the trace elements into the major rock-forming minerals.

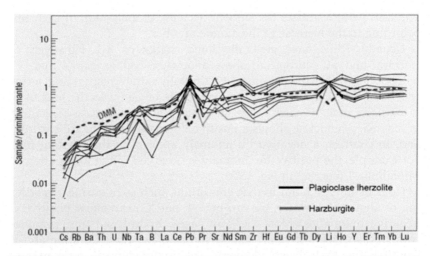

FIGURE 1.38 The primitive mantle-normalized trace element patterns (PM-normalized trace element patterns) of the Horoman peridotite samples. Horoman is a peridotite massif placed in the northern island, Hokkaido, in Japan. The samples are plagioclase lherzolites and harzburgites. DMM means the hypothetical depleted mantle. The plot is from Malaviarachchi et al. (2008). *Source: Copyright: Nature Publishing Group, authors permission from NPG, Malaviarachchi, S.P.K., Makishima, A., Tanimoto, M., Kuritani, T., Nakamura, E., 2008. Nat. Geosci. 1, 859–863.*

The trace element pattern has special features. For example, if plagioclase is crystallized and lost from the magma, Sr and Eu negative anomalies will appear in the pattern of the melt. Because Sr^{2+} and Eu^{2+} have the $+2$ valence, therefore, they substitute Ca^{2+} in plagioclase, resulting in higher distribution coefficients than other neighboring REE^{3+}, resulting in negative Sr and Eu anomaly in the trace element pattern. If zircon crystallizes, Zr and Hf show negative anomalies in the pattern. Alkaline elements, B, Pb, Sr, and Li show special behaviors if the fluid or metasomatism-related processes affect the magma source. If the magma source reacted with the fluids enriched in such elements, the trace element pattern shows a positive anomaly in B, Pb, Sr, and Li. If the fluids enriched in such elements are released from the magma source, such elements should show a negative anomaly. Thus, determination of the bulk trace element pattern is becoming a basic database of the silicate samples.

1.3.10 Age dating by radioactive isotopes

Using radioactive isotopes, cosmochemists can determine age, which can bind astrochemistry with astrophysics. The time-criteria can constrain the theoretical models. Therefore, the age dating is very important in astrophysics or astronomy.

The radioactive decay is expressed as the following equation, where N is the number of a radioactive isotope:

$$\frac{dN}{dt} = -\lambda N \tag{1.19}$$

λ is called as the decay constant. The half-life, $T_{1/2}$ has the relation between λ as:

$$T_{1/2} = \ln 2/\lambda = 0.693/\lambda \tag{1.20}$$

When the Eq. (1.19) is integrated,

$$N_t = N_0 e^{-\lambda t} \tag{1.21}$$

N_0 is the initial number of the radioactive isotope, and t is the elapsed time from the start (the element synthesis). The number of the daughter isotope is:

$$D_t = D_0 + (N_0 - N_t) = D_0 + N_0(1 - e^{-\lambda t}) \tag{1.22}$$

Generally, in astrochemistry, $t = 0$ is present, and t is the age of the sample. Thus, the above equation changes as follows:

$$D_P = D_T + N_P(e^{\lambda T} - 1) \tag{1.23}$$

where "P" indicates the present, T is the age from the present, and D_T is the initial amounts of the daughter isotope at the time T. For example, we consider the decay of ^{147}Sm to ^{143}Nd. As neodymium has a stable isotope, ^{144}Nd, of which number does not change over time. Then Eq. (1.23) is expressed as:

$$\left(^{143}Nd/^{144}Nd\right)_P = \left(^{143}Nd/^{144}Nd\right)_T + \left(^{147}Sm/^{144}Nd\right)_P \left(e^{\lambda T} - 1\right) \tag{1.24}$$

This equation is the basic equation of age dating.

In actual applications, the isochron plot is used (see Fig. 1.39). The $(^{143}Nd/^{144}Nd)_P$ and $(^{147}Sm/^{144}Nd)_P$ values are plotted in x- and y-axes, respectively, for several mineral phases. If all the phases became isotopic equilibria at age T, and did not suffer any disturbances from T to present, the data should form a line which is called as an isochron as shown in Fig. 1.39. If we calculate the y-intercept, it is an initial value of $(^{143}Nd/^{144}Nd)_T$. The slope corresponds to the age. As the age becomes older and older, the slope becomes steeper and steeper. The age calculation can be done by the famous ISOPLOT program by Ludwig (1999), which can be obtained freely from their website.

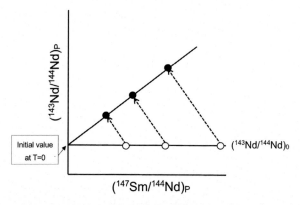

FIGURE 1.39   The isochron plot for the ^{147}Sm–^{143}Nd system. The present data of three minerals are plotted as a solid circle. At age T, each point was in melt (magma) with isotopic equilibrium as the open circle. As time goes on, one ^{147}Sm atom becomes one ^{143}Nd atom one by one. Thus, the slope of the dotted arrow becomes −1. Note that each mineral must have the same ^{143}Nd/^{144}Nd ratio at T = 0, and that each mineral must be closed for Sm and Nd.

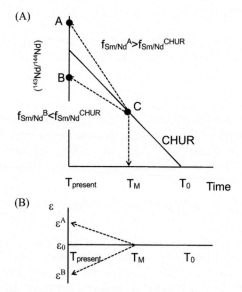

FIGURE 1.40   Model age in ^{147}Sm–^{143}Nd system. (A) Evolution curve of (^{143}Nd/^{144}Nd). (B) Evolution curve in ε-notation.

1.3.11   Model age

We explain the model age using the ^{147}Sm–^{143}Nd isotope system and Fig. 1.40. In this model, only one event occurred at T_M. Before T_M, only one uniform reservoir (UR) existed. Here, we assume that the first UR was CHondritic Uniform Reservoir or CHUR reservoir, which had the

Sm/Nd ratio of the averaged carbonaceous chondrites ($f_{Sm/Nd}^{CHUR}$). In Fig. 1.40A, the horizontal axis is the time, which went from right to left. T_0 is the initial point when the element synthesis was finished.

At T_M, Point C is the time when one reservoir is separated into two reservoirs, A and B by partial melting. The melts (B) evolved into the enriched crust, and the residual phase (A) formed the depleted mantle. After the partial melting event, A and B had $^{147}Sm/^{144}Nd$ ratios of $f_{Sm/Nd}^{A}$ and $f_{Sm/Nd}^{B}$, respectively ($f_{Sm/Nd}^{A} \geq f_{Sm/Nd}^{B}$). $T_{present}$ is the present time, and T_M is called as the model age. Reservoir A has a higher $f_{Sm/Nd}^{A}$ value than that of CHUR, resulting in a higher $(^{143}Nd/^{144}Nd)_T$ value than that of CHUR during $T_{present} < T < T_M$. In contrast, reservoir B has a lower $f_{Sm/Nd}^{B}$ value than that of CHUR (it is also called as an "enriched reservoir" which corresponds to a liquid phase such as the crust), resulting in a lower $(^{143}Nd/^{144}Nd)_T$ value than that of CHUR during $T_{present} < T < T_M$.

The $(^{143}Nd/^{144}Nd)_T^{A}$ and $(^{143}Nd/^{144}Nd)_T^{B}$ values are also defined by the ε value, where

$$\varepsilon_T^i = \left[(^{143}Nd/^{144}Nd)_T^i / (^{143}Nd/^{144}Nd)_T^{CHUR} - 1 \right] \times 10^4 \qquad (1.25)$$

i = A or B, and $T_{present} < T < T_M$. As shown in Fig. 1.40B, when ε-notation is used, the evolution curve of CHUR becomes a horizontal linear line of $\varepsilon_T^{CHUR} = 0$, and the evolution curves of A and B split from T_M into ε^A and ε^B at present by the linear lines. Of course, this is a too simplified two-stage model, however, it is very useful to treat the radiogenic isotope data.

This ε-notation appears when the natural mass fractionation in the stable isotope ratios is described. It should be noted that the meaning of ε-values in the stable isotope astrochemistry completely different from that of the radiogenic isotope astrochemistry.

The model age is defined as T_M, where the present isotope ratio with the present parent–daughter ratio crosses with the CHUR evolution curve. The model age is very convenient, and can be applied to most isotopic systems, such as La-Ce, Sm-Nd, and Lu-Hf.

For the Rb–Sr isotope system, not CHUR but the UR is used, because Rb is an alkaline element and very volatile, therefore, the primitive Earth could have a different Rb/Sr ratio from that of CI chondrites by evaporation loss of Rb. In the case of the Re-Os systematics, CHUR is also not defined because both elements are partitioned into the core, therefore, CHUR cannot be defined assuming CI chondrites.

1.3.12 The extinct nuclides and the fossil isochron

There are some nuclides, which have relatively short half-lives of 0.1–100 My (see Table 1.11). The half-lives of these elements are so short that they are already extinct. Therefore, these elements are called extinct nuclides.

TABLE 1.11 Important extinct nuclides.

Radioactive parent	Decay scheme	Daughter isotope	Horizontal isotope ratio[a]	Half-life (Myr)
^{26}Al	β^-	^{26}Mg	$^{27}Al/^{24}Mg$	0.7
^{60}Fe	$2\beta^-$	^{60}Ni	$^{56}Fe/^{58}Ni$	1.5
^{53}Mn	β^-	^{53}Cr	$^{55}Mn/^{52}Cr$	3.7
^{107}Pd	β^-	^{107}Ag	$^{108}Pd/^{109}Ag$	6.5
^{182}Hf	$2\beta^-$	^{182}W	$^{180}Hf/^{184}W$	9.0
^{129}I	β^-	^{129}Xe	$^{127}I/^{132}Xe$	16
^{146}Sm	α	^{142}Nd	$^{149}Sm/^{144}Nd$	103

[a]The "horizontal isotope ratio" means the horizontal axis in the fossil isochron.

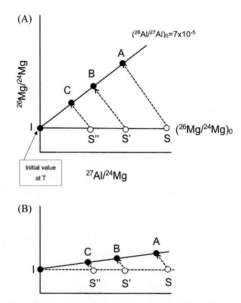

FIGURE 1.41 **The fossil isochron plot for the $^{26}Al-^{26}Mg$ system.** (A) Elemental fractionation occurred at T, and three phases (minerals), S, S', and S" with different $^{27}Al/^{24}Mg$ were formed at T. These phases are found now as A, B and C, and as S, S' and S" at T, respectively. From the slope, T, the initial value of $^{26}Mg/^{24}Mg$, and $(^{26}Al/^{27}Al)_0$ can be obtained. (B) When such elemental fractionation occurred at $3 \times T$, the axis of (B) becomes eight times smaller than that of (A).

However, their inhomogeneous existences remain as isotopic anomalies of daughter nuclides in the materials formed in the early solar system.

The existence of these elements in the early solar system is proved in the fossil isochron, as shown in Fig. 1.41A. Here, we use ^{26}Al for example. The differences from the isochron in Fig. 1.41 are as follows: As ^{26}Al is already extinct, we cannot use ^{26}Al in the horizontal axis in the isochron. The daughter isotope is ^{26}Mg, and Mg in this sample has non-radiogenic isotopes of ^{24}Mg and ^{25}Mg. Thus, the horizontal axis can be converted to Al/Mg or ^{27}Al/^{24}Mg of the original condition. (Al is monoisotopic and ^{27}Al/^{25}Mg can be also used.)

Elemental fractionation occurred at T after the formation of the ^{26}Al synthesis within five times of $T_{1/2}$ ($5 \times T_{1/2} = 3.5$ Myr; $T_{1/2}$ is the half-life of ^{26}Al) since the formation of the element. Three phases (minerals), S, S', and S'' with different ^{27}Al/^{24}Mg were formed at T, and these phases are found now as A, B, and C. When A, B, and C form a line with positive slope and ^{26}Mg/^{24}Mg of S, S' and S'' at T, (see Fig. 1.41A), the line is called as a fossil isochron. From the fossil isochron, $(^{26}$Al/^{27}Al$)_0$, T and the initial value of ^{26}Mg/^{24}Mg can be determined.

In Fig. 1.41B, the case that the elemental fractionation between Al and Mg occurred after $3 \times T$ from the initial elemental synthesis is shown. The vertical variation becomes eight times smaller, because the amount of ^{26}Al is $(^{1}/_{2})^{3} = 1/8$. When T is five times larger than $T_{1/2}$, the fossil isochron becomes flat, and the date cannot be determined.

1.3.13 The ^{182}Hf$-^{182}$W isotope system

The ^{182}Hf decays to ^{182}W through the double β^{-}-decay with a short half-life of 8.9 My. The ^{182}Hf$-^{182}$W isotope system has a special character compared to other extinct nuclides. Hafnium is a lithophile element and never becomes a siderophile element (see Section 1.3.6), therefore, Hf prefers silicates (mantle) in the mantle-core segregation. Tungsten is not only a moderately siderophile element, but also a moderate lithophile element. Therefore, elemental fractionation between Hf and W is expected to occur, when silicate (Hf-rich) and metal phases (W-rich) are separated, especially at the core formation.

After all ^{182}Hf decayed, W isotope ratios are frozen, and Hf–W chemical fractionation should be dated if the fractionation occurred early enough in the solar system history. As five times of a half-life is a limit for detection, so that 45 My from the formation of the solar nebula is a limit for application of the ^{182}Hf$-^{182}$W isotope system. Harper and Jacobsen (1994) and Halliday and Lee (1999) successfully determined the isotopic anomaly of ^{182}W using Multicollector Inductively Coupled Plasma Mass Spectrometry (MC-ICP-MS; see Fig. 1.29).

The schematic fossil isochron for ^{182}Hf$-^{182}$W is shown in Fig. 1.42. For the denominator isotope, a stable isotope, ^{184}W is generally used.

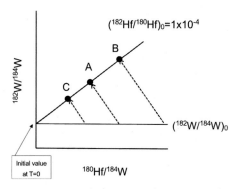

FIGURE 1.42 **The fossil isochron plot for the ^{182}Hf–^{182}W system.** The horizontal axis is ^{180}Hf$/^{184}$W, and the vertical axis is ^{182}W$/^{184}$W.

Most terrestrial samples are plotted at A in Fig. 1.42. However, early differentiated silicate phases are Hf-rich, and plotted at B. In contrast, the metal phase (core) is plotted at C. The W isotope ratios are presented in an ε-unit defined as:

$$\varepsilon^{182}W = \left[\left(^{182}W/^{184}W \right)_{sample} / \left(^{182}W/^{184}W \right)_{standard} - 1 \right] \times 10^4 \qquad (1.26)$$

and this ε-notation is used for the presentation of the degree of fractionation between Hf and W.

The W isotope ratios are presented in an ε-unit as Eq. (1.26), but also in μ-notation defined as:

$$\mu^{182}W = \left[\left(^{182}W/^{184}W \right)_{sample} / \left(^{182}W/^{184}W \right)_{standard} - 1 \right] \times 10^6 \qquad (1.27)$$

As W isotope ratio determination was difficult by the conventional TIMS (see Fig. 1.28) because W and Hf are too refractory to be ionized by thermal ionization on the filament, the development of MC-ICP-MS (see Fig. 1.29) greatly contributed (Harper and Jacobsen, 1994; Halliday and Lee, 1999) because the ICP plasma easily ionized both W and Hf.

To win in the isotope astrochemistry, it is necessary to analyze isotope ratios more precisely than before. This is achieved by getting newer analytical machines and acid-resistant clean rooms, etc. Thus, the isotope astrochemistry always requires a lot of money (a few million dollars). In addition, the cost of maintenance and consumables are not cheap. However, compared to astrochemistry, big science such as launching a spacecraft in astrophysics or building an observatory for astronomy, requires three orders of magnitude larger money.

1.3.14 A single zircon dating by the Pb−Pb method

Zircon is $ZrSiO_4$ and a common mineral in acidic rocks. This mineral is very refractory and stands against metamorphism, erosion, and weathering. As shown in Fig. 1.43A, the zircon crystal is not so homogeneous and contains inclusions (mineral, melt, and fluid). Old zircons are often made of inner core and outer rim. The inner core inherits the older zircon, which means that there was an older source of the zircon such as granites (e.g., Fig. 1.43B). Corroded rims and metamict (amorphous) parts are formed by the radiation of radioactive U and Th. The outer rim forms the euhedral zircon with eroded rims. Our targets for age determination are pure inner core and outer

(A)

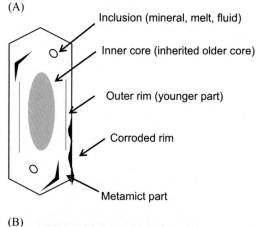

Inclusion (mineral, melt, fluid)

Inner core (inherited older core)

Outer rim (younger part)

Corroded rim

Metamict part

(B)

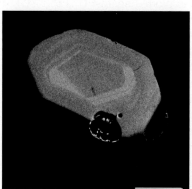

FIGURE 1.43 **Zircon ($ZrSiO_4$) crystal.** (A) A schematic diagram of a zircon crystal. (B) A backscatter electron image of a polished zircon crystal. This image was taken by SEM after the zircon was dated by LA-ICP-MS. The laser pit can be seen at the bottom right. Several growth histories are recorded in this grain. The scale bar at the bottom is 50 μm. *Source: By courtesy Dr. Y. Lu (China University of Geoscience, Beijing) for the photograph (B).*

rim as well. By establishment of HR-SIMS (Fig. 1.34), accurate ages of both inner and outer parts from a single zircon can be obtained.

The zircon crystal contains a high concentration of U but very low Pb in its structure, therefore, Pb in zircon is considered to be only radiogenic Pb. In this dating, two decays of uranium isotopes:

$$^{238}U \rightarrow {}^{206}Pb^*$$

$$\left(^{206}Pb^*/^{204}Pb\right) = \left(^{238}U/^{204}Pb\right) \times \left(e^{T\lambda_{238}} - 1\right) \tag{1.28}$$

$$^{235}U \rightarrow {}^{207}Pb^*$$

$$\left(^{207}Pb^*/^{204}Pb\right) = \left(^{235}U/^{204}Pb\right) \times \left(e^{T\lambda_{235}} - 1\right) \tag{1.29}$$

are used. The asterisks mean the radiogenic lead; T is the crystalized age of the zircon; λ_{238} and λ_{235} are the decay constants of ^{238}U and ^{235}U, respectively. Note that no common leads are assumed to exist in this calculation. The "common lead" means the non-radiogenic Pb incorporated at the zircon formation. From Eqs. (1.28) and (1.29),

$$\left(^{206}Pb^*/^{207}Pb^*\right) = \left(^{238}U/^{235}U\right) \times \left(e^{T\lambda_{238}} - 1\right)/\left(e^{T\lambda_{235}} - 1\right) \tag{1.30}$$

while $\left(^{238}U/^{235}U\right)$ is constant and 137.88. Thus, if $\left(^{206}Pb^*/^{207}Pb^*\right)$ is measured, T is obtained. This is called the Pb—Pb dating method. To determine T precisely, not only accurate determination of the isotope ratio of $\left(^{206}Pb^*/^{207}Pb^*\right)$ but also selection of zircons without common lead are prerequisites. In addition, to choose less disturbed zircons after formation is also important. "Disturb" means Pb or U are lost or added from outside of the zircon crystal by geological processes such as metamorphism or weathering.

When $\left(^{206}Pb^*/^{238}U\right)$ is plotted against $\left(^{207}Pb^*/^{235}U\right)$, one curve that starts from the origin is obtained. This is called a concordia diagram, which is shown in Fig. 1.44. If the data is on this curve, the U—Pb of zircon has not been disturbed from the formation of zircon to the present. If the data is not on this line, the U—Pb of zircon has been disturbed. When the disturbance was only once and the loss of Pb, the data formes a line. One end shows the disturbing age (1 Ga in Fig. 1.44), and the other end shows the formation age (3.5 Ga in Fig. 1.44). This line is called a discordia line. This dating method is called as concordia method.

The age of CAIs (Section 1.4.13.1) or chondrules (see Section 1.4.13.2) are determined by the Pb—Pb age dating method by assuming the common lead isotopic compositions.

1.3.15 Mass discrimination correction laws

In this section, mass fraction laws or mass fractionation correction laws are explained. There is a similar word "mass discrimination." In this book,

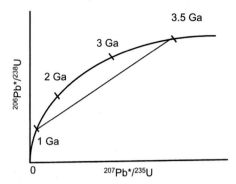

FIGURE 1.44 **The concordia diagram.** The curve indicates the concordia diagram, and the line shows the discordia line. One intercept of the discordia line to the concordia at 3.5 Ga shows the formation age of zircon. Another intercept of the discordia line at 1 Ga indicates the resetting age (the metamorphic age) of zircon.

mass fractionation indicates "the mass-dependent fractionation that occurs in nature," and mass discrimination is "the mass-dependent phenomena that occur in the mass spectrometer or in artificial conditions such as evaporation or ion exchange column chemistry in sample preparation." In mass spectrometry, the mass discrimination inevitably occurs, therefore, its correction is also necessary. These mass discrimination correction laws can be applied to elements with more than three isotopes.

Therefore, we explain them using the three isotopes of Sr as an example. We assume isotopes of ^{86}Sr, ^{87}Sr, and ^{88}Sr are denominator, target, and reference isotopes with the mass of m_{86}, m_{87}, and m_{88}; R_m is a measured isotope ratio of $^{87}Sr/^{86}Sr$; R_c is a mass discrimination corrected ratio of $^{87}Sr/^{86}Sr$; R_r is a reference isotope ratio or a normalizing isotope ratio of $^{88}Sr/^{86}Sr$ (a constant); R_{rm} is a measured reference isotope ratio of $^{88}Sr/^{86}Sr$; and α is called as a mass fractionation correction factor. Then, two isotope ratios of R_c and R_{rm} follow the mass fractionation laws as follows:

- The linear law

$$\alpha = (R_r/R_{rm})/(m_{88} - m_{86}) \tag{1.31}$$

$$R_c = R_m \times [1 + \alpha \times (m_{87} - m_{86})] \tag{1.32}$$

- The power law

$$\alpha = (R_r/R_{rm})^{[1/(m88-m86)]} - 1 \tag{1.33}$$

$$R_c = R_m \times (1+\alpha)^{(m87-m86)} = R_m \times (R_r/R_{rm})^{[(m87-m86)/(m88-m86)]} \tag{1.34}$$

- The exponential law

$$\alpha = \left[\log\left(R_r/R_{rm}\right)/\log\left(m_{88}/m_{86}\right)\right]/m_{86} \qquad (1.35)$$

$$R_c = R_m \times \left(m_{87}/m_{86}\right)^{(\alpha \times m86)} \qquad (1.36)$$

To choose which mass discrimination correction laws are the most appropriate, the raw data of $(R_m, R_c) = ((^{87}Sr/^{86}Sr)_m, (^{87}Sr/^{86}Sr)_c)$ should be x-y-plotted using the above three mass discrimination laws, and examine which law gives the smallest errors, or which law shows the flattest pattern.

There is another law, the Rayleigh fractionation law. When there are two phases and one phase is continuously removed from the system through fractional distillation, it is called the Rayleigh fractionation. If the fractionation factor is constant during the process, the evolution of the isotopic composition in the residual material is described as:

$$\left(\frac{R}{R^0}\right) = \left(\frac{X}{X^0}\right)^{\alpha-1} \qquad (1.37)$$

where R is an isotope ratio (e.g., $^{18}O/^{16}O$); R^0 is an initial ratio; X is the concentration or amount of the more abundant (lighter) isotope (e.g., ^{16}O); and X^0 is the initial concentration. Because $X \gg X_h$ (heavier isotope concentration), X is approximately equal to the amount of original material in the phase, then, $f = X/X^0$ = fraction of material remaining, then,

$$R = R^0 f^{\alpha-1} \qquad (1.38)$$

1.3.16 Isotope dilution method

1.3.16.1 Basics of isotope dilution

The isotope dilution (ID) method is necessary for accurate elemental determination by mass spectrometers which can determine isotope ratios precisely, such as TIMS, ICP-QMS, or MC-ICP-MS. The drawback of the ID method is that it cannot be applied to monoisotopic elements, such as Na, P, Co, Au, etc.

Here, ID is explained using Fig. 1.45. A target element needs to have more than two isotopes, shown as 1 and 2 in the figure. A "spike," which is artificially enriched in one isotope, is required. Isotope-2 is assumed as the spike. Usually, the spike is used in a solution form. p and P are the weights (g) of the target elements in the sample and spike; A, B, and R are isotope-2/isotope-1 ratios of the sample, spike and sample–spike mixture; D_{sample} and D_{spike} are the isotopic abundances of the

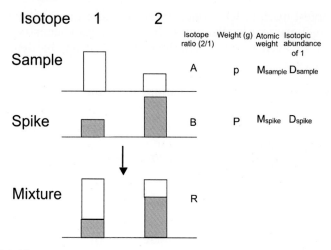

FIGURE 1.45   Isotope dilution method. There are two isotopes 1 and 2. The sample has the isotopic ratio of A, and the enriched spike (artificially made) has the isotopic ratio of B. When p [g] of the sample and P [g] of the spike are perfectly homogenized, the isotope ratio of mixture becomes R.

isotope-1 in the sample and spike; M_{sample} and M_{spike} are the atomic weights of the target element in the sample and spike, respectively.

Then, the target element concentration (C_T^{sample}) is expressed as:

$$C_T^{sample} = [(B - R)/(R - A)] \times Q' \times m_{spike}/m_{sample} \qquad (1.39)$$

$$Q' = (D_{spike}/D_{sample}) \times (M_{sample}/M_{spike}) \times C_{spike} \qquad (1.40)$$

where C_T^{sample} and C_{spike} are the concentrations of the target element in the sample and spike; m_{sample} and m_{spike} are the sample and spike weights; and Q' Eq. (1.38) is called the pseudo-concentration of the spike solution, because Q' has the dimension of concentration. When the target element does not have any radioactive or radiogenic element, neither atomic weights nor atomic abundances change. The pseudo-concentration of the spike solution should be calibrated beforehand using the precisely determined standard solution.

After A, B, and Q' are determined beforehand, and m_{sample} and m_{spike} are measured for each sample, the sample and spike are mixed well by the sample digestion and evaporation processes. After the separation of the target element by solvent extraction or by column chromatography, the isotope ratio of mixture, R is determined using mass spectrometry. Then the concentration of the target element, C_T^{sample} is determined by Eq. (1.39), only by measurement of R. This equation stands so far as the isotope equilibrium for the target element between the sample and the spike is achieved.

The largest merit of ID is, once isotope equilibrium is achieved, losses of the target element by the ion exchange column chemistry or solvent extraction, or even by poor handling, do not affect the determination result or accuracy.

1.3.16.2 Error propagation in isotope dilution

Here, we investigate how the error in the determination of R is propagated to the concentration result (the error propagation or error magnification) in ID. To make the equation simpler, $Q'' = Q' \times m_{sample}/m_{spike}$ is used, and the derivative of Eq. (1.39) for R is:

$$d\, C_T^{sample}/dR = Q'' \times \left[(A - B)/(A-R)^2\right] \qquad (1.41)$$

From Eq. (1.41):

$$d\, C_T^{sample}/C_T^{sample} = R \times (B - A)/[(A - R) \times (B - R)] \times dR/R \qquad (1.42)$$

$$= F(R) \times dR/R \qquad (1.43)$$

where

$$F(R) = R \times (B - A)/[(A - R) \times (B - R)] \qquad (1.44)$$

This F(R) is a function that indicates the deviation of R (dR/R) is magnified to the deviation of the concentration (d $C_T^{sample}/C_T^{sample}$).
If we calculate the derivative of F(R),

$$dF(R)/dR = (B - A) \times (R^2 - A \times B)/\left[(R - B) \times (A-R)^2\right] \qquad (1.45)$$

This function becomes minimal when

$$dF(R)/dR = 0 \ \text{ at } \ R = (A \times B)^{0.5} \qquad (1.46)$$

For example, if we plot F(R) against R when A = 0.9142, B = 74.75 (a case of Sm; $^{149}Sm/^{147}Sm$), the result is shown in Fig. 1.46. As shown in the figure, this function, F(R) takes a minimum value of 1.25 at R = 8.32. F(R) means, for example, when the measurement error is 0.1%, the minimum error using ID is 0.125% (1.25 × 0.1%). If R is smaller, or larger than 8.32, the error in ID is magnified by F(R). When you want 0.2% error after ID, R should be between ∼2 and ∼40. If the R-value is out of this range (under-spiked or over-spiked), the error in ID is >0.2%, therefore, the data should be discarded.

1.4 Introduction to mineralogy and petrology

To understand astrochemical inorganic materials, we must know silicate minerals, rocks, and metals, from which they are made. Thus, in

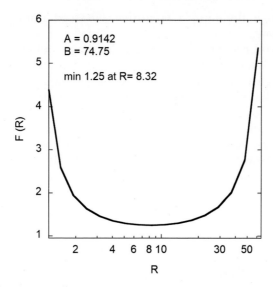

FIGURE 1.46 **Error magnification.** The horizontal axis is the mixture isotope ratio of the sample and the spike. The vertical axis is the error magnification of F(R). This is a curve of A = 0.9142 and B = 74.75. When R = 8.32, the curve takes the minimum value of 1.25.

this section, we learn about minerals that form rocks, and petrology by which the rocks are made. Mineralogy and petrology have been established as the science of natural history, therefore, each mineral or rock has different names irrespective of chemical formulas. The author thinks this is one of the reasons for making especially mineralogy not popular and more tedious. As V.M. Goldschmidt noticed, the crystal (mineral) is controlled by the ionic charge and ionic radius of cations. Thus, we learn basic mineralogy based on chemistry.

1.4.1 Thin section

In mineralogy, petrology, and geology, you first of all need to prepare thin sections of samples such as rocks, minerals, soils, etc. In big institutes, there could be technicians who make thin sections, or you can borrow thin sections of precious samples such as meteorites. The thin section is necessary to observe and analyze samples using a polarizing microscope, scanning electron microscope (SEM), EPMA, or SIMS.

The sample chip is cut, polished, and pasted on a glass slide. Then the sample is cut and polished into ~30 μm thick. The observation of rock samples by the polarizing microscope is done in two steps. One is observation by open Nicol, which is done by a single polarizing light. The other is observation by cross Nicol, in which the polarized light passed through the sample is further observed through the second polarizing plate.

Each mineral has its own characteristic shape and optical properties; thus most rock-forming minerals are identified easily. For example, plagioclase is a clear mineral with twin crystal. Clinopyroxene and orthopyroxene look pale green in the open Nicol. Olivine is colorful when observed in cross Nicol.

When you use SEM-EDX or EPMA, you need to carbon-coat the samples. When you use SIMS, gold-coating of the sample surface is required. In some cases, the lent samples are already coated but the coatings are damaged. When you borrow the thin sections, you must get permission to coat or remove the old coating and re-coat samples. You need to judge whether you need coating again or not case-by-case.

1.4.2 Solid solution and phase diagram

First, we learn solid solution. The solid solution is a solution in which more than two components are compatible and form a single phase. For example, gold (Au) and silver (Ag) dissolve each other at any ratio. The composition of this alloy is written as $Au_x + Ag_{(1-x)}$ $(0 \leq x \leq 1)$ in a chemical formula. This is a solid solution, and the compositions when x is 0 and 1 are called as end members.

There is a case that two end members do not form a perfect solid solution, and an example is shown in Fig. 1.47. Fig. 1.47 is a phase diagram of materials A and B, which do not form perfect solid solutions. We assume the case when the composition is P, which is near A, cools from high temperature. At the beginning, the composition P is a uniform melt (L). As the temperature becomes lower, the phase becomes melt and solid phase a (partial melting), and finally all becomes α, with the composition of P when all solidified. In the case of the composition Q, the melt becomes $L + \alpha$, and finally the solid becomes α, with small amounts of β. In the case of R, the melt becomes directly to $\alpha + \beta$ with a 1:1 mixture, or thin α and β phases are parallelly existing structures (lamella structure).

1.4.3 Rock-forming minerals

Next, we learn rock-forming minerals. In this section, we learn such minerals based on the number of total oxygens in chemical formulas. Minerals can be characterized into groups by the number of oxygen atoms in the formula, of which four, six, eight, and twelve are the main rock-forming minerals. Thus we learn the mineral groups according to

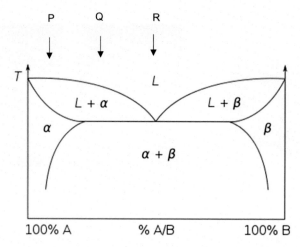

FIGURE 1.47   The phase diagram of a not perfect solid solution made of two materials, A and B. The horizontal axis is the mixing ratio of A and B. The vertical axis is the temperature. L is a liquid phase, and α and β are solid solutions that are mainly composed of A and B, respectively. The materials of A and B form the solid solutions α and β. P, Q, and R indicate compositions of mixtures of A and B. At the composition R where A: B = 1:1, the mixture directly solidified to $\alpha + \beta$. *Source: From Michbich, https://upload.wikimedia.org/wikipedia/commons/thumb/1/1d/Eutektikum_new.svg/369px-Eutektikum_new.svg.png.*

this oxygen number. In astrochemistry and deep-Earth geochemistry, the iron is assumed to be Fe(II), and Fe(III) rarely appears only in very oxidative conditions.

1.4.3.1 Four oxygens: olivine group; $Z^{4+}SiO_4$

A four-oxygen mineral is olivine. The name of olivine came from olive color (see Fig. 1.48A). One end member is Mg_2SiO_4 (forsterite; Fo) and the other end member is Fe_2SiO_4 (fayalite; Fa); and Fa and Fo form a perfect solid solution [$(Mg_xFe_{(1-x)}O)_2SiO_2$] ($0 \leq x \leq 1$).

Olivine changes into wadsleyite at 410 km depth and ringwoodite at 520 km depth, and decomposes into magnesium perovskite [(Mg, Fe)SiO_3; Mg−Pv] and ferropericlase [(Mg, Fe)O] at 660 km depth.

1.4.3.2 Six oxygens: pyroxene group; $Z^{4+}Si_2O_6$

A six-oxygen mineral is pyroxene. Pyroxenes are expressed as XY(Si, Al)$_2O_6$ [X = Ca, Na, Fe^{2+}, Mg, Li; Y = Cr, Al, Fe^{3+}, Mg, Mn, Sc, Ti, V, Fe^{2+}], however, important pyroxenes are mainly without Al, and written as $X_2Si_2O_6$ (X = Ca, Mg, Fe^{2+}), which is shown in Figs. 1.48B and C and 1.49.

Fig. 1.49 is called the ternary diagram, and each corner shows 100 mol% of each mineral. End members of the Ca, Fe, and Mg pyroxene ternary diagram (Fig. 1.49) are wollastonite (Wo; $Ca_2Si_2O_6$), enstatite (En; $Mg_2Si_2O_6$),

(A) (B) (C)

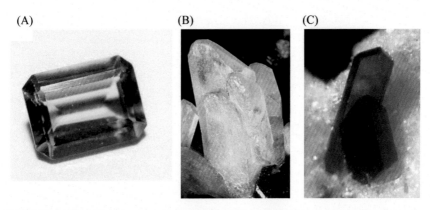

FIGURE 1.48 Photographs of (A) olivine, (B) diopside (pyroxene), and (C) augite
(pyroxene). (A) Olivine. (B) Diopside [$Ca_xMg_yFe_{(2-x-y)}Si_2O_6$; $0.9 < x < 1$, $0.5 < y < 1$]. (C)
Augite [$Ca_xMg_yFe_{(1-x-y)}Si_2O_6$; $0.3 < x < 0.9$] *Credit: (A) Michelle Jo. https://upload.wikimedia.
org/wikipedia/commons/2/2c/Gemperidot.JPG. (B) https://upload.wikimedia.org/wikipedia/. (C)
Didier Descouens. https://upload.wikimedia.org/wikipedia/commons/thumb/9/95/Diopside_Aoste.
jpg/742px-Diopside_AF.*

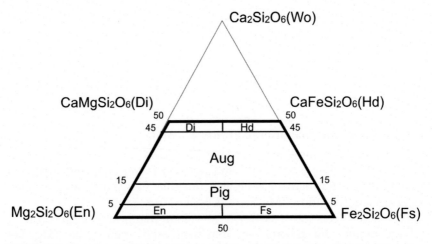

FIGURE 1.49 The Ca-Fe-Mg pyroxene (Ca,Mg,Fe) $_2Si_2O_6$ ternary diagram. The number shows mol %. Three end members are called wollastonite ($Ca_2Si_2O_6$), enstatite
($Mg_2Si_2O_6$), and ferrosilite ($Fe_2Si_2O_6$). *Aug*, Augite; *Pig*, pigeonite; *Di*, diopside; *Hd*,
hedenbergite.

and ferrosilite (Fs; $Fe_2Si_2O_6$). The pyroxene containing $<5\%$ of Ca is called
orthopyroxene (opx), and that of $>5\%$ Ca is clinopyroxene (cpx). The pyroxenes in trapezoids containing $5 < Ca < 15\%$ and $15 < Ca < 45\%$ are called
pigeonite (Pig) and augite (Aug), respectively. The pyroxenes containing
50% Ca of $MgCaSi_2O_6$ and $FeCaSi_2O_6$ are called as diopside (Di) and

hedenbergite (Hd). The trapezoid areas of $45 < Ca < 50\%$ are also named as Di and Hd, respectively, separated at the $Mg:Fe = 1:1$.

NaAl can substitute X_2 in $X_2Si_2O_6$ ($X = Ca$, Mg, Fe^{2+}). When all X_2 are substituted by NaAl, it is called as jadeite ($NaAlSi_2O_6$), and partly substituted one forms the solid solution composed of the diopside (Di)-hedenbergite (Hd)-jadeite triangle, named omphacite at high pressure.

1.4.3.3 Eight oxygens: feldspar group; $Z^{8+}Si_2O_8$

An eight-oxygen mineral is feldspar. The feldspars are expressed as $(Na,K,Ca)Al(Si,Al)Si_2O_8$. There are two important components: $NaAlSi_3O_8$ (albite; Ab; see Fig. 1.50A) and $CaAl_2Si_3O_8$ (anorthite; An; see Fig. 1.50B), which form a perfect solid solution named plagioclase. When K is contained, $KAlSi_3O_8$ (orthoclase; Or; Fig. 1.50C) is one of the end members, and the Or-plagioclase forms a solid solution. In the Martian meteorite, the glassy plagioclase ($NaAlSi_3O_8$) named maskelynite exists.

FIGURE 1.50 Photographs of feldspars and garnets. (A) Albite ($NaAlSi_3O_8$), Crete Island, Greek. A scale bar is 1 in. in length. (B) Anorthite ($CaAl_2Si_3O_8$), Miyake-jima, Japan. $2.4 \times 2.7 \times 1.7$ cm. (C) Orthoclase ($KAlSi_3O_8$), Minas Gerais, Brazil. 27×17 cm. (D) Pyrope ($Mg_3Al_2Si_3O_{12}$), Xinjian-Uygur, China. $9.8 \times 9.0 \times 8.3$ cm. (E) Almandine ($Fe_3Al_2Si_3O_{12}$) on shist, Tyrol, Austria. $19 \times 11 \times 7$ cm. (F) Spessartine ($Mn_3Al_2Si_3O_{12}$) on quartz. Fujian, China. *Credit: (A) Rock Currier, https://commons.wikimedia.org/wiki/File:Albite_-_Crete_(Kriti)_Island, _Greece.jpg. (B) Rob Lavinsky, https://commons.wikimedia.org/wiki/File:Anorthite-221029.jpg. (C) Didier Descouens, https://commons.wikimedia.org/wiki/File:OrthoclaseBresil.jpg. (D) Rob Lavinsky, https://commons.wikimedia.org/wiki/File:Pyrope-260132.jpg. (E) Didier Descouens, https://commons.wikimedia.org/wiki/File:Almandin.jpg. (F) Géry Parent, https://commons.wikimedia.org/wiki/File:Garnet,_quartz,_feldspar_5.jpg.*

1.4.3.4 Twelve oxygens: garnet group; $Z^{6+}Al_2Si_3O_{12}$ and $Z^{12+}Si_2O_{12}$

The garnet group is separated into six end members: $Mg_3Al_2Si_3O_{12}$ (pyrope; see Fig. 1.50D), $Fe_3Al_2Si_3O_{12}$, (almandine; Fig. 1.50E), $Ca_3Al_2Si_3O_{12}$ (andradite), $Mn_3Al_2Si_3O_{12}$ (spessartine; see Fig. 1.50F) and $(Ca_3Cr_2)Si_2O_{12}$ (uvarovite), and their solid solutions. Garnets appear at high pressure.

1.4.4 Oxide minerals

1.4.4.1 Spinels

There are important minerals that are not silicate minerals but oxide minerals such as spinels. Spinels are expressed as $([zMg + (1 - z)Fe] O + [xCr_2O_3 + yAl_2O_3 + (1 - x - y)FeTiO_3])$ $(0 \leq x + y, z \leq 1)$, which is shown in Fig. 1.51. This can be easily understood as the solid solution of $MgO-FeO$ is mixed with the $Cr_2O_3-Al_2O_3$ and $FeO-TiO_2$ solid solutions with a ratio of 1:1.

1.4.4.2 Silica minerals; SiO_2

The most popular mineral of SiO_2 is quartz, which is the second most abundant mineral in the continental crust. (The most abundant one is feldspars.) There are a lot of polymorphs (the same composition with different structures) of SiO_2. A high-temperature phase of SiO_2 is tridymite and cristobalite. Coesite is a high-pressure phase (2−3 GPa) of SiO_2, and appears in some meteorite impact sites. Stishovite (> 10 GPa) is a more denser and higher pressure polymorph.

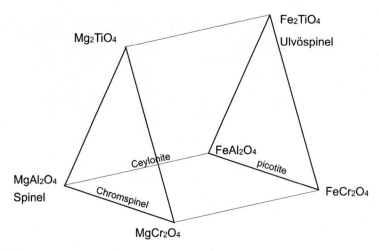

FIGURE 1.51 **The triangular prism for the spinel.** The composition of spinels is shown as $[zMg + (1 - z)Fe]O + [xCr_2O_3 + yAl_2O_3 + (1 - x - y)FeTiO_3]$ $(0 \leqq x + y, z \leqq 1)$.

1.4.4.3 *Titanium-related oxides*

As titanium is $+4$ and the ionic radius is small, Ti behaves solely. The most popular Ti mineral is rutile (TiO_2). When FeO is included, it becomes ilmenite ($FeTiO_3$), which often appears in the lunar mare basalts. As shown in Section 1.4.7, the mare basalts are divided into two groups, high-Ti basalts and low-Ti basalts. High-Ti basalts are thought to have derived from cumulates of ilmenite, which are made from low-Ti basalts.

1.4.4.4 *Perovskites*

Generally, perovskite means $CaTiO_3$. However, as related in Section 1.4.3.1, perovskite in astrochemistry is the high-pressure phase of magnesium silicate, $(Mg,Fe)SiO_3$ (Mg-perovskite; Mg$-$Pv). In the lower mantle, which is deeper than 660 km (~ 24 GPa), the calcium-bearing phase such as pyroxene breaks into calcium perovskite ($CaSiO_3$, Ca$-$Pv), which is the host of trace elements at high pressure.

There are lots of industrious useful materials, which have the perovskite structure and composition of $A^{2+}B^{4+}X^{2-}_3$. $MgCNi_3$ is a metallic perovskite, which attracts attention because it shows superconductivity. Yttrium barium copper oxide (YBCO; $YBa_2Cu_3O_{7-x}$) is a ceramic perovskite, which has also superconductivity. Methylammonium lead triiodide ($CH_3NH_3PbI_3$) also shows a perovskite structure and is expected to be high conversion dye-sensitized solar cells with an efficiency of $>20\%$.

1.4.5 Carbonates

Calcium and magnesium carbonates ($CaCO_3$ and $MgCO_3$) at normal condition is calcite and magnesite. $CaCO_3$ precipitated from hot water is called aragonite.

1.4.6 Nickel$-$iron alloy

In iron meteorites, nickel$-$iron alloy is often observed. In particular, kamacite (Ni-poor α-iron) and taenite (Ni-rich γ-iron) are often observed in iron meteorites showing the Widmanstatten pattern (see Fig. 4.10A).

1.4.7 The phase diagram of water

At high pressure, water becomes mineral, ice. The phase diagram of water is shown in Fig. 1.52. There are many interesting points in these phase diagrams. There are many polymorphs in solids (ice). At least ten polymorphs of ice are found. Many properties (viscosity, self-diffusion,

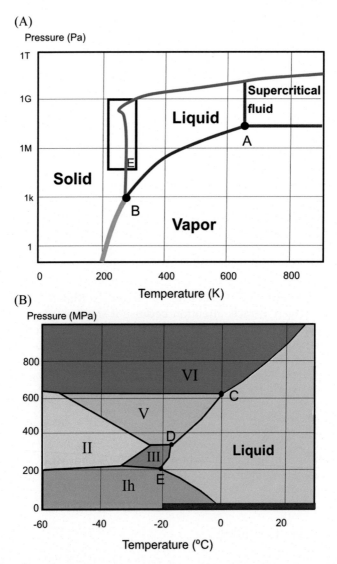

FIGURE 1.52   The phase diagram of water. (A) The horizontal and vertical axes are temperature (K) in linear scale and pressure (Pa) in logarithmic scale, respectively. The blue, green, and red lines show boundaries between solid-liquid, solid-vapor, and liquid-vapor, respectively. The area of right and upper side of dotted purple lines below the blue line is supercritical fluid. Points A and B are the critical and triple points, respectively. The temperature (°C) and pressure (MPa) of A and B are 647 and 22; and 273 and 0.6, respectively. The red "E" indicates the Earth's condition. The area surrounded by the black lines is enlarged in (B). (B) The enlarged phase diagram of the dotted area in (A). The right-side green area indicates liquid water areas. The left-side blue areas indicate the ice (solid) areas. The ice phases of I, II, III, V, and VI are shown. Points C, D, and E are triple points where three phases coexist. The temperature (°C) and pressure (MPa) of C, D, and E are 0.16 and 632; −17 and 350; and −22 and 210, respectively. The red line at the bottom indicates the typical habitable zone of the Earth.

Raman spectra, etc.) of cold liquid water changes around at 200 M Pa (we are living in 0.1 MPa). Pure liquid water with a temperature of $< -22°C$ cannot exist at any pressure.

1.4.8 Characterization of volcanic rocks and ultramafic rocks

In this section, we learn the petrology. Igneous rocks are defined as the rocks in which magmas are solidified. Volcanic rocks are the magma (lava) that is rapidly cooled after eruption. The characteristics of the volcanic rocks are the porphyritic texture which is composed of phenocrysts and ground mass.

The chemical classification of volcanic rocks is often used. The easiest classification is only based on silica (SiO_2 wt.%). The rocks of silica of <45%, 45%−52%, 52%−63%, and >63% are called ultramafic, mafic, intermediate, and felsic rocks, respectively.

There is another chemical classification, which is shown in Fig. 1.53. When the horizontal axis is silica and the vertical axis is total alkaline oxides ($Na_2O + K_2O$ wt.%), the rocks below the dotted line are called total alkaline oxides, and above the dotted line are alkaline rocks. In more detail, the igneous rocks are classified by the solid lines. Especially, basalt, basaltic andesite, andesite, and dacite often appear as the rock name.

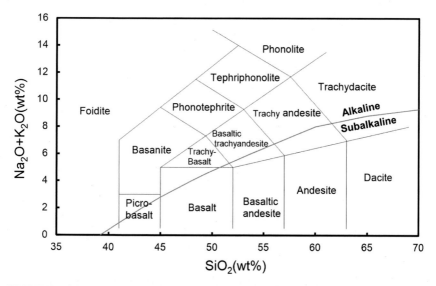

FIGURE 1.53   **The chemical classification of igneous rocks based on the silica versus total alkaline oxides.** The boundary of alkaline/subalkaline rocks is shown in the broken line. The igneous rocks are classified by contents of silica (SiO_2 wt.%) versus those of alkali oxides ($Na_2O + K_2O$ wt.%).

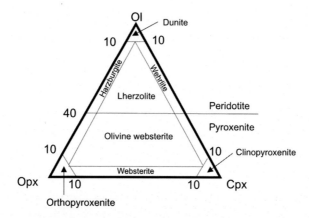

FIGURE 1.54 The ternary diagram of ultramafic rocks. Each corner of the mineral shows 100 wt.% of the mineral. The number shows the wt.% of each mineral. *Ol*, Olivine; *Opx*, orthopyroxene; *Cpx*, clinopyroxene.

The ultramafic rock classification is shown in Fig. 1.54 which forms the upper mantle. The ultramafic rocks are characterized by the ternary plot of olivine-orthopyroxene-clinopyroxene. There are two separation methods. One is the rocks with olivine content of 100%−60% and 60%−0% are named as peridotites and pyroxenites. The other separation method is as follows: Corners of Ol, Opx, and Cpx of >90% are named as dunite, orthopyroxenite, and clinopyroxenite. Ultramafic rocks of triangle sides of Ol, Opx, and Cpx of <5% are websterite, wehrlite, and harzburgite, respectively. Ultramafic rocks of Ol of 90%−60%, Opx and Cpx of >5% are lherzolites, and Ol of 60%−5%, Opx and Cpx of >5% are olivine websterite.

1.4.9 Basic geological knowledge on sedimentary rocks, metamorphism, and deep sea volcanism

"Amphibolites" are metamorphic rocks, which are considered to be volcanic rocks as basalt before the metamorphism. The "pillow structure" is typical to lavas erupted in the deep sea, for example, mid-oceanic ridge basalts (MORBs). Even after metamorphism, the pillow structure can sometimes remain in amphibolites. When the pillow structure is observed, it is clear evidence that the basalt has erupted under the sea as MORBs or the sea existed at the eruption.

"Turbiditic" sediments mean sediments formed the near-continent conditions. Their original grain sizes are relatively large compared to the pelagic sediments. In contrast, "pelagic" sediments are sediments formed far from the continent, therefore, the original grain sizes are fine. The pelagic sediments sometimes form alternate layers with "cherts," whose composition is almost 100% silica. The origin of the "cherts" is considered to be pure-chemical origin or biogenic origin. Recent cherts are biogenic in origin.

"Foliation" is observed in metamorphic rocks especially when the sheet-like minerals like "mica or chlorite" align to one planar or wavy directions. "Schistosity" is a mode of foliation in metamorphism, which is planar (sheet-like) and rather strong. When metamorphic rocks show strong foliation or strong schistosity, the strong parallel force to the direction of the original bedding was supplied to the rocks.

When "garnet" is observed in metamorphic rocks, it means the metamorphic temperature and pressure were relatively high. When the metamorphism is going in the direction of higher temperature and pressure, it is called as "progressive" metamorphism. In contrast, when the metamorphism goes to lower temperature and pressure, it is called as "retrogressive" metamorphism.

1.4.10 Determination of major elements

To characterize volcanic rocks, the major element determination including silica (SiO_2 wt.%) and total alkaline ($Na_2O + K_2O$ wt.%) is required. About 50 years ago, the major elements were analyzed by the CLASSIC method, where the rocks were powdered, melted by fusion, dissolved with HF, and separated into each element which was made into oxide and determined its weight by the balance.

Nowadays, after the rock powder is prepared, the powders are made into glass beads, and major elements are determined automatically by X-ray fluorescence spectrometry (XRF). Details of the XRF are shown in Fig. 1.55. The sample (rock powder) is melted with flux and made into a glass bead. In XRF, intense high-energy X-ray is showered at a glass bead, and lower energy electrons (e.g., K or L shells' electrons) are removed. Then the higher energy electrons in L, M, N, ... shell go down to the K or L shells (L → K, M → L, etc.), and the characteristic X-rays that are specific to each element are emitted. Then the characteristic X-rays of each element are separated for concentration determination of each element.

For XRF, standard silicate powders of various ranges of elemental compositions are required to determine major element compositions precisely.

1.4.11 Partition coefficients

The concepts of partition coefficients and the partial melting are important in astrochemistry. We assume the system is partially molten.

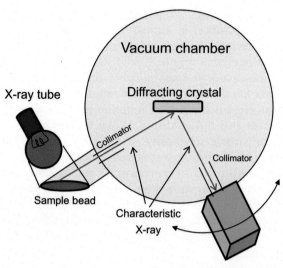

Gas flow proportional counter

FIGURE 1.55 **A schematic diagram of X-ray fluorescence spectrometer (XRF).** The sample is diluted and melted with ten times the weight of flux, which is made of pure $LiBO_2 + Li_2B_4O_7$ that does not absorb characteristic X-rays of heavier elements. X-ray made by the Rh-anode X-ray tube is showered onto a sample bead. The secondary characteristic X-rays are emitted from the sample. The direction of the X-ray is collimated to the diffracting crystal. From the sample to the detector, the chamber is kept in a vacuum. The beams enter the diffracting crystal and are dispersed. The characteristic X-ray of the target element is detected by a gas flow proportional counter, which can be moved along the focal plains and counts the X-ray pulses.

One of the trace elements, T, is partitioned based on the partition coefficient, K_d. The K_d value is defined as:

$$K_d = C_S/C_L \qquad (1.47)$$

where C_S and C_L are concentrations of the element T in solid and liquid, respectively.

1.4.12 Classification of the meteorite

The meteorite is characterized as the fall and the find meteorites. The fall meteorite is the meteorite that was observed from the falling from the sky. The Chelyabinsk meteorite, which fell in Russia in 2013, and the Allende meteorite, which fell in Mexico in 1969, are the typical fall meteorites. The find meteorite is not the fallen meteorite. Initially, it was just a stone, and later it was appraised by a specialist as a meteorite. The fusion crusts, which

■ Iron meteorites

Ni poor	Hexahedrite	IIA, IIG
↕	Octahedrite	IA, IB, IIB, IIC, IID, IIE, IIIAB, IIICD, IIIE, IIIF, IVA
Ni rich	Ataxite	IB, IIICD, IVB
Others		IC

■ Stony-iron meteorites

● Pallasites   Main group pallasites, Eagle station pallasites, pyroxene pallasites

● Mesosiderites

■ Stony meteorites

● Chondrites

• Carbonaceous chondrites

CI chondrites

CM-CO chondrites (mini-chondrule)

CM chondrites (CM1-CM2; Mighei-like), CO chondrites (CO3-CO3.7; Ornans-like)

CV-CK chondrites

CV chondrites (CV2-CV3.3; Vigarano-like), CK chondrites (Karoonda-like)

CR chondrite clan

CR chondrites (Renazzo-like)

CH chondrites (Allan Hills 85085-like)

CB chondrites (Benucubbin-like)

• Ordinary chondrites

H chondrites (H3-H7), L chondrites (L3-L7), LL chondrites (LL3-LL7)

• Enstatite chondrites

E chondrites (E3-E7), EH chondrites EH3-EH7), EL chondrites (EL3-EL7)

• Other chondrites

R chondrites (Rumuruti-like), K chondrites (Kakangari-like)

● Achondrites

• Primitive chondrites   Acapulcoites, Lodranites, Winonaites

• Asteroidal achondrites

HED meteorites (4 Vesta?)   Howardite, Eucrite, Diogenite

Angrites, Aubrites, Ureilites, Brachinites

• Lunar meteorites

• Martian meteorites (SNC meteorites)   Shergottites, Nakhlites, Chassignites, Others (ALH84001)

FIGURE 1.56   Classification of meteorites. Meteorites are classified as iron meteorites, stony-iron meteorites, and stony meteorites. The iron meteorites are thought to be the core of planetesimals. Stony-iron meteorites are made of metal and olivine crystal. Stony meteorites are separated into chondrites and achondrites. Chondrites are classified into carbonaceous chondrites, ordinary chondrites, enstatite chondrites, and other chondrites. Achondrites are separated into primitive chondrites, asteroidal achondrites (HED, etc), and Lunar and Martian meteorites.

FIGURE 1.57 **Photographs of sliced meteorites.** (A) Toluca iron meteorite (octahedrite, IA), width of ~10 cm. (B) Esquel pallasite, width of ~5 cm. (C) Vaca muerta mesosiderite, width of ~3 cm. (D) Murchison CM2 chondrite, width of 1 cm. (E) Allende CV3 chondrite, width of 5 cm. The chondrules (round-shaped particles) and CAI (irregular shaped white materials near the center) can be seen. (F) Tagish Lake C2 chondrite, the height of 5 cm. (G) Chelyabinsk LL5 chondrite, a width of 4 cm. (H) Abee EH4 chondrite, a width of 31 cm. (I) NWA 2989 acapulcoite, width 1.3 cm. (J) NWA5000 lunar meteorite, a width of 16 cm. (K) Millbillillie eucrite, a width of ~3.5 cm. (L) ALH84001 martian meteorite, width 18 cm. Figures of (D) and (F) are the photographs of not sliced but chipped surfaces. Figure (L) is the photo of the surface (the fusion crust). *Credit: (A) H. Raab,*

are melted and oxidized surfaces by heat generated by the compaction of air with the supersonic velocity during falling (not the friction with the air!), can be seen on the surface. The fusion crusts are one of the proofs of the meteorite.

Meteorite hunters are collecting meteorites from Australian and African deserts. Such meteorites are named as NWA XXX (Northwest African Desert; Morocco, Algeria, Western Sahara, and Mali), Dar al Gani XXX (Libya Desert), and Dhofar XXX (Oman Desert) (XXX indicates the number). Lots of meteorites are also found in glaciers in Antarctica, named with the base name, e.g., ALH-YYXXX, Yamato-YYXXX, or Aska-YYXXX, etc. (YY is the last two digits of the dominical year; XXX is the number).

The meteorites are finally characterized by the chemical composition and textures observed by an optical microscope and oxygen isotope ratios of $^{17}O/^{16}O$ and $^{18}O/^{16}O$. As the appraisement of the meteorite is not easy, there are lots of fakes in the internet shops.

The classification of meteorites is summarized in Fig. 1.56. The meteorites are divided into three types: (1) iron meteorites, (2) stony-iron meteorites, and (3) stony meteorites. The iron meteorites are mainly made of FeNi-alloy (kamacite and taenite), and sometimes contain troilite (FeS), graphite (C), schreibersite [(Fe, Ni)$_3$P], and cohenite [(Fe, Ni, Co)$_3$C]. Based on Ni contents, the iron meteorites are divided into (1) hexahedrite, (2) octahedrite (see Fig. 1.57A), (3) ataxite, and (4) other groups. The iron meteorites are further divided by Ga, Ge, and Ir contents into groups of I, II, III, and IV plus A, B, C, D, E, and F.

The stony-iron meteorites are divided into (1) pallasites (metallic iron plus silicates; Fig. 1.57B) and (2) mesosiderites (breccias which show metamorphism, 16 Psyche origin?; Fig. 1.57C). The iron meteorites and the stony-iron meteorites are considered to have come from the core and core-mantle boundary in evolved asteroids, respectively.

The stony meteorite is further divided into (1) chondrite and (2) achondrite (see Fig. 1.56). The chondrites occupy $\sim 86\%$ of meteorites

https://commons.wikimedia.org/wiki/File:TolucaMeteorite.jpg. (B) Doug Bowman, https://upload.wikimedia. org/wikipedia/commons/thumb/0/00/Esquel.jpg/640px-Esquel.jpg. (C) Brocken Inaglory, https://upload.wiki-media.org/wikipedia/commons/thumb/d/d9/Vaca_muerta_mesosiderite.jpg/640px-Vaca_muerta_mesosiderite. jpg. (D) United States Department of Energy; Carl Henderson. http://www.anl.gov/Media_Center/ Image_Library/images/fermi-dust.jpg. (E) Basilicofresco, https://upload.wikimedia.org/wikipedia/commons/ thumb/2/24/Allende_meteorite.jpg/634px-Allende_meteorite.jpg. (F) Mike Zolensky, NASA JSC. https:// upload.wikimedia.org/wikipedia/commons/1/17/Tagish_Lake_meteorite.jpg. (G) Pavel Maltsev. https:// upload.wikimedia.org/wikipedia/commons/a/a1/Meteorit-chebarkul-macro-mix2.jpg. (H) Google. https://art-sandculture.google.com/asset/abee/CAGUO17Jy9tRfA?hl = ja. (I) James St. John, https://commons.wikime-dia.org/wiki/File:NWA_2989_meteorite,_acapulcoite_(14601736517).jpg. (J) Steve Jurvetson. https:// commons.wikimedia.org/wiki/File:NWA5000,_large_slice.jpg. (K) The author's collection. (L) NASA-JSC. https://upload.wikimedia.org/wikipedia/commons/c/c4/ALH84001.jpg.

and are composed of chondrules and matrices. When no chondrules are observed in the meteorite, they are called achondrites. The chondrules are small silicate spherules (0.1 mm—a few cm), once melted at high temperature ($>1800K$) and quenched ($\sim 1000\,K\,h^{-1}$; Hewins, 1997). The matrix is a low-temperature component. According to the thermal heating (metamorphic) grade, the chondrites are numbered from 1 (the least metamorphism) to 7 (melted).

The chondrite is also divided into three types, (1) carbonaceous chondrites (CI, CM-CO, CV-CK, CR; Fig. 1.57D−F), which contain carbon compounds (graphite, carbonates, organic compounds; carbon contents of $<\sim 3\%$), (2) ordinary chondrites (L, LL, and H; Fig. 1.57G), and (3) enstatite chondrites (rich in enstatite; EL and EH; Fig. 1.57H).

The chondrites are called with alphabet and heating grade as CV3, LL6, etc. The CI and CM chondrites contain water of $3\%-20\%$, and are metamorphic grades of $1-3.7$ or without numbers (CI, CR, CH, and CB chondrites). The carbonaceous chondrites are the least affected by thermal events and, thus are considered to be the most primitive material in the meteorites.

The achondrites are a mixture of non-chondrites: (1) the chondritic composition like acapulcoite (the origin is unknown; Fig. 1.57I), (2) the evolved composition like angrites, (3) HED meteorites, (4) lunar meteorites (Fig. 1.57J), and (5) martian meteorites (SNC meteorites). The HED meteorites are the combination of the first letters of Howardites (mixture of eucrites and diogenites), Eucrite (basaltic; see Fig. 1.57K), and Diogenite (peridotitic). The HED meteorites are considered to come from 4 Vesta asteroid, which is the evolved asteroid with the core and mantle. The SNC meteorites are also the combination of the first letters of Shergottites, Nakhlites, and Chassignites, which possibly came from Mars (Fig. 1.57M). Some special meteorites are considered to come from the Moon (e.g., ALHA A81005).

1.4.13 Some basic knowledge on the meteorites and asteroids

1.4.13.1 The calcium−aluminum rich inclusions

In Allende CV3 chondrite, not only the chondrules, but also white irregular shaped inclusions exist (See Fig. 1.57E). These white inclusions, called as the calcium−aluminum rich inclusions (CAIs), have a very refractory composition. These inclusions are so refractory that they are considered to have formed in the very early cooling stage of the solar nebular and survived homogenization and heat in the solar nebula. The age of the CAIs obtained by the Pb−Pb method shows the oldest age in the solar system so far from 4567.4 ± 1.1 to 4567.17 ± 0.70 My (Amelin et al., 2002), and is considered to show the earliest age in the solar system. The formation of the solar nebula is considered to start within 1 My of the shockwave of the supernova, because the CAIs contain the evidence of ^{26}Al, of which half-life is 0.7 My (see Section 1.3.11).

1.4.13.2 Why are chondrules globular shape?

Chondrules are mm-scale, previously melted silicates and cooled-very-fast spherules in chondrites. The ages of the chondrules are significantly 2−3 My younger (after) than those of the CAIs after Section 1.4.13.1. The planetesimal formation and even metallic core segregation in meteorite parent bodies are already finished within 1 My, chondrule formation is the subsequent event.

Many mechanisms are proposed for forming chondrites, such as shock waves in the protoplanetary disk, and magnetic flares, etc.

Johnson et al. (2014) established the idea of impact jetting in which two spherical projectiles impact each other forming the melts. Johnson et al. (2015) further reported that impacts can produce enough chondrules during the first five My of planetary accretion to explain their observed abundance (see Fig. 1.58). They simulated protoplanetary impacts, finding that material is melted and ejected at high speed when the impact velocity exceeds 2.5 km s^{-1}. The jetted melt will form mm-scale droplets and the cooling rates will be fast, 10−1000 K h^{-1}. An impact origin for chondrules implies that meteorites are a byproduct of planet formation rather than leftover building blocks of planets.

1.4.13.3 The evidence of early formation and differentiation of protoplanets: constraints from iron meteorites

The iron meteorites are considered to be the core of the meteorite parent bodies, and good samples for constraining the protoplanet differentiation and accretion in the presolar disk. Kruijer et al. (2014) found variations of 5−20 parts per million in ^{182}W, resulting from the decay of now-extinct ^{182}Hf in iron meteorite groups, IIAB, IID, IIIAB, and IVA. These ^{182}W isotope ratios imply that core formation occurred after ∼1 My, and the iron meteorite parent bodies probably accreted ∼0.1 to 0.3 My after the formation of the CAIs.

1.4.13.4 Presolar grains

From the remaining materials after treatment with hydrofluoric acid, hydrochloric acid, and nitric acid which are called "acid residues," SiC micro-particles with a diameter of one μm were found. The isotopic ratios of carbon and silicon were measured by SIMS (see Fig. 1.34). The observed isotopic ratios were significantly different from those of the Earth, and other materials from other types of meteorites. These particles were considered to be remnants before the establishment of the present Solar System, therefore, they were named as "presolar" grains.

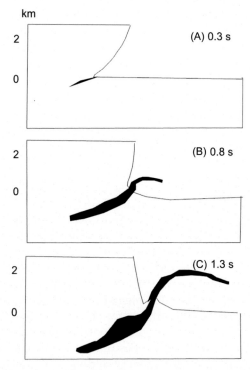

FIGURE 1.58 **Schematic diagrams of jetting of melted material during an accretionary impact.** Chondrules can be formed when two planetesimals collide with each other. A projectile of 10 km in diameter strikes the flat target at 3 km s^{-1}. The black parts show the melt jets with a temperature of >1600K. *Source: Modified from Johnson, B.C., Minton, D.A., Melosh, H.J., Zuber, M.T., 2015. Impact jetting as the origin of chondrules. Nature 517, 339–341.*

The presolar grains were considered to have derived from a carbon-rich AGB star (the mass was 1.5–3 $M_\odot$; $M_\odot$ is the mass of the Sun). This means that the small mass of such an AGB star was added to the large mass of the solar nebular which composed the present solar system.

1.4.13.5 The Yarkovsky effect

When the asteroid is rotating, the asteroid gets heat on the day surface, and emits heat (radiation) on the night surfaces. When the asteroid is not symmetric, the transfer of heat or radiation can be inhomogeneous, and the radiation becomes forced to drift the asteroid from the expected orbit. This is called as the Yarkovsky effect.

1.4.13.6 The Yarkovsky–O'Keefe–Radzievskii–Paddack effect

The spin rate and obliquity of the pole of the small astronomical body can change due to the scattering of solar radiation off its surface and the emission of its own thermal radiation. This is called as the Yarkovsky–O'Keefe–Radzievskii–Paddack (YORP) effect (please do not confuse it with the Yarkovsky effect in Section 1.4.13.5). The YORP effect typically occurs for asteroids with the heliocentric orbit in the solar system.

1.4.13.7 CI chondrites and the trace elemental abundance of the Earth

CI chondrites have never exceeded 200°C, therefore, they contain lots of volatile components such as water and organic materials. Thus, it is believed to contain the materials of the origin of life, the first materials of the solar nebular, or the origin of the Earth. Furthermore, the target asteroids of sample return projects of both Hayabusa 2 and OSIRIS-REx are C-type asteroids, which are made of carbonaceous chondrites.

Refractory elemental abundances of the Earth are assumed to be similar to those of the CI chondrites. This is supported by the coincidence of the compositions of the carbonaceous chondrite to those of the photosphere of the Sun. McDonough and Sun (1995) estimated the elemental abundance of the Earth is 2.75 times that of the CI chondrite for highly refractory elements. The semi-volatile elements and volatile elements of the Earth such as C, N, O, S, halogens, etc. are difficult to estimate. Because, some can go into the core, and some can be lost or evaporated during the formation of the Earth and the Moon.

The elemental abundances of the CI chondrites compared to Si are shown in Fig. 1.59. The relative low abundances of Li, Be, and B are clearly observed. The even-atomic-number elements are more abundant than the odd-atomic-number elements, which are clearly observed as the zig-zag pattern in heavy elements in the figure. This is called as the "Oddo–Harkins" rule. This is caused by the nucleus's relative stability of even-atomic-number elements compared to odd-atomic-number elements.

1.4.13.8 The Palermo scale and the Torino scale

The Palermo scale, or the Palermo Technical Impact Hazard Scale is a logarithmic scale used by astronomers to rate the potential risk of impact of near-Earth objects (NEOs). It combines two types of data—probability of impact and estimated kinetic yield—into a single "hazard" value. A rating of 0 means the hazard is similar to the background hazard. A rating of +2 would indicate the hazard is 100 times as great as a random background event. Scale values less than −2 reflect events for which there are no likely consequences, while Palermo Scale values between −2 and 0 indicate situations that merit careful monitoring.

The Torino Scale is also used to categorize the impact hazard of NEOs, such as asteroids and comets. The Torino Scale is based on the log–log plot of the probability of impact in the horizontal axis, and the kinetic energy (megatons of TNT; Mt) in the vertical scale. The Torino

X/Si x 10^6

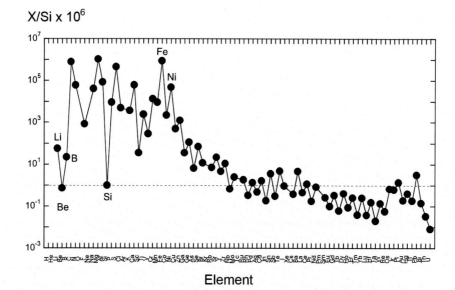

Element

FIGURE 1.59   **The elemental abundances of the CI chondrites compared to Si.** Each elemental abundance in the CI chondrite is divided by the Si abundances times 10^6. The horizontal dotted line shows $X/Si \times 10^6 = 1$. *Source: Data from McDonough, W.F., Sun, S.-s, 1995. The composition of the Earth. Chem. Geol. 120, 223–253.*

scale is an integer scale from 0 to 10. Zero indicates an object has a negligibly small chance of collision with the Earth, or the object is too small to pass through the atmosphere. Ten (the probability of >0.99 and the kinetic energy of $>10^5$ Mt corresponding to >1 km in diameter) indicates that the collision causes a global disaster.

1.5 Introduction to organic astrochemistry

1.5.1 Amino acids

Amino acids, especially 2-, alpha- or α-amino acids are biologically important. These prefixes indicate the type of amino acids that have amine ($-NH_2$) and carboxylic acid ($-COOH$) at the same carbon. Therefore, the general chemical formula of the amino acids becomes $H_2NCHRCOOH$ where R is an organic substituent known as side chain.

In Fig. 1.60, typical amino acids are shown. Histidine, threonine, isoleucine, tryptophan, leucine, lysine, valine, methionine, and phenylalanine are essential amino acids for humans. These amino acids cannot be synthesized inside the human body. The essential amino acids are different for each living thing.

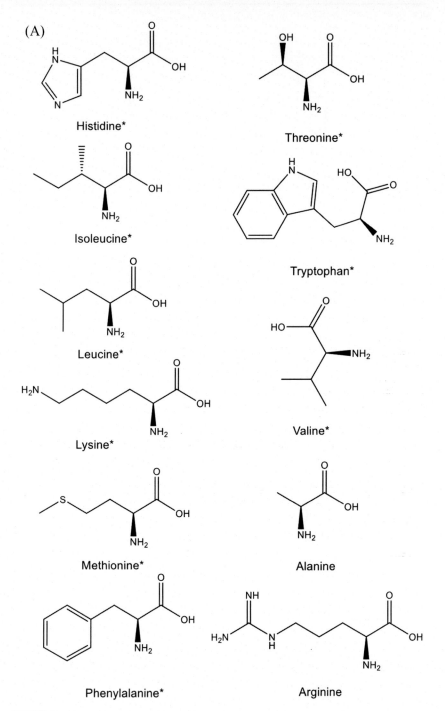

FIGURE 1.60 Amino acids. (A) Amino acids. (*) indicates the essential amino acids for humans, which cannot be synthesized in the human body. (B) Amino acids (continued). The C-NH₂ bond shown in thick line indicates C is the chiral center of the amino acids. See Fig 10.3 in Chapter 10 for chirality.

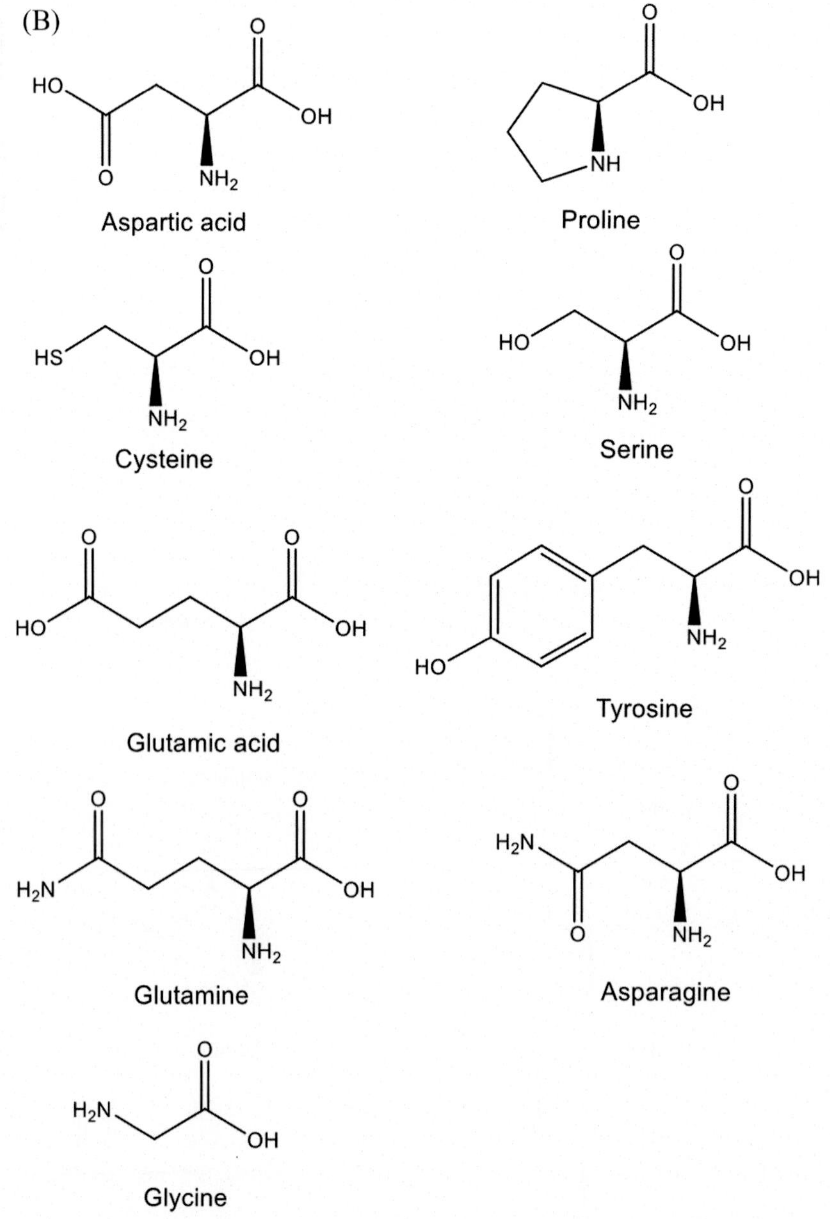

(B)

Aspartic acid

Proline

Cysteine

Serine

Glutamic acid

Tyrosine

Glutamine

Asparagine

Glycine

FIGURE 1.60 Continued.

In Table 1.12, abbreviations of 20 amino acids in proteins are shown. Each amino acid has a three-letter abbreviation and a one-letter abbreviation. There are four kinds of side chains: negative, positive, uncharged polar, and nonpolar side chains.

1.5.2 Proteins

Two amino acids with side chains of R_1 and R_2 can combine as follows:

$$NH_2 - CH(R_1) - COOH + NH_2 - CH(R_2) - COOH$$

$$\rightarrow NH_2 - CH(R_1) - CO - NH - CH(R_2) - COOH + H_2O \qquad (1.48)$$

TABLE 1.12 The 20 amino acids. Each amino acid has a three-letter and a one-letter abbreviation.

Amino acid	Abbreviation	Alphabet	Side chain
Aspartic acid	Asp	D	Negative
Glutamic acid	Glu	E	Negative
Arginine	Arg	R	Positive
Lysine	Lys	K	Positive
Histidine	His	H	Positive
Asparagine	Asn	N	Uncharged polar
Glutamine	Gln	Q	Uncharged polar
Serine	Ser	S	Uncharged polar
Threonine	Thr	T	Uncharged polar
Tyrosine	Tyr	Y	Uncharged polar
Alanine	Ala	A	Nonpolar
Glycine	Gly	G	Nonpolar
Valine	Val	V	Nonpolar
Leucine	Leu	L	Nonpolar
Isoleucine	Ile	I	Nonpolar
Proline	Pro	P	Nonpolar
Phenylalanine	Phe	F	Nonpolar
Methionine	Met	M	Nonpolar
Tryptophan	Trp	W	Nonpolar
Cysteine	Cys	C	Nonpolar

There are negative, positive, uncharged polar and nonpolar side chains.

This $-CO-NH-$ bond is called as a peptide bond. The amino acid can combine infinitely (see Fig. 1.61 for a combination of three amino acids in a 3D image).

Proline is not an amino acid, but an imino acid. However, it is treated as an amino acid. As proline has an unusual ring structure, the peptide bond becomes not as usual (see Fig. 1.62).

The polypeptide polymer molecule is called a protein, and some polypeptides have special biological functions. The end of the protein with a free carboxyl group is called as C-terminus or carboxy terminus (the amino acid with R_3 in Fig. 1.61), and the end with a free amino group is called as N-terminus or amino terminus (the amino acid with R_1 in Fig. 1.62).

There are two secondary structures in proteins. One is a coil structure called alpha-helix (α-helix). There is another motif in proteins, which is called beta-sheet (β-sheet) structure.

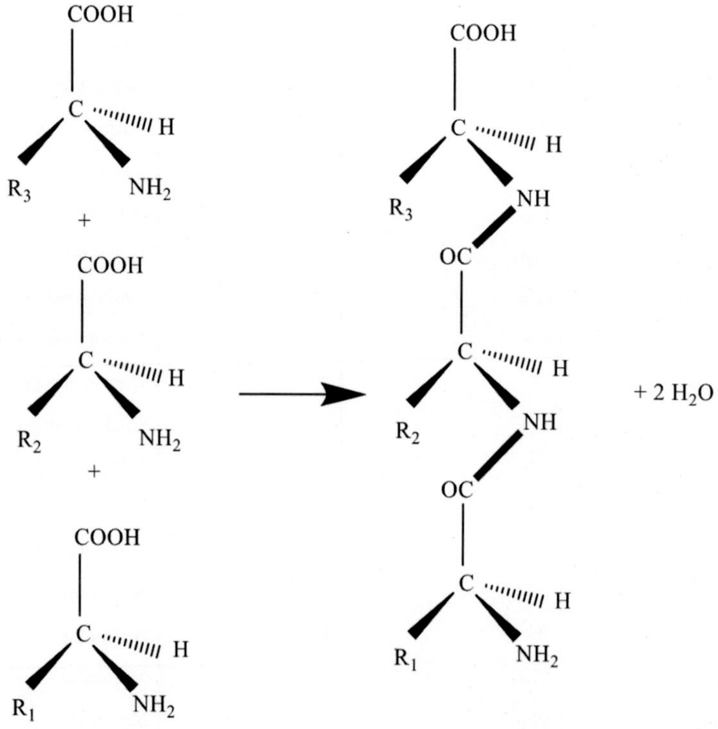

FIGURE 1.61 **Combination of three amino acids by peptide bonds.** When there are three amino acids, NH_2 of the first one reacts with COOH of the second one, forming the NH$-$CO bonding. NH_2 of the second one reacts with COOH of the third one, similarly forming the NH$-$CO bonding. Thus, the amino acids can connect infinitely. *Source: From "Biochemistry for Material Science", Akio Makishima, page 39, Fig. 2.2.1, Copyright Elsevier (2017).*

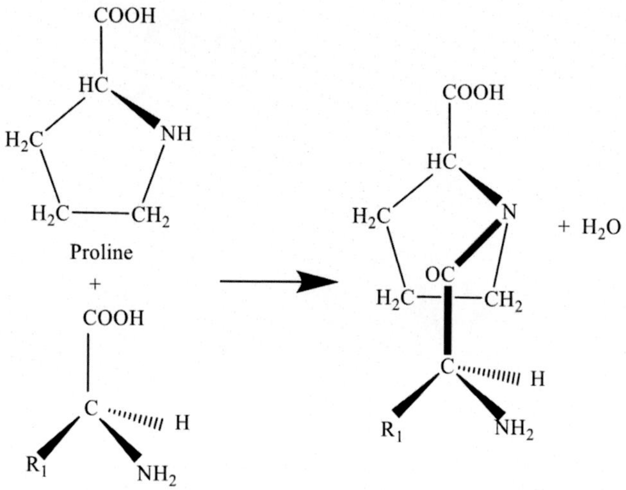

FIGURE 1.62   Combination of amino acid and proline by peptide bonds. As proline has a 5-member ring with NH, COOH of the next amino acid connects forming N—CO bonding, remaining the 5-member ring of proline.

1.5.3 The Protein Data Bank

The name of proteins is registered in the Protein Data Bank (PDB) with PDB ID. When the PDB ID is known, a three-dimensional image of the protein is also registered with the reference(s), which can be obtained by accessing http://www.rcsb.org.

For example, sperm whale myoglobin by Watson (1969) is registered as "1MBN." As many as 5070 sperm whale myoglobins were registered in PDB in February 2023. The 3D structure of 1MBN is shown in Fig. 1.63. This protein is a metalloprotein, which is made of the protein itself and heme (or heam; see Fig. 1.64). The heme is shown as connected balls in Fig. 1.63. Red, orange, blue, and gray balls indicate oxygen, iron, nitrogen, and carbon, respectively. The protein holds the heme, and coils colored in the rainbow is the α-helix, which was explained in the previous section.

1.5.4 Purines, pyrimidines, DNA, and RNA

In Fig. 1.65, purines and pyrimidines are shown. Adenine (A) and guanine (G) are purines, and cytosine (C), thymine (T), and uracil (U) are pyrimidines. These are the most important parts in the nucleic acid, and genetic information is stored in the sequence of these molecules.

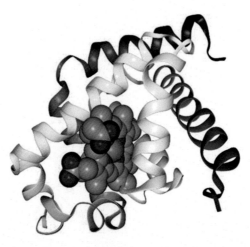

FIGURE 1.63 **The sperm whale myoglobin with heme (1MBN;Watson, 1969).** The balls show heme. Red, orange, blue, and gray balls indicate oxygen, iron, nitrogen, and carbon, respectively. *Source: From "Biochemistry for Material Science", Akio Makishima, page 40, Fig. 2.3.1A, Copyright Elsevier (2017).*

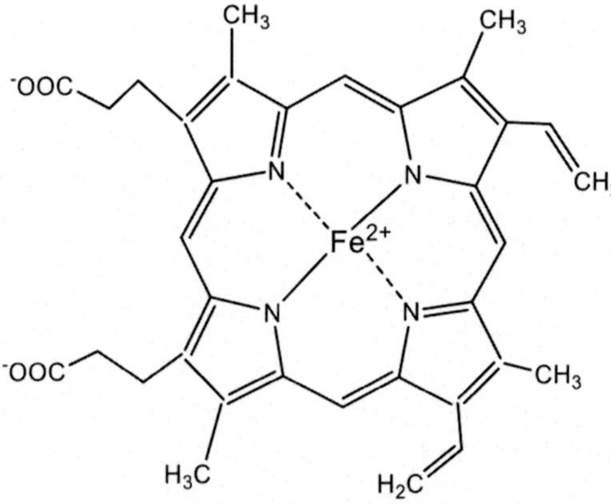

FIGURE 1.64 **The chemical formula of heme (porphyrin IX containing Fe $^{2+}$ ion).** The heme contains iron 2 + in the center of the porphyrin, which is a flat molecule with four nitrogen connecting to the iron ion. *Source: From "Biochemistry for Material Science", Akio Makishima, page 41, Fig. 2.3.2, Copyright Elsevier (2017).*

The red color nitrogen atom in Fig. 1.65 is connected to pentose sugar to form DNA or RNA.

These adenine (A) or T, G, and C are bound with the five-carbon sugar, called deoxyribose, which is connected with one phosphoric acid. This one set of molecules is called a nucleotide. When the five-carbon sugar is

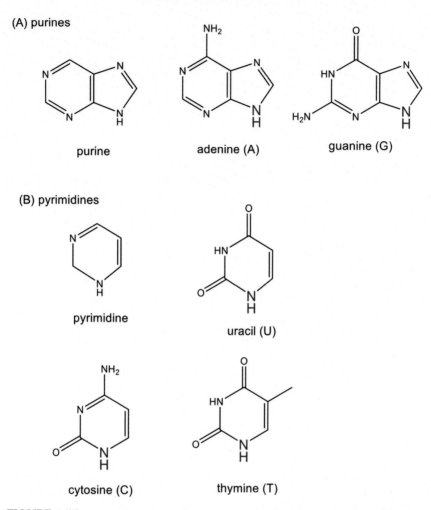

FIGURE 1.65 **Purines and pyrimidines.** (A) Purines. Purine is made of a six-member ring containing two nitrogen atoms fusing with the five-member ring with two nitrogen atoms, totally forming a nine-member ring with four nitrogen atoms. Adenine is one C−H bond of the six-member ring of the purine near the five-member ring substituted with NH_2. Guanine is the one carbon of the adenine's NH_2 replaces C−O, and another CH becomes C−NH_2. (B) Pyrimidines. Pyrimidine is a six-member ring with two nitrogen atoms. Cytosine is two CH of the pyrimidine becomes C−O and C−NH_2. Uracil is both of two CH of the pyrimidine becomes C−O. Thymine is another CH of uracil that becomes C−CH3.

deoxyribose (Fig. 1.66A), the polymer of this molecule is called deoxyribonucleic acid (DNA). When the five-carbon sugar is ribose (Fig. 1.66B), the polymer of this molecule is called ribonucleic acid (RNA).

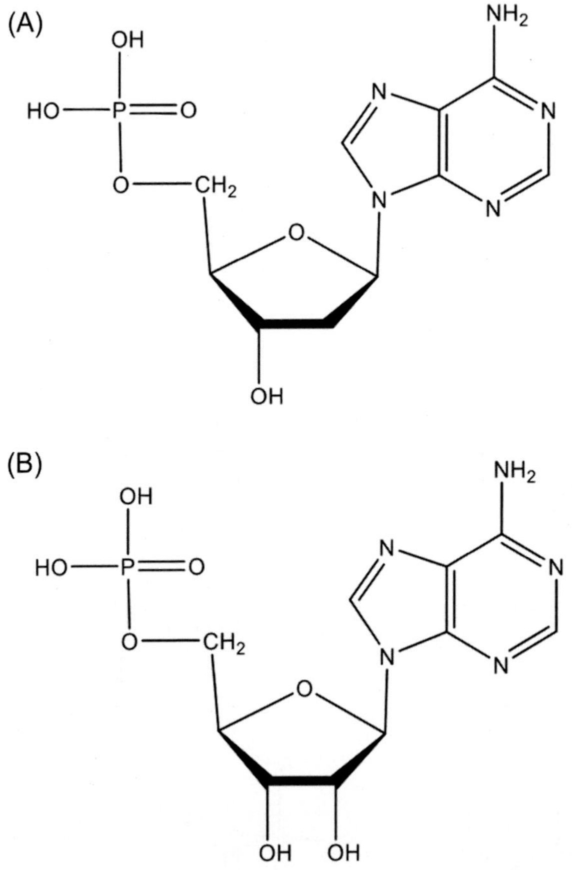

FIGURE 1.66   Nucleotides. (A) Adenine-deoxyribonucleic acid-phosphoric acid (adenine-DNA). (B) Adenine-ribonucleic acid-phosphoric acid (adenine-RNA). The molecule (B) is also called as adenosine monophosphate (AMP). *Source: "Biochemistry for Material Science", Akio Makishima, page 43, Fig. 2.5.2, Copyright Elsevier (2017).*

The abbreviation of A, G, C, T, and U are often used to show the genetic information in deoxyribonucleic acid (DNA, see Fig. 1.66A) and ribonucleic acid (RNA, see Fig. 1.66B). In DNA and RNA, A, G, C, and T, and A, G, C, and U are used, respectively.

1.5.5   Hydrogen bond

The five molecules form a constant combination by the hydrogen bond. Adenine (A) combines with thymine (T) or uracil (U) in DNA or RNA, respectively, and guanine (G) combines with cytosine (C), respectively. The hydrogen bonds are shown as dotted lines in Fig. 1.67.

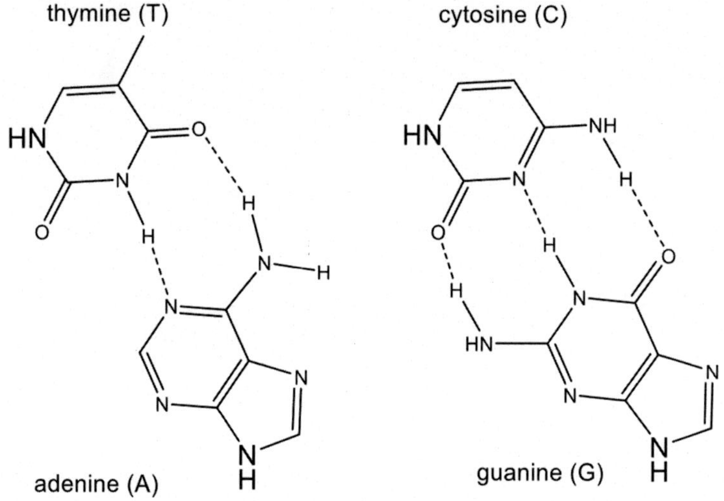

FIGURE 1.67   **Hydrogen bonds.** Adenine and thymine (uracil) combine with two hydrogen bonds, and cytosine and guanine combine with three hydrogen bonds. This is the cause why A−T (U) and C−G pair do not make mistakes in DNA duplication. *Source: From "Biochemistry for Material Science", Akio Makishima, page 44, Fig. 2.6.1, Copyright Elsevier (2017).*

The hydrogen bond is an electrostatic attraction between polar molecules, such as nitrogen or oxygen and hydrogen. The hydrogen bonds appear in various reactions in living things. The 3D structure of proteins, which are one of the hot regions in biochemical research, is determined using hydrogen bonds. Recently, some software have succeeded to reconstruct the protein structure using artificial intelligence (e.g., Jumper et al., 2021).

Astonishingly, living things use hydrogen bonds to store the genetic information only by A−T and G−C pairs. These bonds are not so strong to be cut, but strong enough not to be cut easily. Some carcinogens have similar structures to these molecules, and enter into the A−T or G−C pairs and break the genetic information.

1.5.6 Helix and double helix

Nucleotides make long polymers, as shown on the left side of Fig. 1.68, which is called from 5′ to 3′. The two polymers make the double helix as shown in Fig. 1.68. Two ends of phosphoric acid and deoxyribose are called 5′ end and 3′ end respectively. The direction of 5′→3′ is defined to be the normal direction, and when the base direction of the DNA in Fig. 1.68 is needed to be shown, it is shown along the normal direction as AGT, not as TCA.

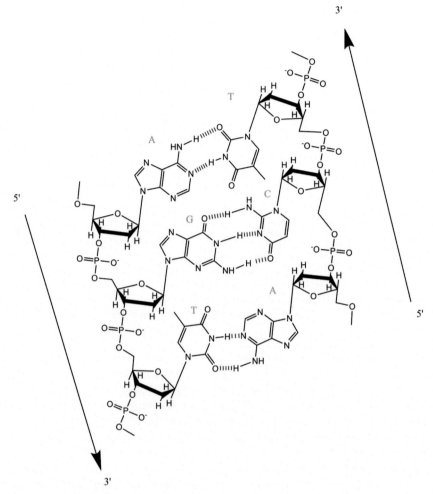

FIGURE 1.68   DNA. Characters of A, G, T, and C indicate adenine, guanine, thymine, and cytosine. Dotted bonds show hydrogen bonds. *Source: From "Biochemistry for Material Science", Akio Makishima, page 45, Fig. 2.7.1, Copyright Elsevier (2017).*

1.5.7 Synthesis of proteins from DNA

For the synthesis of proteins from DNA, the proteins are not directly synthesized from DNA. DNA is copied to RNA, then the protein is synthesized from RNA. The first step is DNA to RNA (step 1), and the other step is RNA to protein (step 2). In this section, these two steps are explained in more detail.

The genetic information of DNA (see Fig. 1.69A) is copied to RNA (Fig. 1.69B), which is called as "transcription." It should be noted that

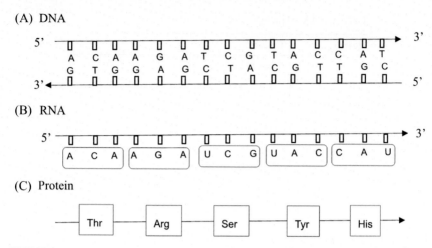

FIGURE 1.69 **Synthesis of DNA to protein.** (A) DNA pairs. (B) RNA. Three nucleotides represent one protein. (C) A protein sequence is formed from the RNA sequence. *Source: From "Biochemistry for Material Science", Akio Makishima, page 46, Fig. 2.8.1, Copyright Elsevier (2017).*

RNA is a chain of ribonucleotides; that T (thymine) in DNA is transcribed into U (uracil) in RNA; and that RNA is not a double-helix molecule, but a single-line molecule. Compared to DNA molecules with 250 million nucleotide pairs long, RNA molecules are no more than a few thousand nucleotides long. Many RNAs are much shorter.

Then the information of RNA is copied to protein, which is called as "translation." Three sets of nucleic acids are called as a codon, which is indicated as dotted squares in Fig. 1.69B. Each codon is translated into one protein following the genetic code (see Fig. 1.69C and Table 1.13). When the stop codon appears, the protein synthesis stops and ends.

1.5.8 Transcription (synthesis of RNA from DNA)

The transcription is performed by an enzyme named RNA polymerase. The RNA polymerase moves stepwise along the DNA, unwinding the DNA helix making an mRNA chain 5'- to 3'- direction. As the RNA is released from the DNA in a short time, the RNA syntheses can start at various positions of the DNA before the RNA syntheses finish.

The genetic information in RNA is not stored permanently, thus, one mistake occurs in every 10^4 nucleotide copy. In contrast, in DNA, one mistake occurs in every 10^7 nucleotides copy. There are several types of RNAs and three main RNAs are shown in Table 1.14. Most RNAs are rRNAs, and mRNAs are 3%–5%.

TABLE 1.13 The genetic codes (codons).

Protein	Codon
Asp	GAU, GAU
Glu	GAA, GAG
Arg	AGA, AGG, CGA, CGC, CGG, CGU
Lys	AAA, AAG
His	CAC, CAU
Asn	AAC, AAU
Gln	CAA, CAG
Ser	AGC, AGU, UCA, UCC, UCG, UCU
Thr	ACA, ACC, ACG, ACU
Tyr	UAC, UAU
Ala	GCA, GCC, GCG, GCU
Gly	CAA, CAG
Val	GUA, GUC, GUG, GUU
Leu	UUA, UUG, CUA, CUC, CUG, CUU
Ile	AUA, AUC, AUU
Pro	CCA, CCC, CCG, CCU
Phe	UUC, UUU
Met	AUG
Trp	UGG
Cys	UGC, UGU
Stop	UAA, UAG, UGA

TABLE 1.14 Three main types of RNA.

Type of RNA	Function
mRNA	Messenger RNAs, code for protein
rRNA	Ribosomal RNAs form the basic structure of the ribosome
tRNA	Transfer RNAs, adapters between mRNA and amino acids

1.5.9 Synthesis from RNA to protein

The translation of the nucleotide sequence of an mRNA molecule into protein takes place in the cytoplasm on a large ribonucleoprotein assembly called a ribosome. The amino acids used for protein synthesis are first attached to a family of tRNA molecules, each of which recognizes, by complementary base-pair interactions, particular sets of three nucleotides in the mRNA (codons). The sequence of nucleotides in the mRNA is then read from one end to the other insets of three according to the genetic code (see Table 1.13).

To initiate translation, a small ribosomal subunit binds to the mRNA molecule at a start codon (AUG, see Table 1.13) that is recognized by a unique initiator tRNA molecule. A large ribosomal subunit binds to complete the ribosome and begin protein synthesis. During this phase, tRNAs, which bear a specific amino acid, bind sequentially to the appropriate codons in mRNA through complementary base pairing between tRNA anticodons and mRNA codons. Each amino acid is added to the C-terminal end of the growing polypeptide in four sequential steps: tRNA binding, followed by peptide bond formation, followed by two ribosome translocation steps. The mRNA molecule progresses codon by codon through the ribosome in the 5′- to 3′- direction until it reaches one of three stop codons (see Table 1.13). A release factor then binds to the ribosome, terminating translation and releasing the completed polypeptide.

Eucaryotic and bacterial ribosomes are closely related, despite differences in the number and size of their rRNA and protein components. The rRNA has the dominant role in translation, determining the overall structure of the ribosome, forming the binding sites for the tRNAs, matching the tRNAs to codons in the mRNA, and creating the active site of the peptidyl transferase enzyme that links amino acids together during translation.

In the final steps of protein synthesis, two distinct types of molecular chaperones guide the folding of polypeptide chains. These chaperones recognize exposed hydrophobic patches on proteins and serve to prevent the protein aggregation that would compete with the folding of newly synthesized proteins into their correct three-dimensional conformations. This protein folding process must also compete with an elaborate quality control mechanism that destroys proteins with abnormally exposed hydrophobic patches. In this case, ubiquitin is covalently added to a misfolded protein by a ubiquitin ligase, and the resulting polyubiquitin chain is recognized by the cap on a proteasome that moves the entire protein to the interior of the proteasome for proteolytic degradation. A closely related proteolytic mechanism, based on special degradation signals recognized by ubiquitin ligases, is used to determine the lifetimes of many normally folded proteins. By this method, selected normal proteins are removed from the cell (Alberts et al., 2008).

1.5.10 Synthesis of proteins with hexahistidine-tag (6xHis-tag)

In biochemistry, proteins are often synthesized with the polyhistidine tag using *Escherichia coli* (abbreviated as *E. coli*), a popular method for biochemists.

For the synthesis of the target protein, the amino acid sequence of the target protein must be known. When the amino acid sequence is known, the sequence of DNA is also determined, because three DNA sequences determine one amino acid. In this sequence, the six-repetition of histidine (histidine is an essential amino acid) called as hexahistidine-tag (or 6xHis-tag or His6 tag) is added.

This recombinant DNA is inserted into that of *E. coli*. The DNA of *E. coli*. is one circular DNA, so that the DNA is cut at one place (the place commonly or preferentially used is called as a "cloning vector"), and connected again at two places as the 6xHis-tag-(DNA of the target protein)-6xHis-tag. When the protein is expressed by *E. coli*, the protein (recombinant protein) has the 6xHis-tag at the beginning and the end. This 6xHis-tag is used for the purification of protein described in the next section.

1.5.11 Purification of protein by immobilized metal affinity chromatography using Ni-NTA resin

xHis-tag has a high affinity with Ni^{2+}, because the polyhistidine easily makes the stable chelate complex with Ni^{2+}. This character is used for the purification of the protein. Nickel ions are immobilized by chelation with nitrilotriacetic acid (Ni-NTA) bound to a solid support of chromatographic resin (Ni-NTA resin; see Fig. 1.70). This resin is commercially available, and the purification method of proteins with the 6xHis-tag is called as immobilized metal affinity chromatography (IMAC).

After the *E. coli* produced the protein based on the recombinant DNA, the protein needs to be purified. The cell membranes of *E. coli* containing the target recombinant protein are broken. Then the solution containing the recombinant protein is passed through the column packed with the Ni-NTA resin at a relatively high pH (e.g., ~8) with ~20 mM imidazole. The tagged protein is passed through the column, and adsorbed on the resin, while other materials, such as recombinant DNA, cytoplasm, and other proteins and enzymes which are not tagged, are washed away. Then the column was washed with ~200 mM imidazole for the target protein to recover. Finally, the polyhistidine tag is removed by endopeptidases, and pure proteins are recovered.

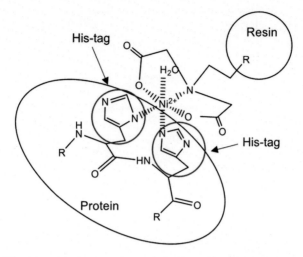

FIGURE 1.70 **Reagents for protein recombination.** As histidine has high affinity to Ni ions, the histidine-tagged protein can be easily concentrated onto the nickel nitrilotriacetic acid (Ni-NTA) bound to a chromatographic resin (Ni-NTA resin). *Source: From "Biochemistry for Material Science", Akio Makishima, page 12, Fig. 2.12.1, Copyright Elsevier (2017).*

1.5.12 CRISPR-Cas9 gene editing

In 2012, J. Doudna and E. Charpentier published the finding that editing genomic DNA becomes possible using CRISPR-Cas9 (Clustered Regularly Interspaced Short Palindromic Repeats-Crispr ASsociated protein 9) with guide RNA (Jinek et al., 2012). Both were awarded the Nobel Prize in Chemistry in 2020.

The CRISPR is a DNA sequence found in the genomes such as bacteria and archaea. In 1987, Y. Ishino first found CRISPR as the strangely repeated sequence in the genome of *E. coli* (Ishino et al., 1987). The Cas9 protein (see Fig. 1.71) targets the DNA domain of a protospacer adjacent motif which is a part of CRISPR. The Cas9 is combined with single-guide RNA (sgRNA or gRNA; made of tracrRNA in hairpin form and crRNA), and cleavages the target DNA. In Cas9, the part of the sequence of crRNA is inserted into the cleavaged DNA.

1.5.13 Fischer projection

Fischer projection was devised by H.E. Fischer, by which a three-dimensional organic molecule is presented in a two-dimensional form. Usually, the Fischer projection is used for representing monosaccharides (simple sugars), of which the general formula is $C_nH_{2n}O_n$. All bonds

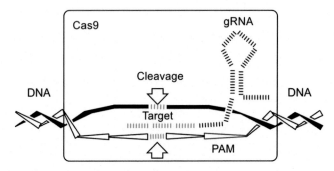

FIGURE 1.71　Schematic diagram showing the work of CRISPR-Cas9. See text for details.

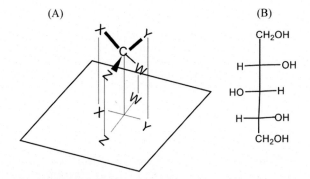

FIGURE 1.72　Fisher Projection. (A) Projection of a tetrahedral molecule on a flat plane. (B) A Fischer projection of xylitol.

except initial and terminal bonds are indicated as horizontal or vertical lines. The carbon chain is shown as the vertical lines, and the center of crossing lines exists carbon atoms. An example of a Fischer projection of a tetrahedral molecule on a flat plane is shown in Fig. 1.72A. A Fischer projection of xylitol, which is a sugar alcohol used as a sweetener, is shown in Fig. 1.72B. As xylitol is positively beneficial for dental health, it is used in chewing gum, lozenges, etc.

1.5.14 Tholin

Tholin means a variety of organic compounds formed by solar ultraviolet or cosmic ray irradiation of carbon-containing compounds, such as carbon dioxide, methane, and ethane. Sometimes they combine with nitrogen or water. Tholin is observed on the surfaces of comets, many icy moons, and Kuiper belt objects. Tholin also exists on the surface of the moons, such as Titan, Europa, Rea, and dwarf planets, such as Pluto, Ceres, Charon, etc.

1.5.15 Fischer−Tropsch reaction and carbon isotopic fractionation

Fischer−Tropsch reaction or process is a chemical reaction that converts a mixture of carbon monoxide and hydrogen into hydrocarbons:

$$n\ CO + (2n + 1)H_2 \rightarrow C_nH_{2n+2} + n\ H_2O \qquad (1.49)$$

The Fischer−Tropsch reaction can also make oxidized carbon:

$$2n\ CO + (n + 1)H_2 \rightarrow C_nH_{2n+2} + n\ CO_2 \qquad (1.50)$$

Lancet and Anders (1970) first measured carbon isotopic fractionation in the Fischer−Tropsch reaction. They observed fractionation of -50 to $-100‰$ at 400K, which was similar to those of carbonaceous chondrites. When hydrated silicates are formed, the reaction of:

$$4(Mg, Fe)_2SiO_4 + 4\ H_2O + 2\ CO_2 \rightarrow 2(Mg, Fe)CO_3 + 2\ (Mg, Fe)_3Si_2O_5(OH)_4$$
$$\text{olivine} \qquad\qquad \text{carbonate} \qquad \text{serpentine}$$

$$(1.51)$$

occurs. This is a serpentinization reaction.

1.5.16 Confocal Raman spectroscopy

When the monochromatic light is lit on a sample molecule, the light is reflected, absorbed, or scattered. In some cases, Raman scattering occurs, in which lights of shorter and longer wavelengths (Stokes and anti-Stokes Raman scattering lights) are emitted from the sample molecule. The Raman scattering is dependent on the vibrational states of the molecule but very weak ($\times 10^{-7}$). To enhance the Raman scattering lights, a larger number of photons in the incident light is required. Thus, laser light is generally used.

The confocal microscopy is an optical imaging technique to enhance resolution and contrast. There is a spatial pinhole at the confocal plane of the lens to remove out-of-focus light of Z-direction. The Raman spectroscopy instrument is used with this confocal microscope, and not only a small X−Y area (>20 nm area) but also the Z-directional image can be analyzed. Therefore, the confocal Raman microscope enables to obtain an X−Y−Z Raman scattering light image, or three-dimensional imaging.

1.5.17 Characterization of the present life

Life is divided into eight hierarchies (life, domain, kingdom, phylum, class, order, family, genus, and species) in the biological classification, which is called as "taxonomic rank." All the lives on the Earth are divided into three domains, Archaea, Bacteria, and Eukaryota (see Fig. 1.73).

Archaea has neither a cell nucleus nor any other membrane-bound organelles in their cell. Bacteria is a few micrometers in length and has various shapes, ranging from spheres to rods and spirals. Both archaea and bacteria have only one kingdom, therefore, life is divided into eight kingdoms. Eukaryota cell contains a nucleus and other organelles enclosed within membranes. For example, any plants or animals are Eukaryotes.

1.5.18 Stromatolites

Stromatolites are layered sedimentary formations (microbialite) which are produced by photosynthetic microorganisms such as cyanobacteria, sulfate-reducing bacteria, and Pseudomonadota (see Fig. 1.74A). These microorganisms make adhesive compounds that cement sand and other rocky materials. Stromatolites occur widely in the fossil record of Precambrian (Fig. 1.74B), but rare today.

1.5.19 Hydroxyl alcohols, hydroxyl aldehydes, and hydroxyl ketones

Hydroxyl alcohols, hydroxyl aldehydes, and hydroxyl ketones with linear and branched structures are shown in Fig. 1.75.

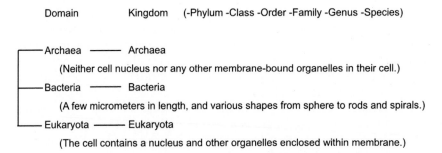

FIGURE 1.73 **Taxonomic rank and characterization of the present life.** The life on the Earth is divided into three domains, Archaea, Bacteria, and Eukaryota. Life is characterized by biological rank. Eukaryotes are animals and plants. The cell contains a nucleus etc. Bacteria and Archaea do not have a nucleus. Bacteria have some membrane-bound organelles, but Archaea do not.

(A)

(B)

FIGURE 1.74 Stromatolites. (A) Modern stromatolites in Shark Bay, Western Australia. (B) Fossilized stromatolites, about 425 Ma, in the Soeginia Beds near Kubassaare, Estonia. *Source: (A) From Paul Harrison (Reading, UK). https://en.wikipedia.org/wiki/Stromatolite#/media/File:Stromatolites_in_Sharkbay.jpg. (B) Wilson44691. https://en.wikipedia.org/wiki/Stromatolite#/media/File:SoegininaStromatolitesEstonia.jpg.*

The chemical formulas are shown by the Fisher projection (see Section 1.5.13). When an aldehyde functional group exists at a carbon atom of position 1, it is called aldoses (Fig. 1.75A). When a ketone functional group resides in a carbon atom at position 2 or 3, it is called ketoses (Fig. 1.75B).

In the case of the five-carbon aldoses, they are called pentoses or monosaccharides. The pentoses are separated into two groups: one

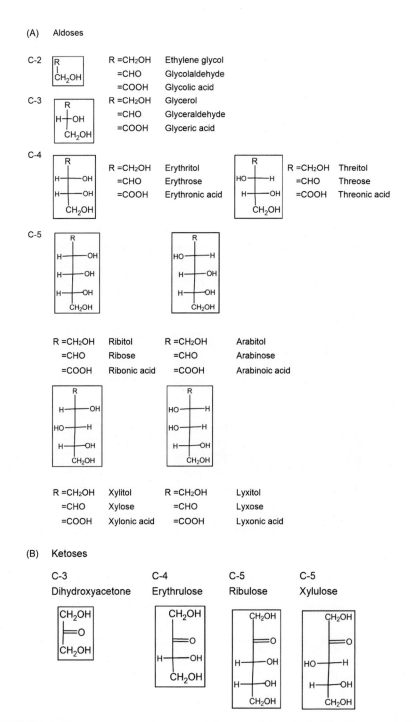

FIGURE 1.75 Structures and names of linear and branched (A) aldoses and (B) ketoses. (A) Aldoses with C-2, C-3, C-4, and C-5 are shown. In the Fisher plot, it is a straight molecule of carbon with the head and tail are CH₂OH, CHO, or COOH, and middles are CHOH, which is chiral carbon. (B) Ketoses with C-3, C-4, and C-5 are shown. In the Fisher plot, the head and tail are CH₂OH, and the middles are C = O and CHOH.

group is aldopentoses, and the other group is ketopentoses. The aldopentoses have an aldehyde functional group at position 1. The ketopentoses have a ketone functional group in position 2 or 3. As the aldopentoses have three chiral centers, eight (2^3) enantiomers are possible including D- and L-riboses, arabinoses, xyloses, and lyxoses. As the ketopentoses have two chiral centers, four (2^2) enantiomers (see Fig. 1.75B) are possible as D- and L-ribuloses and xyluloses.

1.5.20 The schematic structure of the eukaryote cell

The schematic structure of the eukaryote cell is shown in Fig. 1.76. The cell is contained in the cell membrane, which is made of a double layer of phospholipids (schematically depicted in Fig. 1.77A), or called

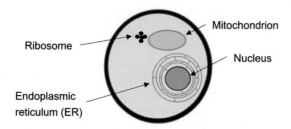

FIGURE 1.76 **The schematic structure of the eukaryote cell.** The eukaryote cell has a nucleus, mitochondrion, ribosome, and endoplasmic reticulum which connects the nucleus to various parts of the cell.

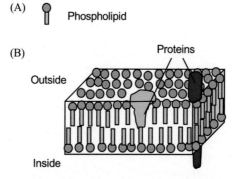

FIGURE 1.77 **The bilayer cell membrane made of phospholipid molecules.** (A) The phospholipid. The phospholipid is composed of a hydrophilic glycerol-phosphate head (depicted as a pink sphere) with long hydrophobic fatty acid tails (depicted as a pale green bar connected to the sphere). (B) Modern cell membranes. The membrane is composed of a bilayer of the phospholipid. Functional proteins are embedded in the membrane. Some proteins penetrate the membrane to transport ions from inside to outside and vice versa. Early life should have a similar membrane but simpler functions.

as phospholipid bilayer (see Fig. 1.77B). This membrane is embedded with many kinds of proteins to protect the cell and cytoplasm from hostile conditions such as high osmotic pressure.

The DNA is contained in chromosomes at the nucleus, which is clearly separated by the phospholipid bilayer from the cytoplasm. In contrast, the archaea and bacteria cells called as prokaryote cells have no nucleus. Mitochondria, which generate energy efficiently, are also separated by the phospholipid bilayer. Mitochondria also do not exist in the prokaryote cells. Endoplasmic reticulum, which is also the bilayer transport network for molecules, exists around the nucleus. Ribosomes, which are large complex of RNA and protein molecules, also exist.

1.5.21 Short story for the origin of life

After the late veneer, the Earth cooled, and the life-related molecules were formed (see Chapter 2, Section 7.1). Then, life is born.

After the life-related molecules of life are synthesized, polymers (nucleic acids and peptides; it is called as the prebiotic world) occur. Then, the formation of supramolecular architectures (membrane and protocells) occurred. Subsequently, the three main hypothetical steps are proposed to replicate or translate complicated information to the next generation. They are:

- Peptide World
- Peptide + Nucleic Acid World
- RNA World
- Protocell.

It is still a long way from the protocell to the modern life. The genetic information transfer system needed to be evolved, and the present RNA/DNA system should be formed and developed.

1.5.21.1 The peptide world

After prebiotic or prebiologic amino acids are synthesized under heating or dehydrating conditions, especially with a catalyst, amino acids are polymerized and form proteins.

However, if the simplest life, which is made only by one peptide sequence (or one protein) of 20 amino acids, was accidentally born, the possibility of the amino acid sequence (or the protein) is $1/20^{20} = 1.0 \times 10^{-26}$. Even if one protein (20 amino acid sequences) can be synthesized in 1 hour (it is still too fast), to synthesize all sequences will take 1.1×10^{15} Gyrs, which is far longer than the age of the universe. Furthermore, there are no chemical methods to duplicate a 20 a-peptide ligation with the same sequence. To duplicate the same sequence, the storage system of the peptide ligation is required.

1.5.21.2 Peptide + nucleic acid world

It is a fundamental question of how nucleic acids first assembled and then started the earliest cellular life about 4 billion years ago. The presence of a phospholipid enhanced the yield of polymeric products because the lipid matrix serves to concentrate and organize the mononucleotides.

To encode amino acids, Peptide Nucleic Acids (PNA see Fig. 1.78A) were proposed with a peptide-like backbone and nucleic bases on the side chains (Bohler et al., 1995). The PNA not only play the role of information carriers but also work as catalysts in a pre-RNA World.

1.5.21.3 RNA world

In the RNA (see Fig. 1.78B) world hypothesis, the nucleic acid played both the role of information storage and the role of the catalysts at the initial stage of evolution. However, the sudden appearance of

FIGURE 1.78 (A) Peptide nucleic acids (PNA) and RNA. (A) PNA (peptide nucleic acid) as a hypothetical information carrier. (B) RNA (ribonucleic acid) is the information carrier of the present life. R is the site where the genetic information is stored.

self-replicating RNA is unlikely. The RNA world should have been preceded by the pre-RNA world, in which another system of replicating molecules was working. Or there is a possibility that life evolved using directly both pre-RNA and RNA systems, where Peptide + Nucleic acid World are working. The RNA world finally evolved into the protein, RNA, and DNA world today.

1.5.21.4 The protocell

The first life might be made of a single cell, and three functional units as follows: (1) compartments, by which the inside is separated from the outside of the cell physically and chemically; (2) cytoplasm, which is the cell content. In the cytoplasm, the chemical reactions called "metabolism" work for the production of energy and materials to keep life; and (3) genetic information, which is stored safely, and passed to the next generation by RNA, or DNA, etc. In addition to these three basic functions, the first life needs to perform (4) reproduction or duplication of the life must be done.

Such first cell is called as the "protocell." Many scientists proposed that the protocell should have compartments (membranes), and the protocell is considered to have a self-reproducing compartment (Morowitz et al., 1988; Varela et al., 1974). Membranes keep their integrity, facilitate the material exchange between the external environments, and protect ion gradients which is advantageous in using energy sources (Pascal et al., 2006).

The protocell membranes should be like the modern cell. The cell membranes are essentially made of phospholipids, amphiphilic molecules composed of a hydrophilic glycerol-phosphate head (depicted as a sphere in Fig. 1.77A) bound to long hydrophobic fatty acid tails (depicted as a bar combined to the sphere in Fig. 1.77A). They form bilayers and carry proteins for energy-producing processes (Fig. 1.77B). The cell membranes define the self-boundary and retain the necessary contents. At early stages, the cell membrane had few functions, and increased functions as long time passed.

When the cell is duplicated, the cell membrane is also duplicated. The protocell is a cell-like compartment enclosed in a lipid vesicle containing nucleic acid and other biomolecules. The important problem is how they acquired the ability of self-replication.

1.5.21.5 Road from protocell to archaea, bacteria, and eukaryotes

If the protocell has evolved into the three present lives, archaea, bacteria, and eukaryota, the common ancestor is named as "cenancestor" (Fitch and Upper, 1987) or the last universal common ancestor (abbreviated as LUCA) (Forterre and Philippe, 1999). The cenancester existed several billion years ago, and its diversification made three domains of life. As nothing is known about the protocell, biologists have an interest in how eukaryotic cells formed. The idea of the cenancestor has changed into the model that the eukaryotic cell was generated by endosymbiosis of a bacteria cell with an archaea cell. The origin of the eukaryote cell (see Fig. 1.76), which contains a nucleus and other membrane-bound compartments, is one of the conundrums in the evolution of life on Earth (Embley and Martin, 2006).

1.5.22 Nuclear magnetic resonance

1.5.22.1 Introduction

Nuclear magnetic resonance, which is abbreviated as NMR, is a phenomenon where nuclei in a magnetic field absorb and re-emit electromagnetic radiation. This electromagnetic radiation (E) or frequency (n) is determined by the strength of the magnetic field ($\mathbf{B}$), the magnetic properties of the isotope of the atom (m):

$$E = -m\mathbf{B} \tag{1.52}$$

and

$$m = g \; m \; \hbar \tag{1.53}$$

where g, m, and $\hbar$ are the z-component of the magnetic moment, the spin status (e.g., $\pm 1/2$), and the reduced Planck constant. In the NMR, the splitting energy (ΔE) between $+m$ and $-m$ is determined at the constant magnetic field ($\mathbf{B}_0$):

$$\Delta E = 2 \; g \; m \; \hbar \; \mathbf{B}_0 \tag{1.54}$$

From Eq. (1.54), NMR does not occur for the isotope of the spin status of zero. All nuclei with even protons and neutrons have no nuclear spin, therefore, the spin status becomes zero, and the NMR is not observed. The resonance energy changes according to the magnetic moment (g) around the target nuclei at the constant magnetic field.

The resonance frequency is VHF and UHF television broadcasts (60−1000 MHz; VHF and UHF are 30~300 and 300~3000 MHz, respectively). NMR allows the observation of specific quantum mechanical magnetic properties of the atomic nucleus. The magnetic property of the nucleus changes according to the shielding effects by the electrons around the target nucleus, because the electron shielding reduces the magnetic field around the nucleus.

Thus NMR spectroscopy is one of the principal techniques to obtain the topology, dynamics, and three-dimensional information of molecules. The proton-NMR is also used routinely in medical imaging techniques, such as magnetic resonance imaging (MRI; The word "nuclear" reminds non-scientists of "atomic bombs or atomic reactions," therefore, the word "nuclear" was avoided to use in MRI).

Commonly used nuclei for NMR analyses are ^{1}H (spin $- 1/2$), ^{11}B (spin 3/2), ^{13}C (spin $- 1/2$), ^{14}N (spin $- 1$), ^{19}F (spin $- 1/2$), ^{31}P (spin $- 1/2$), etc. Therefore, boranes and carboranes are suitable for the application of NMR, resulting in rapid advances of their structural studies.

1.5.22.2 *Proton nuclear magnetic resonance (^{1}H NMR)*

The proton nuclear magnetic resonance (proton-NMR) determines the nuclear resonances of each ^{1}H in the sample, which change according to the electron shielding effects around protons. The sample is dissolved with solvents without ^{1}H, such as deuterated water (D_2O), deuterated methanol (CD_3OD), or carbon tetrachloride (CCl_4). Then the sample is put in the magnetic field and the resonant signals are measured. The resonance frequencies are presented as the chemical shift values represented in δ in ppm. The δ value is affected by the neighboring functional groups. When the electron-withdrawing effect of the functional group is larger, the chemical shift becomes larger. For example, the δ value of C^1H_3- is:

$$CH_2R(0.8) < COOR(2.0) < COOH(2.1) < CHO(2.2) < Cl(3.0) < OH(3.3) < OCOR$$

$$(1.55)$$

(the value in the parentheses is ppm). In addition, the signal intensity (area) is proportional to the number of the ^{1}H. Furthermore, the peak splits by the spin–spin coupling, which causes the splitting three bonds away.

In Fig. 1.79, an example of a ^{1}H NMR spectrum of ethyl acetate is shown. From the chemical shifts, we know that three types of hydrogens exist. From the peak heights (area), the abundance ratio of 800 ($= 100 \times 2 + 300 \times 2$):1200:1200 ($= 300 \times 2 + 600$) = 2:3:3 is expected. The singlet peak with the chemical shift of ~ 2 is expected to be without hydrogens to three bonds away, resulting in C^1H_3-COOR. The triplet peaks with the smallest chemical shift of ~ 1.3 are expected to have two hydrogens to three bonds away, resulting in $C^1H_3-CH_2-$. Finally, the quartet peaks with the largest chemical shift of ~ 4.1 and the peak area ratio are expected with $CH_3-C^1H_2-OCOR$. Thus, all hydrogens are assigned as shown in the colors of Fig. 1.79 of the material is ethyl acetate.

1.6 Statistics

In this section, basic statistics especially required in astrochemistry are explained.

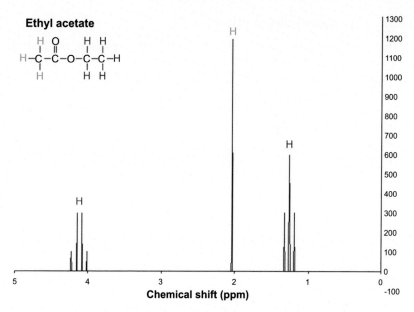

FIGURE 1.79   **The example of ^{1}H NMR spectrum of ethyl acetate.** The intensity (the vertical axis) is plotted versus chemical shift (the horizontal axis). *Credit: T.vanschaik. https:// upload.wikimedia.org/wikipedia/commons/8/80/1H_NMR_Ethyl_Acetate_Coupling_shown.png.*

1.6.1 Average and standard deviation

We assume that there is one test value (the blood pressure or the weight, for example) for one person, which is measured based on the same method and with the same condition. The first value (A_1), the second value (A_2), ..., and the n-th value (A_n) for n persons by the same method with the same condition. Then we obtain the average (μ) as

$$\mu = (A_1 + A_2 + \ldots + A_n)/n \tag{1.56}$$

The standard deviation (sometimes denoted as SD or σ) is obtained as

$$SD = \left[\sum (A_i - \mu)^2 / (n-1) \right]^{1/2} \tag{1.57}$$

which is easily obtained by a "STDEV" function in Microsoft EXCEL worksheet software. We can calculate the relative standard deviation in percentage (RSD%) as

$$RSD\% = SD/\mu \times 100 \tag{1.58}$$

2 RSD% means twofold of RSD%.

1.6.2 The normal distribution

The normal distribution is one of the ideal variations of data. The data scatters as a bell shape. The horizontal axis is the data value, and the vertical axis is the probability. When this curve is integrated from infinite menus to infinite plus, the integrated probability becomes 1. From $-\sigma$ to $+\sigma$, 68.3% of the data are included. From -2σ to $+2\sigma$, 95.4% of the data are included. From -3σ to $+3\sigma$, 99.7% of the data are included.

The probability density function (PDF) is determined as (Fig. 1.80A):

$$f(x) = \frac{1}{\sigma\sqrt{2\pi}} e^{-\frac{1}{2}\left(\frac{x-\mu}{\sigma}\right)^2} \tag{1.59}$$

and the cumulative distribution function (CDF) which is the integrated area of the PDF from minus infinity to x is determined as (Fig. 1.80B):

$$F(x) = \int_{-\infty}^{x} f(t)dt = \frac{1}{2}\left[1 + \mathrm{erf}\left(\frac{x-\mu}{\sigma\sqrt{2}}\right)\right] \tag{1.60}$$

where erf (z) (this is called as error function) is:

$$\mathrm{erf}(x) = \frac{2}{\sqrt{\pi}} \int_{0}^{z} e^{-t^2} dt \tag{1.61}$$

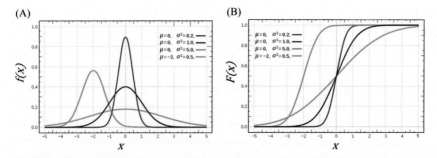

(A)   **(B)**

FIGURE 1.80 **The probability density function and the cumulative distribution function.** (A) indicates the probability density function of the normal distribution. (B) shows the cumulative distribution function of the normal distribution. The normal distribution is the bell-shaped symmetric distribution. From $-\sigma$ to $+\sigma$, 68.3% of the data are included. From -2σ to $+2\sigma$, 95.4% of the data are included. From -3σ to $+3\sigma$, 99.7% are included. *Credit: (A) https://commons.wikimedia.org/wiki/File:Normal_Distribution_PDF.svg. (B) https://upload.wikimedia.org/wikipedia/commons/c/ca/Normal_Distribution_CDF.svg.*

1.6.3 The standard error

We assume a parent population that has a normal distribution of data with an average of μ and a standard deviation of σ. We want to estimate the average and standard deviation of the parent population by picking up a randomly smaller number of the parent population, and we make subpopulation, because the parent population is too large to use all data in the statistics in many cases.

The average and the standard deviation of the subpopulation are μ' and SE, respectively. From the statistics, μ is equal to μ'. The standard deviation of the subpopulation is called a standard error, SE. The standard error (SE) should be:

$$SE = SD/n^{1/2} \tag{1.62}$$

1.6.4 The student's t-distribution

The Student's t-distribution (or the t-distribution) appears in many statistics analyses, such as Student's t-test. If we take n samples from a normal distribution, t-distribution with degrees of freedom of $n = n - 1$ can be defined. It is the same case of the standard error explained in Section 1.6.3. The t-distribution is also symmetric and bell-shaped like the normal distribution, however, the t-distribution has wider tails (Fig. 1.81A).

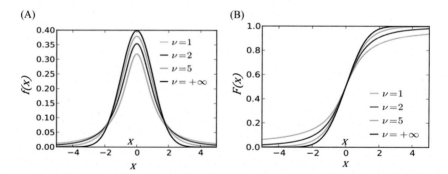

FIGURE 1.81 **The probability density function and the cumulative distribution function.** (A) indicate the probability density function of the Student's t-distribution. (B) shows the cumulative distribution function of the Student's t-distribution, respectively. The t-distribution is also symmetric and bell-shaped like the normal distribution, however, the t-distribution has wider tails. *Credit: (A) IkamusumeFan, https://commons.wikimedia.org/ wiki/File:Student_t_pdf.svg. (B) Hydrox, https://upload.wikimedia.org/wikipedia/commons/thumb/e/ e7/Student_t_cdf.svg/600px-Student_t_cdf.svg.png.*

PDF of the Student's t-distribution is given by:

$$f(x) = \Gamma\left(\frac{\nu+1}{2}\right)\Big/\sqrt{\pi\nu}\,\Gamma\left(\frac{\nu}{2}\right) \times \left(1+\frac{x^2}{\nu}\right)^{-\frac{\nu+1}{2}} \tag{1.63}$$

here

$$\Gamma(z) = \int_0^\infty t^{z-1}e^{-t}dt \tag{1.64}$$

The CDF is skipped because it is too complicated (see Fig. 1.81B).

References

Alberts, B., Johnson, A., Lewis, J., et al., 2008. Molecular Biology of the Cell, sixth ed. Garland Science, New York.

Amelin, Y., Krot, A.N., Hutcheon, I.D., Ulyanov, A.A., 2002. Lead isotopic ages of chondrules and calcium-aluminum-rich inclusions. Science 297, 1678–1683.

Baade, W., 1944. The resolution of Messier 32, NGC 205, and the central region of the Andromeda nebula. Astrophys. J. 100, 137–146.

Bohler, C., Nielsen, P.E., Orgel, L.E., 1995. Template switching between PNA and RNA oligonucleotides. Nature 376, 578–581.

Bus, S.J., Binzel, R.P., 2002. Phase II of the small main-belt asteroid spectroscopic survey. A feature-based taxonomy. Icarus 158, 146–177.

Embley, T.M., Martin, W., 2006. Eukaryotic evolution changes and challenges. Nature 440, 623–630.

Fitch, W.M., Upper, K., 1987. The phylogeny of transfer-RNA sequences provides evidence for ambiguity reduction in the origin of the genetic-code. Cold Spring Harb. Symp. Quanti. Biol. 52, 759–767.

Forterre, P., Philippe, H., 1999. The last Universal Common Ancestor (LUCA) simple or complex? Biol. Bull. 196, 373–375.

Halliday, A.N., Lee, D.C., 1999. Tungsten isotopes and the early development of the Earth and Moon. Geochim. Cosmochim. Acta 63, 4157–4179.

Harper, C.L., Jacobsen, S.B., 1994. [182]Hf-[182]W: a new cosmochronometer and method for dating planetary core formation events. Sixth Meeting of the European Union of Geosciences, p. 451.

Hewins, R.H., 1997. Chondrules. Annu. Rev. Earth Planet. Sci. 25, 61–83.

Ishino, Y., Shinagawa, H., Makino, K., Amemura, M., Nakata, A., 1987. Nucleotide sequence of the iap gene, responsible for alkaline phosphatase isozyme conversion in Escherichia coli, and identification of the gene product. J. Bacteriol. 169, 5429–5433.

Jinek, M., Chylinski, K., Fonfara, I., Hauer, M., Doudna, J.A., Charpentier, E., 2012. A programmable dual-RNA–guided DNA endonuclease in adaptive bacterial immunity. Science 337, 816–821.

Johnson, B.C., Bowling, T.J., Melosh, H.J., 2014. Jetting during vertical impacts of spherical projectiles. Icarus 238, 13–22.

Johnson, B.C., Minton, D.A., Melosh, H.J., Zuber, M.T., 2015. Impact jetting as the origin of chondrules. Nature 517, 339–341.

Jumper, J., Evans, R., Pritzel, A., Green, T., Figurnov, M., Ronneberger, O., et al., 2021. Highly accurate protein structure prediction with AlphaFold. Nature 596, 583–589.

Krasznahorkay, A.J., Csatlos, M., Csige, L., Gacsi, Z., Gulyas, J., et al., 2016. Observation of anomalous internal pair creation in Be-8: a possible indication of a light, neutral boson. Phys. Rev. Lett. 116, 042501.

Krasznahorkay, A.J., Krasznahorkay, A., Csatlos, M., Csige, L., Timar, J., Begala, M., et al., 2023. Observation of anomalous internal pair creation in Be-8: a possible indication of a light, neutral boson. Available from: https://doi.org/10.48550/arXiv.2308.06473.

Kruijer, T.S., Toubout, M., Gischer-Gödde, M., Berminghan, K.R., Walker, R.J., Kleine, T., 2014. Protracted core formation and rapid accretion of protoplanets. Science 344, 1150−1154.

Küppers, M., O'Rourke, L., Bockelee-Morvan, D., et al., 2014. Localized sources of water vapour on the dwarf planet (1) Ceres. Nature 505, 525−527.

Lancet, M.S., Anders, E., 1970. Carbon isotope fractionation in the Fischer-Tropsch Synthesis and in meteorites. Science 170, 980−982.

Lu, Y.H., Makishima, A., Nakamura, E., 2007. Coprecipitation of Ti, Mo, Sn and Sb with fluorides and application to determination of B, Ti, Zr, Nb, Mo, Sn, Sb, Hf and Ta by ICP-MS. Chem. Geol. 236, 13−26.

Makishima, A., Masuda, A., 1993. Primordial Ce isotopic composition of the solar system. Chem. Geol. 106, 197−205.

Makishima, A., Nakamura, E., 2006. Determination of major, minor and trace elements in silicate samples by ICP-QMS and ICP-SFMS applying isotope dilution-internal standardization (ID-IS) and multi-stage internal standardization. Geostand. Geoanal. Res. 30, 245−271.

Malaviarachchi, S.P.K., Makishima, A., Tanimoto, M., Kuritani, T., Nakamura, E., 2008. Highly unradiogenic lead isotope ratios from the Horoman peridotite in Japan. Nat. Geosci. 1, 859−863.

McDonough, W.F., Sun, S.-s, 1995. The composition of the Earth. Chem. Geol. 120, 223−253.

Morowitz, H.J., Heinz, B., Deamer, D.W., 1988. The chemical logic of a minimum protocell. Orig. Life Evol. Biosphere 18, 281−287.

Pascal, R., Boiteau, L., Forterre, P., Gargaud, M., Lazcano, A., et al., 2006. Prebiotic chemistry-: Biochemistry-: Emergenceo of life (4.4-2 Ga). Earth Moon Planets 98, 153−203.

Schauer, A.T.P., Bromm, V., Drory, N., Boylan-Kolchin, M., 2022. On the probability of the extremely lensed z = 6.2 earendel source being a population III star. Astrophys. J. Lett. 934, L6.

Tholen, D.J., 1989. Asteroid Taxonomic Classifications. Asteroids II. University of Arizona Press, Tucson, pp. 1139−1150.

Varela, F.G., Maturana, H.R., Uribe, R., 1974. Autopoiesis: the organization of living systems its characterization and a model. Biosystems 5, 187−196.

Watson, H.C., 1969. The stereochemistry of the protein myoglobin. Prog. Stereochem. 4, 299.

Welch, B., Coe, D., Diego, J.M., Zitrin, A., Zackrisson, E., Dimauro, P., et al., 2022. A highly magnified star at redshift 6.2. Nature 603, 815−818.

2

Origin of elements

2.1 The origin of the universe

2.1.1 Cosmic microwave background radiation

To understand the origin of our solar system and galaxy, we must know the origin and evolution of the universe and the elements. Penzias and Wilson (1965) found the cosmic microwave background radiation (CMBR). They got the Nobel Prize in 1978. CMBR (see Fig. 2.1), which was like black body radiation at 3K (2.72548 ± 0.00057K; Gawiser and Silk, 2000) was considered to be the remnant of the Big Bang (BB). CMBR has the largest redshift of z = 1089, corresponding to 13.812 giga-light-years (1.31×10^{26} m).

The Cosmic Background Explorer spacecraft (1989−96; Fig. 2.2A), the WMAP (Wilkinson Microwave Anisotropy Probe) spacecraft (2001−10; Fig. 2.2B), and the Planck spacecraft (2009−13), which were launched by NASA, NASA and ESA, respectively, played key roles to observe CMBR precisely, and to establish the current standard model of cosmology: the Lambda-Cold Dark Matter model. This model includes: (1) the age of the universe is 13.799 ± 0.021 giga-years; (2) the current expansion rate of the universe (the Hubble constant) is 67.8 ± 0.9 km/s/Mpc^{-1}; and (3) the current universe consists of 4.9% ordinary matter, 26.8% dark matter (neither absorbs nor emits lights), and 68.3% dark energy (accelerates the expansion of the universe). The WMAP contributed to obtain not only these data but also to prove the existence of a cosmic neutrino background and the cosmic inflation paradigm.

2.1.2 Origin and evolution of the universe: the Big Bang theory

The BB started 13.8 Gyr ago. There is no accurate theory to explain what occurred in the universe at the beginning of BB. However, at the early stage of the universe, it is generally believed that the universe experienced several

Introductory Astrochemistry
DOI: https://doi.org/10.1016/B978-0-443-23938-0.00002-9

FIGURE 2.1   **Nine-year Wilkinson Microwave Anisotropy Probe (WMAP) image of cosmic microwave background radiation (2012).** This detailed all-sky picture of CMBR was obtained by 9 years of WMAP (which is a spacecraft launched by NASA in 2001) data at the Lagrange-2 point (see Section 1.1.3). The colors are artificial and show small temperature variations of ± 200 microKelvin. The image reveals 13.77-billion-year-old temperature fluctuations that correspond to the seeds of the galaxies. *Credit: NASA/WMAP Science Team. http://map.gsfc.nasa.gov/media/121238/ilc_9yr_moll4096.png.*

(A)   (B)

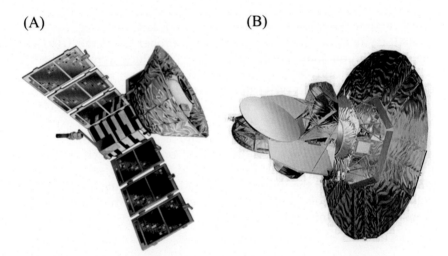

FIGURE 2.2   **Artist illustrations of spacecraft for observation of CMBR.** (A) Cosmic Background Explorer placed on the Earth orbit to observe CMBR. (B) Wilkinson Microwave Anisotropy Probe, placed on the Lagrange-2 point to observe CMBR. *Credit: (A) NASA. https://commons.wikimedia.org/wiki/File:Cosmic_Background_Explorer_spacecraft_model.png. (B) NASA. https://commons.wikimedia.org/wiki/File:WMAP_spacecraft.jpg.*

"phase changes" as ice changes into water. Each phase is called an "epoch." In this section, the evolution of the universe (from 0 to present) is briefly related according to the timeline shown in Fig. 2.3 and Table 2.1.

- **Planck epoch:** The universe started from an extremely high-temperature and high-density state. Such a hot-dense universe might have started from quantum fluctuations. The four fundamental forces of the present universe: (1) gravity, (2) electromagnetism, (3) the weak nuclear force, and (4) the strong nuclear force were only one fundamental force.
- **Grand unification epoch:** By 10^{-36} s from the start of BB, as the universe was expanded and cooled, the phase transition occurred, in which the gravity was separated from one fundamental force and the remained force is called the electrostrong interaction.
- **Inflationary epoch:** By 10^{-32} s, the universe swelled at least 10^{78} times in volume (10^{26} times in each spatial dimension), which is known as inflation, and the inflation ended. However, there is no model for why and how the inflation occurred. The universe was a hot mixture of quarks, antiquarks, and gluons.
- **Electroweak epoch:** In the inflationary stage, the high temperature of 10^{28}K became lower to 10^{22}K, and the electroweak interaction separated

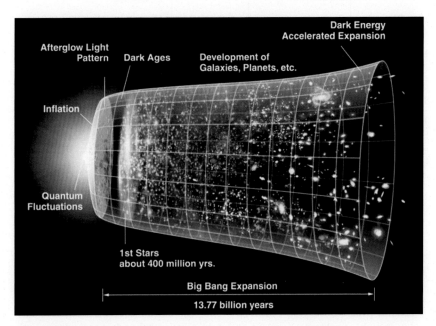

FIGURE 2.3 **Evolution of the universe**. The horizontal and vertical axes indicate the time and size, respectively. Both the axes are not to scale. *Credit: NASA. http://map.gsfc. nasa.gov/media/060915/index.html.*

TABLE 2.1   The timeline of the universe.

Time	Epoch	Temperature (K)	Main events
$0-10^{-43}$ s	Planck epoch	$>10^{32}$	One fundamental force.
$<10^{-36}$ s	Grand unification epoch	$>10^{29}$	Gravity and electrostrong interaction.
$<10^{-32}$ s	Inflationary epoch Electroweak epoch	$10^{28}-10^{22}$	Inflation and electrostrong interaction separated.
10^{-12} s	End of Electroweak epoch	10^{15}	Four fundamental forces like today appeared.
$10^{-12}-10^{-5}$ s	Quark epoch	$10^{15}-10^{12}$	Quark-gluon plasma.
$10^{-5}-1$ s	Hadron epoch	$10^{12}-10^{10}$	Baryons (protons and neutrons) appeared.
1 s	Neutrino decoupling	10^{10}	Neutrinos travel freely. CnB formed.
$1-10$ s	Lepton epoch	$10^{10}-10^{9}$	Leptons (electrons and neutrinos) were formed.
$10-10^{3}$ s	Big Bang nucleosynthesis	$10^{9}-10^{7}$	^1H, ^4He and trace ^2D, ^3He, and ^7Li formed.
10 s–370 kyr	Photon epoch	$10^{9}-4000$	^2D was photodisintegrated. Photons travel freely.
18–370 kyr	Recombination	4000	Clearing up the universe, filled with forming neutral hydrogen. CMBR.
370–150 kyr	Dark ages	4000–60	No stars.
300–400 Myr First stars >1–10 Gyr Modern galaxies	Star and galaxy formation and evolution	~60	The first stars appeared. The first clusters of stars and galaxies appeared.
200 Myr–1 Gyr	Reionization	60–19	The universe is filled with ionized plasma.
13.8 Gyr	Present time	2.7	The universe became 46 billion light-years.

kyr, Kilo years; *Myr*, million years; *Gyr*, 10^9 years.

into the electromagnetism and the weak nuclear interaction. Next, the second phase transition occurred where the electrostrong interaction separated into the strong nuclear interaction and the electroweak interaction (end of Electroweak epoch). Thus, the fundamental four forces began to work. The universe continued to expand at a much slower rate.

About 4 Gyr, the expansion began to increase by the dark matter. The universe is still expanding today. After 10^{-12} s from BB, the early universe can be better understood by physical experiments.

- **Quark epoch:** From 10^{-12} to 10^{-5} s, the universe is filled with the quark-gluon plasma.
- **Hadron epoch:** From 10^{-5} to 1 s, as the temperature lowered, baryons, such as protons (hydrogen) and neutrons which are composed of three quarks appeared. Theories predict that the proton: neutron ratio was 7:1. The equal numbers of antibaryons should have formed, however, this did not occur. The reason is not clear.
- **Neutrino decoupling:** At around 1 s after BB, the BB neutrinos began to travel freely through space. As neutrinos seldom interact with the matter, such neutrinos exist even today as cosmic neutrino background (CnB) like CMBR.
- **Lepton epoch:** The universe is filled with leptons (electron, electron neutrino, muon, muon neutrino, tauon, tauon neutrino) and antileptons, which are thermally equilibrated.
- **BB nucleosynthesis:** From 10 to 10^3 s, the BB nucleosynthesis occurred, and protons (^{1}H) and helium (^{4}He) with trace amounts of deuterium (^{2}D), ^{3}He and lithium-7 (^{7}Li) were formed.
- **Photon epoch:** From 10 s to 370 ky, the universe consisted of plasma made of nuclei, electrons, and photons.
- **Recombination:** From 370 to 150 ky, the universe became cool enough for the hydrogen ions and electrons to bound each other forming hydrogen atoms at ~ 3000K after ~ 370 ky from BB. The photon (light) can pass freely within the universe. It is called as "recombination" or "clearing up the universe."
- CMBR is the remnant of the electromagnetic field at this time. The reason why 3000K is observed as 2.7K in CMBR is the redshift or Doppler effect (see Section 1.1.4). The recombination occurred 370 ky after BB, which corresponds to a redshift of z = 1089, and the universe was filled with neutral hydrogens.
- **Dark ages:** Although the light can pass freely, there were no stars at this time. It is called "dark ages" as shown in Fig. 7.3.
- **Star and galaxy formation and evolution:** Inhomogeneous distribution of hydrogen formed gas clouds, and first stars appeared by the accumulation of hydrogen, and finally galaxies formed in the universe.
- **Reionization:** The universe changed from neutral to ionized plasma again (cosmic reionization). This process started from z = 20−10 (200−500 Ma after BB) and ended in z = 6 (~ 1 Gyr after BB). However, the hydrogen ions and electrons were too sparse to interact, thus the universe was kept clear.
- **Present time:** The universe became a sphere with a radius of 46 billion light-years. The dark energy is accelerating the expansion rate.

To find and observe the most distant (the oldest) galaxies and stars in deep space is one of the big themes of cosmosciences to know the early stage of BB. By the combination of the Hubble Space Telescope and Spitzer Space Telescope (2003−20), (Oesch et al., 2016) found a bright galaxy named GN-z11 with z = 11.1, which meant 13.4 billion years in the past, just after 400 Myr from BB (Fig. 2.4A). It was surprising that a galaxy so massive existed only 400 Myr after the very first stars started to form, which implied that stars were produced at a huge rate forming a galaxy.

In July 2022, by the James Webb Space Telescope, Naidu et al. (2022) discovered GLASS-z13 with z = ∼13.1 (Fig. 2.4B). It is currently the oldest galaxy ever found, dating back to just 300−400 million years after the BB. Another similarly old galaxy, GN-z11 was found near GLASS-z13.

2.2 Origin of elements

As over 99.8% mass of the solar system is that of the sun which is made of 71% of hydrogen (H) and 27% of helium (He), therefore, the solar system as well as the sun is mainly composed of these two elements. The rest of 2% elements are mainly composed of the Earth and

(A)   (B)

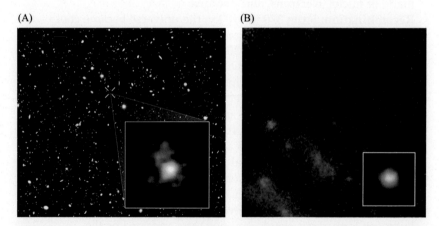

FIGURE 2.4 **The most distant galaxies in deep space.** (A) Hubble Space Telescope surveyed fields containing tens of thousands of old galaxies. The inset is Galaxy GN-z11 with z = 11.1, just 400 Myr after BB, which looks red by the strong redshift. (B) Color composite of JWST-NIRCam image showing GLASS-z13 as a red dot among other galaxies The inset shows the GLASS-z13, with z = ∼13.1. *Credit: (A) NASA, ESA, P. Oesch (Yale University), G. Brammer (STScI), P. van Dokkum (Yale University), and G. Illingworth (University of California, Santa Cruz). https://en.wikipedia.org/wiki/GN-z11#/media/File:Distant_galaxy_GN-z11_in_GOODS-N_image_by_HST.jpg. (B) NASA/STScI/GLASS-JWST program: R. Naidu, G. Brammer, T. Treu. https://commons.wikimedia.org/wiki/File:NASA-GLASS-z13-Context-JWST-20220722.jpg. The inset: https://commons.wikimedia.org/wiki/File:NASA-GLASS-z13-Closeup-JWST-20220722.jpg.*

Moon. These elements are synthesized in the star and the supernova which is the last flash of the dying star. Recently, the neutron star merger which is the collision of two neutron stars is considered to be the main origin of neutron-rich isotopes (Kobayashi et al., 2020).

2.2.1 Three forces are sustaining a star in a fine balance

A star is sustained by three forces: (1) energy produced by nuclear reaction, which makes the star swell; (2) gravity by the huge body, which makes the star contract; and (3) repulsion by electron degeneration, which prohibits a proton and an electron to be a neutron at the center of the star (see Fig. 2.5). It is easy to imagine (1) and (2), however, (3) may be difficult to image. Suppose that there were no (3), the core of the star would become neutrons at once and the star ceases to glare.

2.2.2 Elemental synthesis in the Big Bang and the first stars

The BB produced only ^{1}H and ^{4}He with traces of ^{2}D, ^{3}He, and ^{7}Li. The large amounts of ^{1}H and ^{4}He gases gathered and formed the first-generation star. Therefore, the first stars have very low metallicity (see Section 1.2.4). Theoretically, the first stars should have appeared after 0.3−0.4 Gyr of the BB.

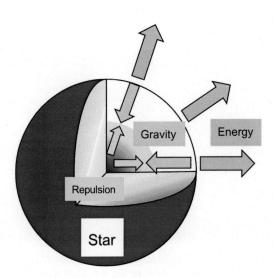

FIGURE 2.5 **The three forces sustaining a star.** The energy produced by nuclear fusion is trying to swell the star. The gravity of huge gases is contracting the star. The most inner place, repulsion by electron degeneration is holding the pressure of the gravity.

Search for the first stars is one of the hot topics in astronomy. J1808-5104 is an ultra metal-poor star that has a Fe/H ratio of $<10^{-4}$ of that in the Sun (Schlaufman et al., 2018). It is a single-lined spectroscopic binary (SB1), but the companion star is invisible. It resides in our Milky Way.

The optical spectrum of SMSS J031300.36-670839.3 in our Milky Way shows almost no evidence of iron (Keller et al., 2014; see Fig. 2.6). Its Fe/H ratio was $<10^{-7}$ of that in the sun, and the original mass of supernova was ~ 60 times heavier than that of the Sun. From chemical and physical investigation, a star that afterward became this supernova could be a first-generation star of ~ 13.6 Gyr ago.

2.2.3 Elemental synthesis of light elements up to carbon

In this section, the evolution of stars of mass of <8 $M_\odot$ ($M_\odot$ is the mass of the sun) is discussed. These stars can synthesize up to the atomic number below C. To synthesize heavier elements than these, massive stars of mass of >8 $M_\odot$ are required.

At the beginning of a star, ^{3}He and ^{1}H burns into ^{4}He. Then, the proton$-$proton chain (p$-$p chain) reaction occurs,

$$^1H + {^1H} \rightarrow {^2D} + e^+ + \nu + \gamma$$

and the reaction of

$$e^+ + e^- \rightarrow 1.0\,\text{MeV}$$

FIGURE 2.6 **The photograph of the oldest star (a supernova), SMSS J031300.36−670839.3.** The oldest star was found in February 2014, by the Sky Mapper Telescope at Siding Spring Observatory in Australia. *Credit: NASA/STSCI. https://commons.wikimedia.org/wiki/File:Oldest Star-SM0313-SMSSJ031300366708393-20140210.jpg.*

occurs and the energy is produced. Then the reaction

$$^2D(^1H, \gamma)^3He$$

occurs. γ is a positive energetic photon. Then the temperature of the star increases and becomes 10−14 million K, and the next p−p chain reaction continues:

$$^3He + {}^3He \rightarrow {}^4He + 2^1H + 12.9 \, MeV \qquad (2.1)$$

When the mass of the star is <3 $M_\odot$, the nuclear reaction ends here, and the star dies leaving a white dwarf with the He nuclear. As the temperature of the star cools, it becomes a brown dwarf, and finally a black dwarf.

When the mass of the star is >3−8 $M_\odot$, the temperature increases to 14−23 million K, and the p−p chain reaction continues:

$$^3He\left(^4He, \gamma\right)^7Be \qquad (2.2)$$

$$^7Be(e^-, \nu)^7Li$$

$$^7Li\left(p, {}^4He\right)^4He$$

and when the temperature becomes >23 million K, the next p−p reaction occurs:

$$^3He(^4He, \gamma)^7Be \qquad (2.3)$$

$$^7Be(p, \gamma)^8B$$

$$^8B \rightarrow {}^8Be + e^+ + \nu + \gamma$$

$$^8Be \rightarrow 2^4He + \gamma$$

Another reaction path is the CNO cycle,

$$^{12}C(p, \gamma)^{13}N$$

$$^{13}N \rightarrow {}^{13}C + e^+ + \nu + \gamma$$

$$^{13}C(p, \gamma)^{14}N(p, \gamma)^{15}O$$

$$^{15}O \rightarrow {}^{15}N + e^+ + \nu + \gamma$$

$$^{15}N + p \rightarrow {}^{12}C + {}^4He + \gamma$$

Total: $4p \rightarrow {}^4He + 2e^+ + 2\nu + 25.10 \, MeV$

In the core of the old star, ^{4}He becomes rich, and the temperature becomes >100 million K. In this condition, the triple-alpha process can occur:

$$^4\text{He} + {}^4\text{He} \rightarrow {}^8\text{Be} + \gamma$$

$$^8\text{Be} + {}^4\text{He} \rightarrow {}^{12}\text{C} + \gamma$$

Totally, 7.275 MeV is produced. ^{8}Be is very unstable, and returns to two ^{4}He in 2.6×10^{-16} s. However, in the condition where He burning occurs, this reaction becomes an equilibrium reaction. In addition, the energy of the second reaction is almost the same as the exited state of ^{12}C. Therefore, these two rare reactions can happen. In case the mass of a star is 3–8 $M_\odot$, the nuclear reaction ends when all He was burnt leaving a white dwarf with a carbon core.

2.2.4 Elemental synthesis of heavy elements up to Fe

In 1957, Burbidge, Burbidge, Fouler, and Hoyle published a famous B^{2}FH theory (Burbidge et al., 1957). In this chapter, the elemental synthesis of heavy elements from oxygen to nickel was explained to occur in massive stars (>8 $M_\odot$). As there are small modifications, the theory survived over half a century. They proposed C-burning, Ne-burning, O-burning, Si-burning, etc. In this section, the elemental synthesis of heavier elements over carbon is explained based on this theory.

The synthesized heavy elements with ^{1}H and ^{4}He gas formed the star again, and when the star dies with forming the elements again. This process was repeated again and again, and the present universe, our galaxy, and our solar system were formed.

When the mass of the star is >8 $M_\odot$, nuclear reactions go further, and oxygen atoms are formed:

$$^{12}\text{C} + {}^4\text{He} \rightarrow {}^{16}\text{O} + \gamma$$

And then ^{20}Ne forms (E means positive energy):

$$^{16}\text{O} + {}^4\text{He} \rightarrow {}^{20}\text{Ne} + \text{E} + \gamma$$

In the C-burning, the following reactions occur (E means positive energy).

$$^{12}\text{C} + {}^{12}\text{C} \rightarrow {}^{20}\text{Ne} + {}^4\text{He} + \text{E}$$

$$^{12}\text{C} + {}^{12}\text{C} \rightarrow {}^{23}\text{Na} + {}^1\text{H} + \text{E}$$

After the C-burning, the Ne-burning occurs:
$$^{20}\text{Ne} + {}^{4}\text{He} \rightarrow {}^{24}\text{Mg} + \text{E}$$

$$^{20}\text{Ne} + \text{g} \rightarrow {}^{16}\text{O} + {}^{4}\text{He} - \text{E}$$

For the Ne-burning, >120 MK (million K) and 4 GPa are required, therefore, Ne-burning occurs only in the star with a mass of >8−11 M$_{\odot}$. In such high temperatures, photodisintegration cannot be ignored, and some Ne nuclei decompose into O. In a few years, the core becomes unstable and is made of O and Mg. Then the O-burning process begins at > 1.5 GK (giga K) and 40 GPa:

$$^{16}\text{O} + {}^{16}\text{O} \rightarrow {}^{28}\text{Si} + {}^{4}\text{He} + \text{E}$$

$$^{16}\text{O} + {}^{16}\text{O} \rightarrow {}^{31}\text{P} + {}^{1}\text{H} + \text{E}$$

$$^{16}\text{O} + {}^{16}\text{O} \rightarrow {}^{31}\text{S} + \text{n} + \text{E}$$

$$^{16}\text{O} + {}^{16}\text{O} \rightarrow {}^{30}\text{Si} + 2{}^{1}\text{H} + \text{E}$$

$$^{16}\text{O} + {}^{16}\text{O} \rightarrow {}^{30}\text{P} + {}^{2}\text{H} - \text{E}$$

This process lasts from half to one year, and the core of the star becomes Si-rich and finally becomes Si. However, the temperature is not high enough to ignite Si. At this time, H-burning shell, He-shell, C-shell, Ne-shell, O-shell, and Si-core exist from outside to inside of the star-like onion shells (see Fig. 2.7).

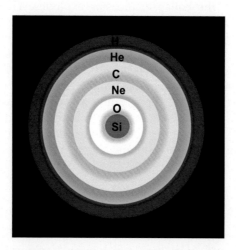

FIGURE 2.7 **Onion shell structure in a red giant star**. Onion shell structure in an O-burning star. H-burning shell, He-shell, C-shell, Ne-shell, O-shell, and Si-core exist from outside to inside.

The Si-burning process occurs only in a massive giant star with $>8-11$ $M_\odot$. This process is a 2-week-long process, and then the star becomes a type-II supernova. In the Si-burning processes, new elements are produced in sequence and end up making ^{56}Ni within a day.

$$^{28}Si + {}^4He \rightarrow {}^{32}S$$

$$^{32}S + {}^4He \rightarrow {}^{36}Ar$$

$$^{36}Ar + {}^4He \rightarrow {}^{40}Ca$$

$$^{40}Ca + {}^4He \rightarrow {}^{44}Ti$$

$$^{44}Ti + {}^4He \rightarrow {}^{48}Cr$$

$$^{48}Cr + {}^4He \rightarrow {}^{52}Fe$$

$$^{52}Fe + {}^4He \rightarrow {}^{56}Ni$$

Thus, the elements lighter than Fe are formed. This reaction stops, and does not go to ^{60}Zn, because ^{56}Ni is the most stable nuclei. ^{56}Ni decays to ^{56}Co by β^+-decay, and ^{56}Co decays to ^{56}Fe by β^+-decay. Then the core of the star suddenly collapses by gravity because the following reaction occurs:

$$^{56}Fe \rightarrow 13{}^4He + 4n - E$$

This reaction is endothermic, therefore, the core collapses at once. By the shockwave of the collapse, the star explodes. This is the type II supernova (see Fig. 2.8). In case, the mass of the star is $11-20$ $M_\odot$, the repulsion of electrons is overwhelmed by the shock of explosion, and protons in nuclei become neutron, and the core becomes a neutron star.

When the mass of the star is between 20 and $40-50$ $M_\odot$, the core can be not the neutron star but the black hole. The larger mass of the star ($>40-50$ $M_\odot$) can collapse directly into a black hole without a supernova. The story is easy to explain, but theoretical treatments are beyond the ability of the author. For those who are interested in the theories, the author nominates two famous classic books by Chandrasekhar (1967) and Clayton (1983).

2.2.5 Elemental synthesis heavier than Fe and supernova

In the B^2FH theory, heavier elements than Fe were synthesized by "r-process," "s-process," and "p-process" which mean "rapid-process," "slow-process," and "proton-rich-isotope synthesis process," respectively.

The r-process has been believed to occur in the type II supernova. When the type II supernova explodes, neutron-rich condition is

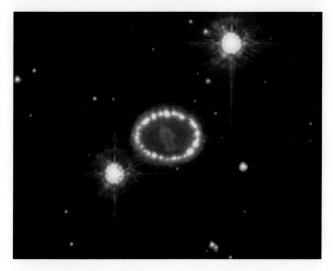

FIGURE 2.8 **Remnant of SN 1987A, type II supernova**. There is a remnant of the supernova at the center. *Credit: NASA, ESA, P. Challis, and R. Kirshner (Harvard-Smithsonian Center for Astrophysics). https://en.wikipedia.org/wiki/File:HST_SN_1987A_20th_anniversary.jpg.*

achieved, and neutron-rich isotopes were synthesized at once. Afterward, unstable neutron-rich isotopes were decayed through β^--decay, and heavy elements formed. Currently, this process is considered to be hard to happen, and the process is replaced by the neutron star merger (see Section 2.2.8).

The s-process is considered to form neutron-rich nuclei "slowly" in thousands of years, and occur in asymptotic giant branch stars (AGB stars) or stars with low metallicity. The neutrons are supplied by the following reactions:

$$^{13}C + {}^4He \rightarrow {}^{16}O + n$$

$$^{22}Ne + {}^4He \rightarrow {}^{25}Mg + n$$

One neutron is added to the seed element (Z, N), and the isotopes, (Z, N + 1), (Z, N + 2),..., (Z, N + n) are formed one by one. When a short half-life isotope of (Z, N + n + 1) appears, the reaction ends by β^--decay. Then the neutron addition reaction starts from the next seed element (Z + 1, N + n), and the isotopes of (Z + 1, N + n + 1), (Z + 1, N + n + 2),... are formed until an unstable isotope appears. This reaction is repeated, and this process continues up to ^{209}Bi. An example of Xe to Eu is shown in Fig. 2.9.

The B^{2}FH theory could not explain how proton-rich isotopes could be generated by neither r-process nor s-process, and named them as p-process. Later, the very strong light (photons) generated in the type II supernova can break the nuclei, and this process is named as

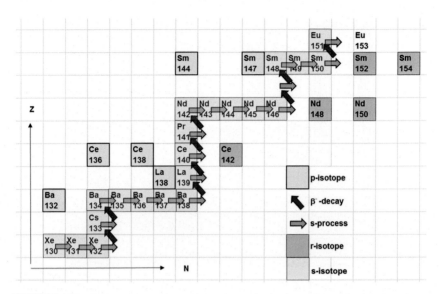

FIGURE 2.9 **Elemental synthesis by the s-process from Xe to Eu.** The horizontal axis is the neutron number, and the vertical axis is the atomic number. The s-process is shown as the blue horizontal arrow, and the β^--decay is shown in the slanted red arrow. The right side of the s-process isotope lines are the r-isotopes, which are neutron-rich orange isotopes, and the left side of the s-process isotopes are p-isotopes, which are neutron-poor yellow isotopes.

photodisintegration process (Cameron, 1957). Thus, this photodisintegration process was also called p-process. In the early theory, "p" meant "proton-rich," but now it means "photodisintegration." The number of p-isotopes is limited.

2.2.6 Synthesis of Li, Be, and B

When the elements are plotted by the elemental abundances compared to Si to the atomic number (see Section 1.4), Li, Be, and B show a large depletion in the plot. The B^2FH theory could not explain the low abundances of Li, Be, and B, and they named x-process. Nowadays, these elements are considered to be synthesized by the spallation of high-energy galactic cosmic rays (Viola and Mathews, 1987). This is the reason why the solar abundances of Li, Be, and B are very low. Tajitsu et al. (2015) and Hernanz (2015) proposed that ^7Li can also be produced in a supernova.

2.2.7 Type Ia supernova

Chandrasekhar (1931) calculated the limit of the mass of a white dwarf. He concluded that if the mass of a white dwarf becomes 1.4 $M_\odot$

(it is called as the Chandrasekhar limit), the electron degeneration cannot hold the gravity, and the white dwarf collapses and explodes as a supernova (type Ia supernova), and becomes a neutron star.

In Section 2.2.3, the star with a mass of <3 $M_\odot$ becomes a white dwarf after H and He are consumed. The gravity balances with the repulsion of electron degeneration because of the small masses. The star becomes the white dwarf and cooler.

However, there is a possibility that the white dwarf is in a binary system. This possibility is not so low. In such a case, one star becomes a white dwarf, and the other star becomes a red giant. The gases of the red giant star with low gravity are attracted by the white dwarf with high gravity (see Fig. 2.10). The gases from the red giant gather on the white dwarf, and explosive nuclear fusion can occur on the white dwarf. This is considered to be the real identity of the AGB star. If the white dwarf continues to absorb gases from the companion giant star, the mass of the white dwarf can reach the Chandrasekhar limit. Then the white dwarf explodes as type Ia supernova (see Fig. 2.11), where r- and s-processes could occur.

2.2.8 Neutron star merger

As related in Sections 2.2.5 and 2.2.7, the r-process could happen in supernova. To detect the evidence of the r-process, astronomers have long focused on Eu because its spectral lines appear in the visible part of the spectrum. However, no one has ever seen a supernova produce

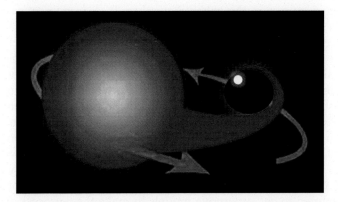

FIGURE 2.10   **The aged binary system.** One star becomes a white dwarf, and the companion star becomes a giant star which starts swelling. The mass is transferred onto the white dwarf. Soon the white dwarf will reach a Chandrasekhar limit, and explode as a type Ia supernova. *Source: Modified from NASA, ESA, and A. Feild (STScI); vectorization by Chris 論: https://commons.wikimedia.org/wiki/File:Progenitor_IA_supernova.svg.*

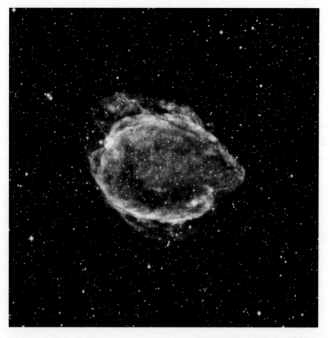

FIGURE 2.11   **G299, a remnant of type Ia supernova**. See text for type Ia supernova. The photograph was taken by the Chandra spacecraft. Red, yellow, green, and blue indicate low to high-energy X-rays. *Credit: NASA. https://commons.wikimedia.org/wiki/File:G299-Remnants-SuperNova-TypeIa-20150218.jpg.*

r-process elements, except Ji et al. (2016), who found the generation of Eu by the r-process in the galaxy Reticulum II. Astronomers saw heavy elements actually formed when two dense neutron stars spiraled into each other, which implied that not supernova but neutron star mergers were the main source of the heavy elements. Lattimer and Schramm (1974) calculated that mergers between neutron stars and black holes can make all of Au, Pt, and other r-process elements in the universe. Symbalisty and Schramm (1982) confirmed the concept that the merger of two neutron stars (see Fig. 2.12) can also make r-process elements.

On August 17, 2017, astronomers detected signals from the merger of two neutron stars in a galaxy 130 million ly away. The gravitational radiation reached the Earth, together with a burst of gamma rays, and a flash of visible light (Abbott et al., 2017). S. Woosley, a theorist, had long struggled to get supernova models to make the r-process elements, and had already concluded that neutron star mergers were the solution. By observing the visible light of the 2017 merger, astronomers deduced that the amount of material was ~5% of a solar mass, which was consistent with the calculation of Lattimer and Schramm (1974).

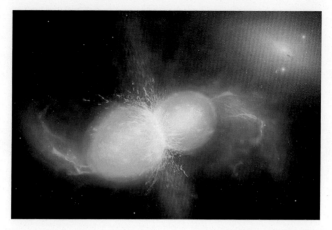

FIGURE 2.12   **The neutron star merger**. The artist's impression of two neutron stars is merging and exploding as a nova. Such a rare event is expected to produce gravitational waves, a short gamma-ray burst, and heavy elements. *Credit: University of Warwick/Mark Garlick.   https://commons.wikimedia.org/wiki/File:Eso1733s_Artist%27s_impression_of_merging_ neutron_stars.jpg.*

The r-process is rare. First, even in our galaxy, only two supernovae in a century, but r-process events occur less often of <1/1000. Second, neutron star mergers made gold and platinum on the Earth and throughout the universe. Third, most supernova (>99%) explosions make none of the r-process elements.

In summary, the probable origin of the elements in the solar system is shown in Fig. 2.13. Heavier elements than Nb are mainly produced by the neutron star merger in high ratio (purple colored), especially for the platinum group elements (Ru, Rh, Pd, Os, Ir, and Pt), Ag, Re, and Au.

2.3 Formation of the proto-Earth in the solar nebular

In Section 2.3, the formation of the solar system from the gas nebular is discussed. In geophysical models, our solar system is considered to start from the gas cloud. Then it began to gather and rotate, and the sun and the planetary system were formed. The Nice and Grand Tack models are built considering the resonance of Jupiter and Saturn. All other planets are governed by these two gas giants in geophysical models. Observation of other planetary systems helped to establish models including late heavy bombardment (LHB). Cosmochemical methods using radioactive isotopes and stable isotopic fractionation became effective in building scenarios after the formation of small particles and silicate planets including the proto-Earth, which was the Earth before the giant impact.

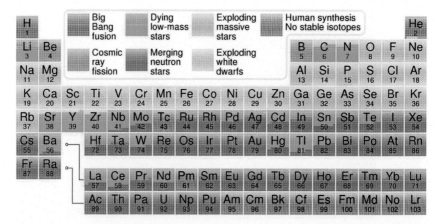

FIGURE 2.13 **Periodic Table indicating probable origins of the elements in the solar system**. Periodic table showing the origin of elements in the solar system. Color codes (origins) are blue (BB), green (s-process), yellow (type II supernova), brown (unstable elements), red (spallation), purple (neutron star merger), and silver (type I supernova). This figure can be changed because of the progress of the studies. *Credit: Jennifer Johnson, Ohio State University. https:// commons.wikimedia.org/wiki/File:Nucleosynthesis_periodic_table.svg.*

2.3.1 Evolution of molecular clouds to the solar nebula

The evolution of early star and surrounding molecular cloud (nebula) are controlled by five components: (1) the centrifugal force, (2) mass ejection (mass and angular momentum loss as jet), (3) mass accretion (mass gain from the disk surrounding the star), (4) a magnetic field (the mass ejection and the accretion follow the magnetic field), and (5) conservation of angular momentum. These five components are schematically depicted in Fig. 2.14 Montmerle et al. (2006).

The star has two strong magnetic fields. One magnetic field extends vertically along the rotation axis of the star. Another magnetic field stretches horizontally. In the accretion-ejection model, these two magnetic fields meet along the X-ring (or at the X-point if plotted two-dimensionally). The accreting materials from the horizontal direction are ejected at the X-ring. Many researchers think that the silicate materials that were heated at the X-ring, melted into a sphere shape, and ejected, then cooled very fast are the origin of the chondrules in the chondrite meteorites (see also Section 1.4.10).

Our solar system is considered to be formed from cosmic gas clouds as shown in Fig. 2.14. Such gas cloud is made of hydrogen, helium, and remnants of supernova (heavy elements), which are already synthesized in Section 2.2.8. The shockwave by a supernova near the place where our sun would exist made inhomogeneity of the gas cloud, and the

FIGURE 2.14   Closeup of young stars in the Orion Nebula. This image is protoplanetary disks, which were taken by the Hubble Space Telescope. *Credit: NASA and ESA. http:// archive.seds.org/hst/OrionProplyds.html.*

cloud of about three ly began to gather and form a dense gas disk (10^3–10^4 au). At the center of the gas disk, there was a huge gas ball (protostar) mainly made of hydrogen.

The various stellar phases with timescales and sizes are shown in Fig. 2.15 after Feigelson and Montmerle (1999). These models are established according to the advancement of observation techniques using infrared and X-ray telescopes.

- The initial stage, which is previously explained, is called as the infalling protostar stage. The accretion and ejection of the molecular gases occur. Although the gas ball size is very large, it cannot be seen by the surrounding clouds. This stage lasts for $<10^4$ y.
- The next stage is the evolved protostar stage. The accretion and ejection of the gases are still vigorously occurring. However, the size became 10 times smaller. This stage lasts for 0.1 Myr. The gas disk as well as the protostar glow with the heat of accretion.
- Then the classical T Tauri stage comes. T Tauri is a variable star in the constellation Taurus. This star is the prototype of T Tauri stars. The star exists near the molecular clouds and is variable in visible light. The variability is caused by the screening of the star light by the gas clouds, proto-planets, and planetesimals. The gas clouds are ejected vigorously. This stage is the ejection stage. The ejection jet is clearly observed. This stage lasts for 1–10 Myr, and the thick disk shrinks to about 100 au. In this stage, the planetary system would be formed.

2. Origin of elements

Infalling protostar $(10^4 \text{ y}, 10^3{\sim}10^4 \text{ au})$

Evolved protostar $(0.1 \text{ Myr}, 5{\times}10^2{\sim}10^3 \text{ au})$

Classical T Tauri star $(1{-}10 \text{ Myr}, 10^2 \text{ au})$

Weak-lined T Tauri star $(1{-}10 \text{ Myr})$

Main sequence star $({\sim}10 \text{ Myr})$

FIGURE 2.15 **Various stellar phases with timescales and sizes (each stage cartoon is not in scale).** See text for details. *Source: Based on Feigelson, E.D., Montmerle, T., 1999. High-energy processes in young stellar objects. Ann. Rev. Astron. Astrophys. 37, 363–408.*

- The next stage is the weak-lined T Tauri stage. The "weak-lined" means that a "weak-Hα" line of ionized hydrogen is observed in the spectrum of the star.

- Finally, the clouds dissipate and the star can be seen. The star becomes a main sequence star in the H-R diagram (see Section 1.2.2).

The evolution model of the star described here is established from the actual observation of stars and theories. However, when the masses of the pre-solar nebula or the planetary system are too small or too large compared to the sun, the evolution of the pre-solar nebula requires different views and theories.

The establishment of such models highly depends on space telescopes, infrared telescopes, and X-ray telescopes with the developments of observation technologies for the gas clouds, which was described in Section 1.2.

2.3.2 From the pre-solar nebula to the solar system

The surrounding gas of the proto-sun becomes hotter and flatter by the effects of gravity, magnetic field, and angular momentum. As related in Section 2.3.1, this gas disk is called nebula or pre-solar disk. As the temperature of the gas clouds decreases, silicates, iron, iron oxides, and carbon compounds, of which sizes are sub-μm to μm, form. Such dust grains collide and build up larger particles in the turbulent circumstellar disk, and finally, in most models of the planet formation, planets grow by the accumulation of smaller bodies called planetesimals about 10−100 km in diameter (e.g., Chambers, 2004).

Each process sounds to proceed without problems. However, the theories cannot solve two problems (Montmerle et al., 2006). Thus, the pre-solar nebula cannot evolve smoothly and continuously into the solar system. One problem is what was the condition of the collision velocity and the density of the gas? If the dust grains have a typical possible velocity in theories, they do not accrete but disrupt each other. Thus, the dust does not grow to the size of pebbles. Although the gas density is different, you can imagine this problem as the problem of space debris. The space debris collides with each other and becomes finer space debris and increases their number, without accreting each other.

Another problem is that the radial migration of planetesimals is unknown during growth. The gases easily drag the planetesimals into the central star within 100−1000 y. Therefore, all planetesimals fall onto the star.

As far as the author investigated, the first problem is not solved yet. To make the meter-sized particles, physical sticking and inelastic collision where the kinetic energy is changed into other energy like heat are required. However, such a condition has not been determined so far.

However, the second problem seems to be solved by the model calculation by Johansen et al. (2007), for example. They started from the meter-sized
boulders and could make an asteroid like a 900 km-sized Ceres considering
magneto-rotational instability in the model calculation.

2.3.3 Formation of gas giant planets and asteroids: the Nice model

The researchers in Observatoire de la Cote d'Azur in Nice, France
(Tsiganis et al., 2005; Morbidelli et al., 2005; Gomes et al., 2005) established
the comprehensive models explaining the evolution of the solar system of
0−0.6 and 700−1000 Myr (the ages indicate the time from the beginning of
the solar system) and origins of the giant planets, the main asteroid belt,
the Trojans asteroids and the Kuiper belt objects, based on the N-body simulation calculation of 3500 planetesimals with a mass of 1/100 of M_E
(M_E is the mass of the Earth), Jupiter, Saturn, Uranus, and Neptune truncated at 30 au. The results of the simulation is shown in Fig. 2.16.

At T = 0, the planetesimals of red (S-type; rocky), gray (C-type; rocky
with ice), and blue (ice and other volatiles such as NH_3, CH_4, etc.) are
distributed systematically from inside to outside (see Fig. 2.17). Inside
the snow line (the distance that the ice starts to exist in the solar system), the red planetesimals exist, and outside of the snow line, gray
ones exist surrounded by the blue ones.

During 0−0.6 and 700−1000 Myr, large evolutionary events of the
solar system are considered to have occurred. In Fig. 2.17, the evolution
of the solar system is depicted after Demeo and Carry (2014). This
model is called as "the Nice model." "Nice" is the place of the research
observatory as well as the nice (put-intended) model.

The model starts just after the giant planets (Jupiter, Saturn, Uranus,
and Neptune) are formed, and the surrounding gas is dissipated. The solar
system is composed of the Sun, the giant planets, and a debris disk of
small planetesimals. The planets then start to erode the disk by accreting
or scattering away the planetesimals. The planets migrate because of the
exchange of angular momentum with the disk particles in this process.
The simulation shows that the fully formed Jupiter starts at 300 M_E at 3.5
au which is favorable for giant planet formation because it is outside of the
snow line. The Saturn's core is initially 35 M_E and ∼4.5 au, and remains
around ∼4.5 au, with increasing the mass to 60 M_E. The cores of Uranus
and Neptune begin at ∼6 and ∼8 au and start from ∼5 M_E, respectively.
Their masses increase just slightly (see Fig. 2.17).

When the mass of Saturn reaches 60 M_E, Saturn suddenly begins migration inwards to 2 au. Jupiter is forced to move inward to ∼1.5 au within
0.1 Myr. Uranus and Neptune migrate slightly inwards with a slow delay
to Jupiter's migration. Then on catching Jupiter, Saturn is trapped in the 1:2
resonance with Jupiter. The migration is reversed, and Saturn and Jupiter

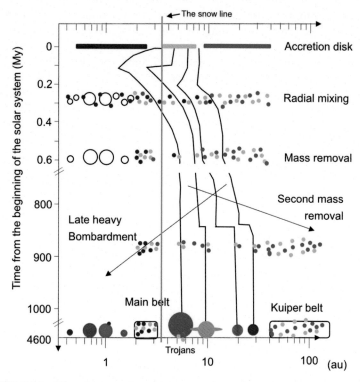

FIGURE 2.16 **The schematic diagram for the evolution of the solar system.** The vertical axis is the time (My) from the beginning of the solar system. Three scales of time (My), 0−0.6, 700−1000, and the present (4567) are shown in one axis. The horizontal axis is the semimajor axis in the au unit. The gray and blue small dots indicate C-type and outer-disk planetesimals, and the red ones indicate S-type (rocky) planetesimals, respectively. The blue ones contain lots of water, which could be a source of water of the Earth. *Source: Based on DeMeo, F.E., Carry, B., 2014. Solar System evolution from compositional mapping of the asteroid belt. Nature 505, 629−634.*

migrate outwards together. They capture Uranus and Neptune in resonance, and migrate outwards as well. Saturn, Uranus, and Neptune reach their full mass at the end of the migration when Jupiter reaches at 5.4 au. The migration rate decreased exponentially as the gas disk dissipated. The final orbital configuration of the giant planets is consistent with their current orbital. The terrestrial planets are formed from this disk within 30−50 Myr after the migration of Jupiter and other planets is finished.

2.3.4 The late heavy bombardment in the Nice model

The LHB was recorded in the Moon craters. The frequency of the Moon crater was strangely very high in 700−1000 Myr (the ages indicate the time from the beginning of the solar system). This phenominon was explained LHB by the Nice model, in which Jupiter and Saturn become 1:2 resonance (Gomes et al., 2005).

Initial of the 1:2 resonance, the Saturn's orbital period was less twice than that of Jupiter. When the two planets crossed the 1:2 mean motion resonance (MMR) their orbits became eccentric. This abrupt transition destabilized the giant planets, leading to a short phase of close encounters among Saturn, Uranus, and Neptune. As a result of the interactions of the ice giants with the disk, Uranus and Neptune reached their current distances and Jupiter and Saturn evolved to their current orbital eccentricities. The main idea is that Jupiter and Saturn crossed the 1:2 resonance at $\sim 700\,\text{Myr}$.

Initially, the giant planets migrated slowly owing to the leakage of particles from the disk. This phase lasted 880 Myr. After the resonance crossing event, the orbits of the ice giants became unstable and they were scattered into the disk by Saturn. They disrupted the disk and scattered objects all over the solar system, including the inner regions. A total of $9 \times 10^{18}\,\text{kg}$ struck the Moon after the resonance crossing.

As Jupiter and Saturn moved from 1:2 resonance toward their current positions, secular resonances swept across the entire asteroid belt. These resonances can drive asteroids onto orbit with eccentricities and inclinations large enough to allow them to evolve into the inner solar system and hit the Moon.

The asteroid objects slowly leak out of the asteroid belt and can evolve into the inner solar system. Roughly the particles arrive in the first 10–150 Myr. It was estimated that $(3-8) \times 10^{18}\,\text{kg}$ of asteroids hit the Moon. This amount is comparable to the amount of comets. So, their model predicts that the LHB impactors should have been a mixture of comets and asteroids. Within the first 30 Myr comets dominated, the last impactors were asteroids. Their results support a cataclysmic model for the lunar LHB. Although many of the LHB are not well known, their simulations reproduce two of the main characteristics attributed to this episode: (1) the 700 Myr delay between the LHB and terrestrial planet formation and (2) the overall intensity of lunar impacts. Their model predicts a sharp increase in the impact rate at the beginning of the LHB.

2.3.5 The Grand Tack model: the resonance of Jupiter and Saturn

The explanation of the large-scale mixing of reddish and bluish material in the asteroid belt (see Fig. 2.17) is lacking in the Nice model. To overcome this problem, the "Grand Tack" model (Walsh et al., 2011) appeared. "Tack" in the Grand Tack model means the migration of the planetesimals by Jupiter, using the sailing analogy. In this model, the

Jupiter and Saturn resonance of 2:3 became important. The 2:3 resonance is proposed by Masset and Snellgrove (2001) as the type II migration, which is now widely accepted.

In this model, during the time of terrestrial planet formation (before the events of the Nice model), Jupiter could have migrated as close to the Sun as Mars is today. Jupiter would have moved through the primordial asteroid belt, emptying it and then repopulating it with scrambled material from both the inner and outer solar system. Then Jupiter reversed and headed back towards the outer solar system. Once the details of the resulting distribution in the Grand Tack model have been compared to the emerging observational picture, it will become clear whether this model can crack the asteroid belt's compositional order. Planetary migration ends within the first billion years of our solar system's 4.5-billion-year history.

The asteroid belt, however, is still dynamic today. Collisions between asteroids are continuously making the bodies to smaller and smaller sizes. The smaller ones are then subject to the Yarkovsky effect, in which uneven heating and cooling of the asteroid alters its orbit (see Section 1.4.10.5). The Yarkovsky effect thoroughly mixes small bodies within each section of the main belt, but once they reach a major resonance such as the 3:1 and 5:2 MMR at the locations, they are swiftly ejected from the main belt (see Table 2.1).

Current observations and models indicate that the strong resonances with Jupiter inhibit the crossing of material from one region to another. These processes continue to make the asteroid belt, erasing the past history and creating new structures in the asteroid belt. New observational evidence that reveals a greater mixing of bodies supports the idea of a solar system is evolving.

2.3.6 Formation of inner planets in the Grand Tack model

In this section, we consider the terrestrial planets that can be produced by the Grand Tack model by Walsh et al. (2011). The terrestrial planets are best formed when the planetesimal disk is truncated with an outer edge at 1 au. These conditions are created naturally if Jupiter tacked at ~1.5 au.

However, before concluding that Jupiter tacked at ~1.5 au, it is necessary to consider that the asteroid belts between 2 and 3.2 au can survive the passage of Jupiter. Volatile-poor asteroids (S-types) are dominant in the inner asteroid belt, while volatile-rich asteroids (C-types) are dominant in the outer asteroid belt. These two main classes of asteroids have partly overlapping semi-major axis distributions, although the number of the C-type is more than that of the S-type beyond ~ 2.8 au. The planetesimals from the inner disk are considered to be "S-type" and those from the outer regions are "C-type."

From hydrodynamic simulations, the inward migration of the giant planets shepherds much of the S-type material inward by resonant trapping, eccentricity excitation, and gas drag. The mass of the disk inside 1 au doubles, reaching ~ 2 M_E. This reshaped inner disk constitutes the initial condition for a terrestrial planet. However, a fraction of the inner disk ($\sim 14\%$) is scattered outward, ending up beyond 3 au. During the subsequent outward migration of the giant planets, this scattered disk of S-type materials is encountered again. A small fraction ($\sim 0.5\%$) of this material is scattered inward and left decoupled from Jupiter into the asteroid belt region.

The giant planets then come across the materials in the Jupiter–Neptune formation region, some of which ($\sim 0.5\%$) are also scattered into the asteroid belt. Finally, the giant planets meet the disk of material beyond Neptune (within 13 au) of which only $\sim 0.025\%$ reaches a final orbit in the asteroid belt. When the giant planets have finished their migration, the asteroid belt population is in place, however, the terrestrial planets require ~ 30 Myr to complete accretion.

The asteroid belt is composed of two separate populations: the S-type planetesimals within 3.0 au (left-overs from the giant planet accretion process: ~ 0.8 M_E of materials between the giant planets and ~ 16 M_E of planetesimals from the 8.0–13 au region). The other is C-type material placed onto orbits crossing the still-forming terrestrial planets. Every C-type planetesimal from beyond 8 au was implanted in the outer asteroid belt. The 11–28 au C-type planetesimals ended up on high-eccentricity orbits that enter the terrestrial-planet-forming region (<1.0–1.5 au), and may represent a source of water for the Earth.

For the Jupiter–Uranus region, this ratio is 15–20; and for the Uranus–Neptune region, it is 8–15. Thus, it is expected that $(3-11) \times 10^{-2}$ M_E of C-type material entered the terrestrial planet region. This exceeds by a factor of 6–22, the minimal mass required to bring the current amount of water to the Earth ($\sim 5 \times 10^{-4}$ M_E; Abe et al., 2000).

The migration of Jupiter creates a truncated inner disk matching the initial conditions of previously successful simulations of terrestrial planet formation. As there is a slight build-up of dynamically excited planetary embryos at 1.0 au, simulations of the accretion needed 150 Myr. Earth and Venus grow within the 0.7–1 au, accreting most of the mass, while Mars is formed from embryos scattered out beyond the edge of the truncated disk. The final distribution of planet mass versus distance quantitatively reproduces the large mass ratio existing between the Earth and Mars, and also matches the quantitative metrics of orbital excitation.

2.3.7 Inconsistency between our solar system and exosolar systems found by the Kepler program

The Kepler program (see Section 1.2.7) found over thousands of exoplanets in about four hundred stellar systems. Thus, more than half of the Sun-like stars have planet(s) in low-eccentricity orbits. However, the periods of the orbits were days to months, and the masses of the planets were $1 M_E < M_p < 50 M_E$, where M_E and M_p are the masses of the Earth and the planet, respectively (Mayor et al., 2011; Batalha et al., 2013). These are called "super-Earths" or "hot Jupiters," because the planets had larger mass but inner orbits than the Earth orbit, resulting in far hotter conditions than that in Earth. Thus, the exosolar planet studies revealed the specialty of our solar system.

In addition, the first planetesimals have formed at ~ 1 Myr after the Sun's birth, while the final accretion of terrestrial planets was in the timescale of 100−200 Myr, well after the dispersion of the nebular gases. In contrast, the exosolar super-Earths are inferred to be gaseous planets, indicating their early formation.

To overcome these problems, Batygin and Laughlin (2015) challenged the calculation following the Grand Tack model (Walsh et al., 2011; Masset and Snellgrove, 2001), where Jupiter moves inward followed by outward migration. The developed model based on calculation is summarized in Fig. 2.17.

In Fig. 2.17A, the left yellow half ball is the sun. The super-Earths, which are common in the Kepler observation, are indicated in blue balls near the sun. In this case, two gas planets are depicted in the figure. These are formed in the gaseous disk, which is indicated by the trapezoid from the sun. The small dark-green circles indicate terrestrial planetesimals of a radius of 10−1000 km. The right-side vertical arrow shows the time. The Jupiter begins to move from ~ 3 au after the Grand Tack model.

As Jupiter migrates inward, the planetesimals collide with each other (Fig. 2.17B), and break and ground into <100 m size. Then in Fig. 2.17C, Jupiter migrates to ~ 1.5 au, and becomes resonant with Saturn, and begins to go on an outside orbit. However, the super-Earths are carried to the sun by inward-drifting debris of planetesimals of <100 m in size. Once the planetesimals become <100 m, aerodynamic drag induces rapid orbital decay.

Thus, the solar system is reset by the movement of Jupiter. After super-Earths are cleared up, the "second-generation planets" are formed from the survived gas and materials (Fig. 2.18D). The terrestrial planets can be formed, instead of the observed gas-rich short-period planets.

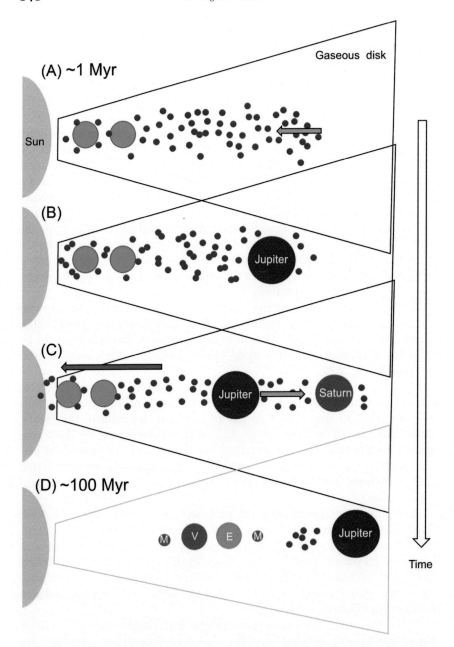

FIGURE 2.17 **The model of** Batygin and Laughlin (2015) **of the early solar system evolution**. See text for details. (Naoz, 2015). *Source: Modified from Naoz, S., 2015. Jupiter's role in sculpting the early Solar System. Nature 112, 4189–4190, by the author.*

This model clearly explains why the short-period super-Earths do not exist in our solar system, and why it took a long time (~ 100 Myr) for the final formation of the terrestrial planets, compared to the early formation (~ 1 Myr) of the sun.

References

Abbott, B.P., Abbott, R., Abbott, T.D., Acernese, F., Ackley, K., 2017. Multi-messenger observations of a binary neutron star merger. Astrophys. J. Lett. 848, L12.

Abe, Y., Ohtani, E., Okuchi, T., Righter, K., Drake, M., 2000. Water in the early Earth. In: Canup, R.M., Righter, K. (Eds.), Origin of the Earth and Moon, Tucson. University Arizona Press, pp. 413–433.

Batalha, N., Rowe, M.J.F., Bryson, S.T., et al., 2013. Planetary candidates observed by Kepler III. Analysis of the first 16 months of data. Astrophys. J. Suppl. Ser. 204 (24), 21.

Batygin, K., Laughlin, G., 2015. Jupiter's decisive role in the inner Solar System's early evolution. Proc. Natl. Acad. Sci. U. S. A. 112, 4214–4217.

Burbidge, E.M., Burbidge, G.R., Fowle, W.A., Hoyle, F., 1957. Synthesis of the elements in stars. Rev. Mod. Phys. 29, 547–650.

Cameron, A.G.W., 1957. Nuclear reactions in stars and nucleogenesis. Pub Astron. Soc. Pac. 69, 201–222.

Chambers, J.E., 2004. Planetary accretion in the inner solar system. Earth Planet. Sci. Lett. 223, 241–252.

Chandrasekhar, S., 1967. An Introduction to the Study of Stellar Structure. Dover Publications, Inc, New York.

Chandrasekhar, S., 1931. The maximum mass of ideal white dwarfs. Astrophys. J. 74, 81–82.

Clayton, D.D., 1983. Principles of Stellar Evolution and Nucleosynthesis. McGraw-Hill, New York.

DeMeo, F.E., Carry, B., 2014. Solar System evolution from compositional mapping of the asteroid belt. Nature 505, 629–634.

Feigelson, E.D., Montmerle, T., 1999. High-energy processes in young stellar objects. Ann. Rev. Astron. Astrophys. 37, 363–408.

Gawiser, E., Silk, J., 2000. The cosmic microwave background radiation. Phys. Rep. Rev. Sect. Phys. Lett. 333, 245–267.

Gomes, R., Levison, H.F., Tsiganis, K., Morbidelli, A., 2005. Origin of the cataclysmic Late Heavy Bombardment period of the terrestrial planets. Nature 435, 466–469.

Hernanz, M., 2015. A lithium-rich stellar explosion. Nature 518, 307–308.

Ji, A.P., Frebel, A., Chiti, A., Simon, J.D., 2016. R-process enrichment from a single event in an ancient dwarf galaxy. Nature 531, 610–613.

Johansen, A., Oishi, J.S., Low, M.-M.M., Klahr, H., Henning, T., 2007. Rapid planetesimal formation in turbulent circumstellar disks. Nature 448, 1022–1025.

Keller, S.C., Bessell, M.S., Frebel, A., Casey, A.R., Asplund, M., et al., 2014. A single low-energy, iron-poor supernova as the source of metals in the star SMSS J031300.36 − 670839.3. Nature 506, 463–466.

Kobayashi, C., Karakas, A.I., Lugaro, M., 2020. The origin of elements from carbon to uranium. Astrophys. J. 900, 179.

Lattimer, J.M., Schramm, D.N., 1974. Black-hole-neutron-star collisions. Astrophys. J. Lett. 192, L145–L147.

Masset, F., Snellgrove, M., 2001. Reversing type II migration: resonance trapping of a lighter giant protoplanet. Monthly Not. R. Astronomical Soc. 320, L55–L59.

Mayor, M., Marmier, M., Lovis, C., et al., 2011. The HARPS search for southern extra-solar planets IV. Occurrence, mass distribution and orbital properties of super-Earths and Neptune-mass planets. Astron. Astrophys.

Montmerle, T., Augereau, J.-C., Chaussidon, M., et al., 2006. 3. Solar system formation and early evolution: the first 100 million years. Earth, Moon, Planets 98, 39−95.

Morbidelli, A., Levison, H.F., Tsiganis, K., Gomes, R., 2005. Chaotic capture of Jupiter's Trojan asteroids in the early Solar System. Nature 435, 462−465.

Naidu, R.P., Oesch, P.A., van Dokkum, P., Nelson, E.J., Suess, K.A., et al., 2022. Two remarkably luminous galaxy candidates at $z \approx 11 - 13$ revealed by JWST. arXiv:2207.09434.

Naoz, S., 2015. Jupiter's role in sculpting the early Solar System. Nature 112, 4189−4190.

Oesch, P.A., Brammer, G., van Dokkum, P.G., Illingworth, G.D., Bouwens, R.J,, Labbe, I., et al., 2016. A remarkably luminous galaxy at $Z = 11.1$ measured with Hubble Space Telescope Grism Spectroscopy. Astrophys. J. 819, 129. Available from: https://doi.org/10.3847/0004-637X/819/2/129.

Penzias, A.A., Wilson, R.W., 1965. A measurement of excess antenna temperature at 4080 Mc/s. Astrophys. J. 142, 419−421.

Schlaufman, K.C., Thompson, I.B., Casey, A.R., 2018. An ultra metal-poor star near the hydrogen-burning limit. Astrophys. J. 867, 98.

Symbalisty, E., Schramm, D.N., 1982. Neutron star collisions and the r-process. Astrophys. Lett. 22, 143−145.

Tajitsu, A., Sadakane, K., Naito, H., Arai, A., Aoki, W., 2015. Explosive lithium production in the classical nova V339 Del (Nova Delphini 2013). Nature 518, 381−384.

Tsiganis, K., Gomes, R., Morbidelli, A., Levison, H.F., 2005. Origin of the orbital architecture of the giant planets of the Solar System. Nature 435, 459−461.

Viola, V.E., Mathews, G.J., 1987. The cosmic synthesis of lithium, beryllium and boron. Sci. Am. 256, 38−45.

Walsh, K.J., Morbidelli, A., Raymond, S.N., O'Brien, D.P., 2011. Mandell: a low mass for Mars from Jupiter's early gas-driven migration. Nature 475, 206−209.

Materials on the Moon

3.1 Characteristics of the materials of the Moon

3.1.1 Sample-return programs of the Moon rocks

Three sample-return programs have succeeded to fetch samples of the Moon. The landing sites are summarized in Fig. 3.1. The first one was the Apollo program, which was the human spaceflight program of NASA. The program succeeded in carrying humans on the Moon from 1969 to 1972 (Fig. 3.2). This program was also sample-return missions by humans and collected totally 380.95 kg of samples (see Table 3.1) from various places on the Moon.

Robotic missions of the Soviet Union (now the Russian Federation), Luna 16, 20, and 24 also succeeded to land on the Moon (see Fig. 3.1), and collected soils of the Moon (101, 30, and 170 g, respectively) and return to the Earth in 1970, 1972, and 1976. However, most samples seem to have scattered and lost at the collapse of the Soviet Union.

In 2020 after 44 years since the last Apollo program, the Chinese robotic sample-return mission, Cheng'e-5 (嫦娥5号) succeeded to collect 1.7 kg samples including the drilled core of the soils.

3.1.2 Rocks collected from the Moon

The Moon rocks from the Apollo program are classified into two main species: the lunar highland rocks and the mare rocks (summarized in Fig. 3.3). The highland rocks are mainly composed of mafic plutonic rocks and regolith breccias. The highland breccias are formed by impacts of the highland igneous rocks. The highland igneous rocks are composed of three types: the ferroan anorthosite suite, the magnesian suite, and the alkali suite. Warren (1993) compiled non-mare Moon rocks. The ferroan anorthosite suite is composed of anorthosite

Introductory Astrochemistry
DOI: https://doi.org/10.1016/B978-0-443-23938-0.00003-0
151

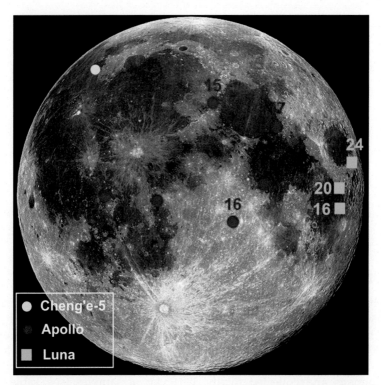

FIGURE 3.1   Landing sites of Apollos, Lunas, and Cheng'e-5. The program numbers of Apollo and Luna are shown. *Source: NASA. http://science.nasa.gov/media/medialibrary/2005/ 07/01/11jul_lroc_resources/landingsites_600.jpg.*

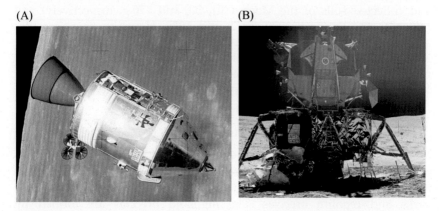

FIGURE 3.2   Apollo spacecrafts for sample-return mission of the Moon. (A) A photo of the command and service module of Apollo 16. (B) A photo of the lunar module of Apollo 16. *Source: (A) NASA. https://upload.wikimedia.org/wikipedia/commons/c/c0/Apollo_ CSM_lunar_orbit.jpg. (B) NASA. https://upload.wikimedia.org/wikipedia/commons/thumb/2/2a/ Apollo16LM.jpg/1280px-Apollo16LM.jpg.*

TABLE 3.1   The Apollo programs and collected sample amounts.

Program	Year	Amounts (kg)
Apollo 11	1969	21.55
Apollo 12	1969	34.30
Apollo 14	1971	42.80
Apollo 15	1971	76.70
Apollo 16	1972	95.20
Apollo 17	1972	110.40

◆ Highland igneous rocks
- Mafic plutonic rocks
  - Ferroan anorthosite   (mafic minerals have low Mg/Fe ratios)
    - Anorthosite (>90 % calcic plagioclase; floated plagioclase cumulates)
  - Magnesian suits (relatively high Mg/Fe ratios)
    - Dunite
    - Troctolites (olivine-plagioclase)
    - Gabbro (plagioclase-pyroxene)
  - Alkali suits
    - Sodic plagioclase
    - Norites (plagioclase-orthopyroxene)
    - Gabbronorites (plagioclase-clinopyroxene-orthopyroxene)
- Regolith breccias   (made by impacts of the highland igneous rocks)

◆ Mare rocks (basalts)
- High-Ti basalts
- Low-Ti basalts
- Very-low-Ti (VLT) basalts
- Very-high-K (VHK) basalts

FIGURE 3.3   Rocks of the Moon. Rocks of the Moon are classified into Highland igneous rocks and Mare rocks (basalts). The former is further divided into Mafic plutonic rocks and regolith breccias. Mafic plutonic rocks are divided into Ferroan anorthosite, Magnesian suits, and alkali suits. Mare rocks are divided into high-Ti basalts, low-Ti basalts, very-low Ti (VLT) basalts, and very high K (VHK) basalts.

($>90\%$ calcic plagioclase). These rocks are considered to represent plagioclase cumulates floated during the lunar magma ocean. The plagioclase is extremely calcic (An $94 \sim 96$), indicating the extreme depletion of alkalis (Na and K) and other volatile elements.

The magnesian suite (Mg suite) is composed of dunites ($>90\%$ olivine), troctolites (olivine-plagioclase), and Gabbro (plagioclase-pyroxene) with relatively high Mg/Fe ratios. The plagioclase is still calcic (An $86 \sim 93$). The trace element contents in plagioclase require equilibrium with a KREEP-rich (KREEP-rich is coined from "K, REE, and P rich") magma, which is inconsistent with their Mg-rich character.

The alkali suite has high alkali contents compared to other Moon rocks. The alkali suite is composed of alkali anorthosites with sodic plagioclase (An $70 \sim 85$), norites (plagioclase-orthopyroxene), and gabbronorites (plagioclase-clinopyroxene-orthopyroxene). The trace element contents in plagioclase also indicate a KREEP-rich parent magma.

The mare rocks are mainly composed of four types of basalts; high-Ti basalts, low-Ti basalts, very low-Ti (VLT) basalts, and very high-K (VHK) basalts. They show a large negative Eu anomaly in the REE pattern. Extremely K-rich rocks were found in the so-called "Very High K (VHK)" basalt (Neal and Taylor, 1992).

The Chinese spacecraft, Chang'e-5 landed on one of the youngest mare basalt units northeast of Mons Rümker in northern Oceanus Procellarum, and collected the basalts from the regoliths. The crater-counting chronology showed the basalts were $3 \sim 1$ Ga. About 800 lithic clasts (> 0.25 mm) were randomly picked from the soil samples, of which about 45% were basalts and about 55% were breccias. The breccias of >80% were basalt fragments. Forty-seven representative basalt clasts with various textures were dated by CAMECA IMS1280 HR-SIMS (see Section 1.3.13), and 2030 ± 4 My were obtained. This age can be a pivotal calibration point for crater-counting chronology (Li et al., 2021).

3.1.3 Similarity between chemical compositions of the Moon and the Earth's mantle

Astrochemistry of the Moon was started to be discussed since the Moon samples were obtained by the Apollo spacecrafts. Dreibus and Wänke (1979) discussed that the moderately to highly siderophile elements (including Mn, V, and Cr) are depleted in the Earth's mantle as well as the Moon relative to those of the CI chondrite composition (see Fig. 3.4A). They also found that the compositions of the Earth's mantle and the Moon are different from those of the eucrite parent body (EPB), of which Mn, Cr, and V are plotted horizontally (see red solid circles in Fig. 3.4B). Therefore, the Moon should have been formed from neither the CI chondrite nor EPB.

Ringwood (1959, 1966) had strong interest in the origin of the Moon. He first noticed that the depletion of V, Cr, and Mn of the Moon to CI chondrites, but that they were similar to those of the Earth's mantle (Ringwood, 1979). He inferred that the origin of the Moon was related to that of the Earth's mantle.

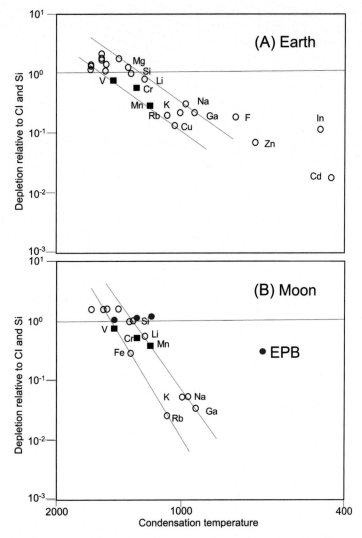

FIGURE 3.4 **Condensation temperature versus depletion of elemental concentration.** (A) The Earth and (B) the Moon. Both horizontal and vertical scales are logarithmic, respectively. The horizontal axis is the condensation temperature. The vertical axis is the concentration ratio of each element in the Earth (A) and the Moon (B) normalized to those of the CI chondrite and silicon. The horizontal line shows the ratio $= 1$. The two slanted lines in each figure show the approximate tendency of moderately refractory to moderately volatile elements of the condensation temperature of $1600°C \sim 800°C$. Black solid squares in figures (A) and (B) show V, Cr, and Mn in each body. Red solid circles in (B) indicate V, Cr, and Mn of the eucrite parent body (EPB) (Wänke et al., 1977). *Source: Data are from Wänke, H., Baddenhausen, Blum, K., et al., 1977. On the chemistry of lunar samples and achondrites. Primary matter in the lunar highlands: a re-evaluation. Proc. Lunar Sci. Conf. 8th, 2191–2213 and Drake, M.J., Newsom, C.J., Capobianco, C.J., 1989. V, Cr and Mn in the Earth, Moon, EPB and SPB and the origin of the Moon: experimental studies. Geochim. Cosmochim. Acta 53, 2101–2111.*

As shown in Fig. 3.4, the Moon's mantle was more depleted in volatile elements and metals compared to those in the Earth's mantle. The mean densities of the Moon and the Earth's mantle were 3.34 and 4.4 g cm^{-3}, respectively. Drake (1983) discussed the origin of the Moon based on the siderophile element abundances (W, Re, Mo, P, Ga, and Ge).

Ringwood and Kesson (1977) concluded that the average abundances of Fe, Co, Ni, W, P, S, and Se of the Moon were similar to those of the Earth's mantle within a factor of ~2. Thus they concluded again that the similarity in siderophile elements between the Moon's and the Earth's mantles implies that the Moon was derived from the Earth's mantle after the Earth's core had segregated. This model was consistent with the model of Wänke and colleagues (Dreibus and Wänke, 1979). Ringwood (1986) discussed the origin of the Moon again. He emphasized that high depletion of metallic iron is a remarkable feature of the Moon. The metallic core of the Moon is < 2%, while the core of the Earth is 32%.

Fig. 3.5 is a plot of abundance ratios of siderophile elements in the Moon versus those of the Earth's mantle against their metal/silicate partition coefficients (Ringwood, 1986). Less siderophile elements than Ni (Mn, V, Cr, Fe, W, P, Co, and S) are similar in both mantles. More siderophile elements than Ni (e.g., Cu, Mo, Re, and Au) are depleted to degrees which correspond well with their metal/silicate partition coefficients. If a small metallic core of <1% of the lunar mass was separated after accretion of the Moon, depletion of highly siderophile elements in Fig. 3.5 can be explained.

Ringwood (1986) favored a fission model. The problems of his model were: (1) the Earth must be formed in the short timescale (~1 My); and (2) the Earth's mantle should have completely melted and degassed for CO_2, N_2, Cl, and noble gases which contradict to the cosmochemical observations. However, supporters of fission models have sometimes appeared (e.g., Ćuk and Stewart, 2012).

Drake et al. (1989) published high temperature partitioning experimental results between solid iron metal, S-rich iron melt, and silicate basaltic melt for the moderately compatible elements of V, Cr, and Mn. Compatibility to metallic phases were Cr > V > Mn at high oxygen fugacity, and V > Cr > Mn at low oxygen fugacity. Solubilities into liquid metal were always larger than those in solid metal. These results suggest that the abundances of V, Cr, and Mn do not reflect core formation in the Earth. Instead, they are consistent with the relative volatilities of these elements (see Fig. 3.4). The similarity in the depletion patterns of V, Cr, and Mn inferred for the Earth's and Moon's mantles is NOT a sufficient reason for the Moon to have been derived from the Earth's mantle.

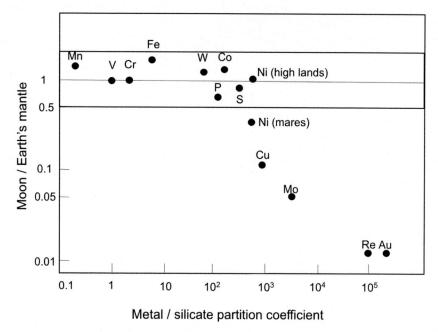

FIGURE 3.5 **The metal/silicate partition coefficients of siderophile elements versus their concentration ratios of the Moon/the Earth's mantle.** Both vertical and horizontal scales are in logarithmic ones. The vertical axis is the concentration ratios of the Moon/the Earth's mantle. The horizontal lines indicates ×2, ×1 and ×0.5 of the elemental abundances of the Moon/the Earth's mantle, respectively. *Source: Data are after Ringwood, A.E., 1986. Terrestrial origin of the Moon. Nature 322, 323–328.*

Ringwood et al. (1990) found that lithophile characters of Cr, V, and Mn are different at pressures, temperatures, and oxygen fugacities in the Earth's upper mantle and in the Moon condition. Their partitioning behaviors into the molten iron preferentially increase because of increase of solubility of oxygen into the molten iron. Therefore, these elements' depletions in the Earth's mantle can be attributed by their siderophilic behavior during formation of the Earth's core. Thus, a large proportion of the Moon could be derived from the Earth's mantle after the Earth's core was segregated.

Ringwood et al. (1991) also conducted the partitioning experiments of Cr, V, and Mn at 1500°C—2000°C and at 3—25 GPa using the Kawai-type apparatus (see Section 1.1.5.2) between metallic iron and silicates, simulating the elemental partitioning in Theia (see Fig. 3.6). The partitionings of Cr, V, and Mn were always $D_{silicate/metal}$ >1, indicating they always remained lithophile. Thus, there were no depletions of Cr, V, and Mn in the Theia's mantle. Accordingly, depletions of Cr, V, and Mn in the Moon were not inherited from the mantle of Theia. The similarity between the

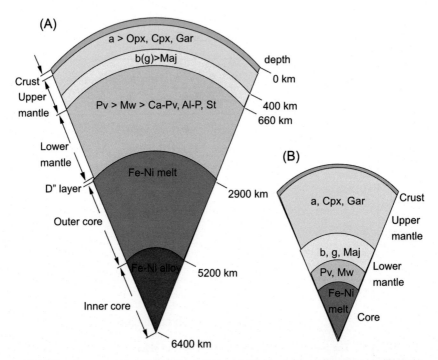

FIGURE 3.6 **Comparison of cross-section of (A) the present Earth and (B) Theia.** Theia is a Martian-size planetesimal with 10%−20% of the Earth's mass (0.46−0.58 of the Earth's radius). The total size between (A) and (B) is proportional, but the thickness of each layer is not. *a*, Olivine; *Opx*, orthopyroxene; *Cpx*, clinopyroxene; *Gar*, garnet; *b*, modified spinel; *g*, spinel; *Maj*, majorite (complicated garnet); *Pv*, perovskite; *Mw*, magnesiowüstite; *Ca-Pv*, Ca-perovskite; *Al-P*, Al rich phase; *St*, stishovite. *Source: Data are from Ringwood, A.E., Kato, T., Hibberson, W., Ware, N., 1991. Partitioning of Cr, V, and Mn between mantles and cores of differentiated planetesimals: implications for giant impact hypotheses of lunar origin. Icarus 89, 122−128.*

Earth's mantle and the Moon's mantle suggests that the Moon's mantle should have come from that of the Earth after core formation.

Gressmann and Rubie (2000) further investigated the partitioning of V, Cr, Mn, Ni, and Co between liquid metal and magnesiowüstite (MgO polymorph at high pressure in the lower mantle) and observed that the partitionings into the metal increased very weakly with increasing pressure (2200°C and 5−23 GPa), indicating slight increase of siderophile behavior, however rather large effects of temperature to the partitioning. If deep magma ocean of 3300°C and 35 GPa and oxygen fugacity with the presence of FeO, the mantle abundance of V, Cr, Mn, Ni, and Co would be explained. Therefore, the Moon is likely to have formed largely from materials that were ejected either from the mantle of a large impactor or from the Earth's mantle.

3.1.4 Constraints from high field strength element (HFSE)

Zirconium, Hf, Nb, and Ta are called high field strength elements (HFSE). These elements easily hydrolyze in popular acids such as hydrochloric acid or nitric acid, which interferes precise concentration determination, when the simple calibration curve method is used in determination of these elements.

As HFSE dissolves only in hydrofluoric acid (HF) forming fluoro-complex or oxo-fluoro-complex (Makishima et al., 1999), HFSE are called "fluorophile" elements (e.g., Makishima et al., 2009; Makishima, 2016), and HF is required to achieve isotopic equilibria.

The Nb/Ta ratio is useful to know the contributions of the proto-Earth and proto-Earth (the impactor). At high pressure, Nb partly becomes a siderophile element, and goes into the core (Wade and Wood, 2001). Therefore, the Nb/Ta ratio becomes lower than that of the chondrite. If the lunar materials are composed of the Earth's mantle, the lunar samples should also show the lower Nb/Ta ratio than that of the chondrite.

Münker (2010) developed a precise analytical measurement method of Zr/Hf and Nb/Ta using isotope dilution-multicollector ICP mass spectrometry (ID-MC-ICP-MS), and obtained precise Zr/Hf and Nb/Ta ratios(to understand the isotope dilution, see Section 1.3.16). He used ^{183}W-^{180}Ta-^{180}Hf-^{176}Lu-^{94}Zr spikes combined with the anion exchange chromatography with intermediate precision (2 RSD) of $< \pm$ 0.6% and $< \pm$ 4% for Zr/Hf and Nb/Ta, respectively. He analyzed 22 lunar rocks, three lunar soils, and some meteorites. For understanding errors (RSD, 2SD, 2SE, etc.), see Section 1.3.6.

The analytical results are schematically shown in Fig. 3.7. The bulk silicate Earth (BSE) value is also plotted in the figure for comparison. The average Nb/Ta ratio of the lunar samples was 17.0 ± 0.8, significantly lower than that of the chondrite of 19.9 ± 0.6. The bulk silicate Earth value of 14 ± 0.3, which is also far lower than that of the chondrite. As Nebel et al. (2010) argue the D″ layer as the Nb deposit, the lower Nb/Ta ratio of the BSE than those of the lunar samples is one of the evidences of the relation between the Moon's and the Earth's mantle.

3.1.5 Recent findings on the Moon's surface

3.1.5.1 Molecular water on the Moon

A characteristic absorption feature at $3 \mu m$ by three independent spacecraft observations has indicated widespread hydration on the lunar surface. However, whether the hydration is molecular water (H_2O) or other hydroxyl (OH) compounds is unknown, because there

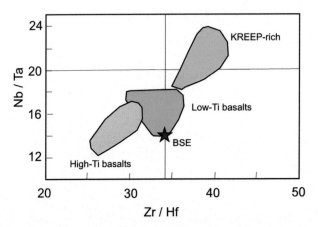

FIGURE 3.7   **A schematic plot of Zr/Hf versus Nb/Ta ratios of lunar samples.** KREEP-rich means KREEP-rich rocks. BSE means the bulk silicate Earth. The horizontal and vertical dotted lines indicate the averages of the CI chondrites. *Source: Data from Münker, C., 2010. A high field strength element perspective on early lunar differentiation. Geochim. Cosmochim. Acta 74, 7340–7361.*

are no established methods to distinguish the two using the $3\,\mu m$ band. However, a molecular water produces a peculiar spectral signature at $6\,\mu m$ that is not shared by other hydroxyl compounds.

Honniball et al. (2021) observed $6\,\mu m$ emission at high lunar latitudes using the NASA/DLR Stratospheric Observatory for Infrared Astronomy (SOFIA). This $6\,\mu m$ emission indicated the presence of molecular water on the lunar surface. The strength of the $6\,\mu m$ band indicated the H_2O abundance of about 100 to $400\,\mu g\,g^{-1}$. The distribution of water is at the small latitude range, which is considered to be a result of local geology, and is probably not a global phenomenon. They also suggested that a majority of the detected water must be stored within glasses or in spaces between grains sheltered from the harsh lunar environment, allowing the water to remain on the lunar surface. When the day for the people to stay on the Moon comes, such water will be a great help to get water on the Moon's surface.

3.1.5.2 Oxygen in hematite on the Moon could come from the Earth

Hematite (Fe_2O_3) is commonly observed as oxidizing products on the Earth, Mars, and some asteroids. Although oxidizing processes are speculated to operate on the lunar surface forming ferric iron–bearing minerals, detections of ferric minerals forming under highly reducing conditions on Moon have remained strange.

Li et al. (2020) analyzed the Moon Mineralogy Mapper data on the *Chandrayaan-1* mission, and found that hematite, a ferric mineral, is present at high latitudes on the Moon, mostly associated with east- and equator-facing sides of topographic highs, and is more prevalent on the nearside than the farside. They proposed that oxygen was delivered from Earth's upper atmosphere, and could be the major oxidant that forms lunar hematite. Hematite at craters of different ages may have preserved the oxygen isotopes of Earth's atmosphere in the past billions of years. Future oxygen isotope measurements can test this hypothesis. In addition, this may reveal the evolution of Earth's atmosphere.

3.2 The age of the Moon

3.2.1 Introduction

By astrophysics, Jacobsen et al. (2014) determined the Moon-forming age to be 95 ± 32 My after condensation by an N-body simulation. If we assume the condensation age to be 4.56 Ga, the Moon was formed in 4.46 Ga. They showed that earlier formation is ruled out at a 99.9% confidence level.

Carlson et al. (2014) divided lunar rocks into five types by the astrochemistry and astro-geology according to time (see Table 3.2).

1. Initial crystallization: In the magma ocean model for the Moon, an initially extensively molten Moon first crystallized mafic silicates that sank into the mantle to form the source regions of much later mare basalt magmatism (Walker et al., 1975; Warren, 1985).
2. FAN and floating plagioclase: After 70%−80% crystallization of the magma ocean, plagioclase began to crystallize from a dense iron-rich differentiated magma leading the plagioclase to float to form the ferroan anorthosite (FAN) series to lunar highland rocks (Dowty et al., 1974). From the model calculation (Elkins-Tanton et al., 2011), the floatation of the crust may occur in a thousand years, then solidification of the magma ocean ends in a few tens of million years. It is strange that the magma ocean continued to ~ 4300 Ma. Because to keep the Moon to be molten, some heat source is required. Tidal heating by the Earth is one of the possibilities.
3. Mafic cumulates with Eu anomaly. The extraction of plagioclase from the magma ocean left the mafic cumulates of the lunar interior with a deficiency in Eu relative to neighboring rare earth elements (REEs). This is reflected as the negative Eu anomaly (see Section 1.3.7) in some mare basalts.
4. Crystallization of KREEP: Further crystallization made residual liquid to be strongly enriched in incompatible elements named as

TABLE 3.2   Age estimates for early lunar differentiation events.

Events	Age (Ga)	References
Giant impact by heating model	4.47	Bottke et al. (2015)
FAN ages	4.360 ± 0.003	Borg et al. (2011)
	4.31 ± 0.07	Nyquist et al. (2010)
Peak in lunar zircon age distribution	4.320	Grange et al. (2013)
The second peak	4.200	Grange et al. (2013)
Oldest point on lunar zircon	4.417 ± 0.006	Nemchin et al. (2009)
Zircon Hf model ages	4.38−4.48	Taylor et al. (2009)
Mare basalt ^{146}Sm-^{142}Nd source age	4.32	Nyquist et al. (1995)
	4.35	Rankenburg et al. (2006)
	4.45	Boyet and Carlson (2007)
	4.33	Brandon et al. (2009)
Pb model age for lunar highlands	4.42	Terra and Wasserburg (1974)
KREEP Sm-Nd and Lu-Hf model ages	4.36 ± 0.04	Gaffney and Borg (2014)
	4.36 ± 0.04	Sprung et al. (2013)
	4.47 ± 0.07	Nyquist et al. (2010)
	∼ 4.26	Lugmair and Carlson (1978)
^{182}Hf-^{182}W lunar model age	<4.50	Touboul et al. (2007)
Mg-suites	4.283, 4.421	Carlson et al. (2014)
urKREEP	4.368 ± 0.029	Gaffney and Borg (2013)

Data are from Carlson, R.W., Borg, L.E., Gaffney, A.M., Boyet, M., 2014. Rb-Sr, Sm-Nd and Lu-Hf isotope systematics of the lunar Mg-suite: the age of the lunar crust and its relation to the time of Moon formation. Phil. Trans. R. Soc. A 372, 20130246.

KREEP for its enrichment in potassium, REE and phosphorus, and other incompatible elements (Warren and Wasson, 1979).

5. Mg-suite cumulate rocks: The final liquid component became the lunar highland crusts, which are plagioclase-rich rocks but are distinguished from FANs by their higher Mg/Fe ratios and the presence of abundant mafic phases. The Mg-suite cumulate rocks are usually to be partial melts of cumulates in the lunar interior (Shearer and Papike, 2005), or the parental magmas originated from large impacts (Hess, 1994).

Thus, at least five ages can be obtained as the age of the Moon. The researchers must understand what age you are discussing, and you are retrieving information.

3.2.2 Age of FANs

As discussed in Section 1.5.1, the ferroan anorthosite (FAN) suite of the lunar crustal rocks is considered to be the primary lunar flotated-cumulated crust that crystallized in the second stages of the magma ocean solidification. It should be remembered that the first solidified silicates in the first stage of the magma ocean solidification sank into the deep mantle.

According to this model, FANs represent the oldest age in the lunar crustal rock types. Attempts to date this rock suite precisely have been failed, because individual isochron measurements are not typically matched to the cosmochemical history of the samples, and have not been confirmed by each isotopic system (Hanan and Tilton, 1987; Carlson and Lugmair, 1988; Alibert et al., 1994; Borg et al., 1999; Norman et al., 2003).

Nyquist et al. (2010) obtained a Sm-Nd mineral isochron age of 4.47 ± 0.07 Ga for FAN-67075. Nyquist et al. (2010) also obtained the Sm-Nd isochron age of 4.31 ± 0.07 Ga for anorthositic clasts of lunar meteorites of Y86032 and Dho 908.

By making improvements to the standard isotopic techniques, Borg et al. (2011) obtained the age of crystallization of FAN 60025 to be 4.360 ± 0.003 Ga using the $^{207}Pb-^{206}Pb$ isotopic system of 4.3592 ± 0.0024 Ga, the $^{147}Sm-^{143}Nd$ isotopic system of 4.367 ± 0.011 Ga and the $^{146}Sm-^{142}Nd$ isotopic system of $4.318_{-0.038}^{+0.030}$ Ga which is the model dependent. These extraordinarily young ages require that either the Moon solidified significantly later (younger) than most previous estimates or the standard model that FANs are floated cumulates of a primordial magma ocean is incorrect. This problem has not been solved yet; thus further studies are required.

3.2.3 Zircons ages

Zircon crystallizes from the melt that saturated in zirconium. Zircon is formed in felsic samples, but sometimes found in mafic cumulates. Most lunar zircons are found in KREEP, but some zircons are found in impact melt breccias. Hafnium, which is contained $\sim 1\%$ in zircon, has high neutron cross-section with thermal neutrons, therefore, elements in zircon are protected against the neutron cosmic rays, which change isotope ratios or cause fission of uranium.

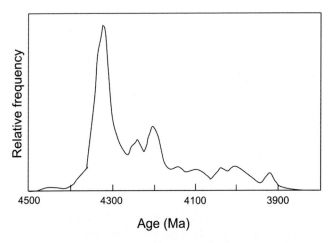

FIGURE 3.8 **Relative frequency of zircons on the Moon.** See the text for details. *Source: After Grange, M.L., Nemchin, A.A., Pidgeon, R.T., Merle, R.E., Timms, N.E. What lunar zircon ages can tell? Lunar Planet Sci 44: 1884, 2013.*

Grange et al. (2013) made an age distribution of zircon (see Fig. 3.8). They found the peaks at 4.320, 4.240, 4.200, and 3.920 Ga between 4.4 and 3.9 Ga. The peak of 4.320 Ga was the most prominent, and the 4.200 Ga was the second. They attributed the periodicity to the radioactive decay heat of the KREEP reservoir, which is enriched in incompatible elements such as K, U, and Th.

Taylor et al. (2009) obtained 4.38−4.48 Ga from the Hf model ages of zircons. Nemchin et al. (2009) found the oldest zircon of 4.417 ± 0.006 Ga, which is considered to be the oldest magmatic activity or magma ocean on the Moon.

3.2.4 Age of KREEP rocks

Lugmair and Carlson (1978) observed that lunar KREEP samples, 12034, 14307, 15382, 65015, 15426, and 75075 are aligned on a line of ∼4.26 Ga of the Sm-Nd isochron. However, the variations of Sm/Nd were not large, all data did not form a clear age result.

Gaffney and Borg (2013) obtained Lu-Hf data for KREEP samples of 15386, 72275,383, 77215, and 78238, and the model age of 4.36 ± 0.04 Ga. Sprung et al. (2013) obtained Lu-Hf data of KREEP-rich rocks of 12034, 14310, 65015, 62235, 68115, and 68815, and the model age of 4.36 ± 0.04 Ga.

Terra and Wasserburg (1974) obtained Pb model age of 4.42 Ga for lunar highlands using U-Th-Pb systematics. Touboul et al. (2007) obtained the ^{182}Hf-^{182}W lunar model age of <4.50 Ga.

3.2.5 Age of Mg-suite rocks

Carlson et al. (2014) determined Rb-Sr, ^{146,147}Sm-^{142,143}Nd, and Lu-Hf isotopic analyses of Mg-suite lunar crustal rocks 67667, 76335, 77215 and 78238, including internal isochron for norite 77215. Isochron ages determined by their study for 77215 were: Rb-Sr = 4.450 ± 0.270 Ga, ^{147}Sm-^{143}Nd = 4.283 ± 0.023 Ga, and Lu-Hf = 4.421 ± 0.068 Ga. The initial Nd and Hf isotopic compositions of all samples indicate that a source region was slightly enriched for incompatible elements, which were previously suggested that the Mg-suite crustal rocks contain a component of KREEP. The Sm/Nd-^{142}Nd/^{144}Nd correlation shown by both a ferroan anorthosite and Mg-suite rocks is consistent with the trend defined by mare and KREEP basalts, the slope of which corresponds to ages between 4.35 and 4.45 Ga. And the ages are in good agreements with the model of lunar formation by the giant impact into the Earth at c. 4.4 Ga.

Gaffney and Borg (2014) named final formed product of magma ocean crystallization as urKREEP. They obtained 4.368 ± 0.029 Ma from the model ages of the Hf and Nd isotope systematics.

3.2.6 Age of the Moon mantle differentiation by the ^{146}Sm-^{142}Nd method

The ^{146}Sm-^{142}Nd method, which uses an α-decay of ($T_{1/2}$ = 103 Ma), is very useful for estimation of the formation age of mantle. The method requires neutron-correction, and the obtained age is the model age, however, it gives important information. Furthermore, it can combine with the ^{147}Sm-^{143}Nd method.

Nyquist et al. (1995) obtained 4.32 Ga as the lunar initial differentiation age which corresponds to the initial crystallization age. Rankenburg et al. (2006) obtained the lunar mantle formation was 215_{-21}^{+23} Myrs from forming the solar system, which corresponds to 4.35 Ga. Boyet and Carlson (2007) obtained the early lunar differentiation to be 4.45 Ga by the ^{146}Sm-^{142}Nd and ^{147}Sm-^{143}Nd methods.

Brandon et al. (2009) re-measured six lunar basalt samples, from Hi-Ti basalts, Lo-Ti basalts, and KREEP, for Sm-Nd and Lu-Hf isotope compositions. They also reevaluated the Moon evolution models. The model that the bulk Moon has a superchondritic ^{147}Sm/^{144}Nd ratio of 6%−8.8% (which means Sm/Nd ratios are higher than that of the chondrite) explains best the Nd−Hf isotope compositions of the measured lunar mare basalts. The ^{142}Nd−^{143}Nd isotope systematics is best interpreted with an age of 229_{-20}^{+24} Ma after the nebular condensation. The mare basalt sources were closed around 150−190 Ma after accretion of the Moon. A long cooling history of a magma ocean is consistent with U−Pb age distributions in zircons (Nemchin et al., 2009; Grange et al., 2013).

3.3 The giant impact model for formation of the Earth-Moon system

3.3.1 Introduction

In this section, the formation of the Earth-Moon system is presented. It is generally accepted that the Earth and Moon are formed by the giant impact, in which a Martian-size object hit the proto-Earth resulting in making the Earth-Moon system. Chemical as well as physical simulations played important roles to establish the giant impact model. The important observations required for the giant impact of the Earth-Moon system from elemental abundances are described in Section 3.3.

Fig. 3.9 is a cartoon showing the relation among the proto-Earth, the impactor named Theia, the giant impact, and the late veneer, in the order of the time, to help you understand these various events. We learn about the giant impact, elemental astrochemistry, and isotope astrochemistry of the Moon in Section 3.4, and the late veneer in Section 4.1.

3.3.2 Peculiarities of the Moon

First of all, peculiarities of the Moon in the solar system are shown (Stevenson and Halliday, 2014). Any model for the origin of the Moon must explain these peculiarities of the Moon.

- The Moon has lower density than the Earth.
- The size of the moon is very large compared to the size of the host planet.
- The Moon is moving away from the Earth.
- The Moon carries most angular momentum of the Earth-Moon system.
- The Moon formed >30 Ma after the solar system formation.
- The start of the Moon was a vigorous magma ocean.
- The Moon is depleted to moderately volatile elements compared to the Earth.
- Isotopic compositions of nonvolatile elements of the Moon are very similar to those of the Earth.

3.3.3 A history of a giant impact model

Darwin (1879) discussed the origin of the Earth-Moon system, and suggested that the Moon should be torn out from the molten Earth mantle by the tidally induced "fission process." This is like a fission process of a heavy nucleus, in which one sphere is teared into two spheres.

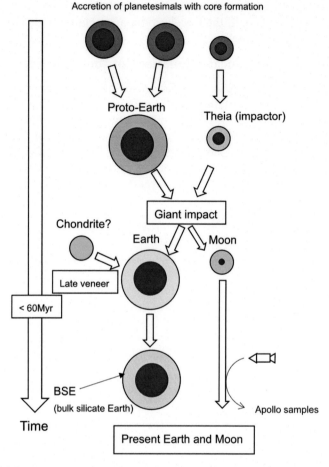

FIGURE 3.9 **A cartoon showing relation among the proto-Earth, Theia, giant impact, and late veneer.** Time proceeds from the top to the bottom.

Later, Daly (1946) proposed that the Moon was created by an impact. However, this impact origin of the Moon was not paid attention until Hartman and Davis (1975) and Cameron and Ward (1976) brought out the idea of the giant impact again. In the initial giant impact model of Daly, some important conclusions were derived:

- The Martian size object (0.1 M_E) hit the proto-Earth (Cameron, 1985).
- The collision made a silicate gas disk with low volatile elements.
- The gas accreted and formed the Moon in short time.
- A large angular momentum of the Earth-Moon system could be explicable.

The impacting body was named as "Theia," who gave birth to the Moon goddess, Selene, in the Greek myth, after Halliday (2000).

3.3.4 Basic astrophysical simulations of the giant impact

Fig. 3.10 shows the basic calculation of the giant impact performed by Agnor and Asphaug (2004) and Asphaug (2010). Some of the giant impact scenarios for the accretion efficiency (ξ) are indicated as functions of v_{rel}/v_{esc}, the colliding masses $M_2:M_1$ and the impact angle θ. The accretion efficiency (ξ) is expressed as $\xi = [(M_F - M_1)/M_2]$ where M_1 and M_2 correspond to the masses of the proto-Earth and the impactor, respectively. M_F is the mass of the largest final body (almost similar to the mass of the present Earth). The v_{rel} and v_{esc} values are the relative velocity of initial two bodies and the escaping velocity from the final body.

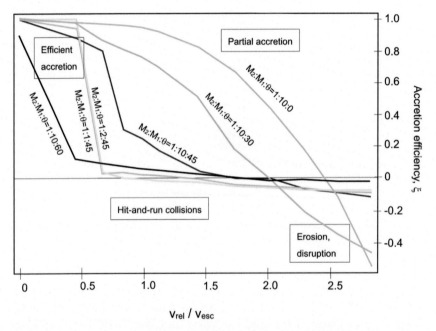

FIGURE 3.10   Model calculation results for accretion efficiency (ξ) in giant impacts as a function of v_{rel}/v_{esc} for colliding masses $M_2:M_1$ and for varying impact angles θ. M_1 and M_2 are masses of proto-Earth and the impactor, respectively. The accretion efficiency (ξ) is expressed as $\xi = [(M_F - M_1)/M_2]$ where M_F is the largest final body. The v_{rel} value is the relative velocity of initial two bodies. The v_{esc} value is the escape velocity from the final body. $M_2:M_1 = 10:1$ corresponds to the mass ratio between the proto-Earth and Mars. Relations of $M_2:M_1:\theta = 10:2:45$ degrees (yellow) and $M_2:M_1:\theta = 10:1:45$ degrees (orange) overlaps, so that it is difficult to discern. *Source: These calculations are based on Agnor and Asphaug (2004) and Asphaug (2010).*

It is easy to understand Fig. 3.10 by assuming extreme cases at $M_2:M_1 = 1:10$ (the case of the Mars and the Earth). When the relative velocities before and after the collision are similar ($v_{rel}/v_{esc} < 0.5$), the two bodies will merge into one body at the impact angles $\theta = 0$, 30, and 45 degrees. Thus the accretion efficiency (ξ) will be ~ 1, which is shown as "efficient accretion" in the figure. In contrast, [(the relative velocity)/(the escape velocity)] (v_{rel}/v_{esc}) is >2.5, two bodies do not accrete irrespective of the impact angles (θ), and the proto-Earth is eroded or even disrupted ($x < 0$). These cases are indicated as "erosion, disruption" in the figure.

When $M_2:M_1$ is 1:10, as the hitting angle increases, the accretion efficiency (ξ) decreases. When $M_2:M_1 = 1:10$, the accretion efficiency (ξ) drops ~ 0.9 to ~ 0 in the (v_{rel}/v_{esc}) range from 0.6 to 1.5 or the impact angle drops from 0 to 60 degrees. The high accretion efficiency can be obtained at the shallow angle resulting in the partial accretion in the figure. As the hitting angle increases from 45 to 60 degrees, or the mass difference decreases from $M_2:M_1 = 1:10$ to 1:1 at the same hitting angle ($\theta = 45$ degrees), the accretion efficiency stays around 0 at the velocity difference from 0.6 to 2.5. This range is named as "hit-and-run collisions." In this range, two bodies hit, but separate into the similar two bodies.

3.3.5 The standard model

The impact models have been converged to the standard model over a quarter of a century. Ida et al. (1997) calculated hundreds of particles after the giant impact. Canup and Asphaug (2001) estimated $M_2 = 0.1 - 0.15\ M_1$, $v_{rel} = \sim v_{esc}$, and $\theta = \sim 45$ degrees (indicated as the red line in Fig. 3.10). Such Mars size impact, v_{esc} is supersonic, therefore, materials are vaporized. The protolunar disk which is made by melts and vapors are formed after the giant impact, then they accrete to form the Moon. Canup (2004) estimated the temperature of the disk to be $\sim 3000K$. The massive diffusion occurred between the Earth and protolunar isotopic system (Pahlevan and Stevenson, 2007). The melt-vapor protolunar disk could explain efficient differentiation of $\sim 4\%$ iron fraction and magma ocean formation of the Moon. From the magma ocean, the rafting crust was formed by plagioclase-feldspar, which became lunar highlands of predominantly anorthosite composition (Warren, 1985; Shearer et al., 2006).

Belbruno and Gott (2005) proposed that Theia was formed at one of the Trojan points near one au. The iron cores of the two planets agglomerate without much mixing of silicate with iron (Nimmo and Agnor, 2006). Dahl and Stevenson (2010) found only a fraction of the Earth's core equilibrates with silicate during a giant impact. The impact-triggered fission (Ćuk and Stewart, 2012) and hit-and-run collision

(Reufer et al., 2012), Theia's mantle either escapes or becomes silicate atmosphere of the Earth. Canup (2012) makes the Moon and the Earth from colliding of equal proto-Earths to solve the isotopic problem.

3.3.6 A latest model proving the similarity between the proto-Earth and Theia (the impactor)

Mastrobuono-Battisti et al. (2015) performed extensive N-body simulations of terrestrial planet formation. To study the compositions of planets and their impactors they analyzed 40 dynamical simulations of the late stages. Each simulation started from a disk of 85–90 planetary embryos and 1000–2000 planetesimals extending from 0.5 au to 4.5 au. Within 100–200 Myrs, each simulation typically produced 3–4 rocky planets formed from collisions between embryos and planetesimals. They also calculated the oxygen isotope ratios and $\Delta^{17}O$ of the Earth and the Moon. Using the large simulation data, they compared the composition of each surviving planet with that of its last giant impactor, which is the last planetary embryo that impacted the planet (the final impact).

Compared to the simulation of Pahlevan and Stevenson (2007), they analyzed a single statistically limited simulation, which included only 150 particles, and compared the compositions of any impactors on any planets during the simulation (not only giant impacts, due to small number statistics) to the compositions of the planets.

As a result, they found that a significant fraction of all planetary impactors has compositions similar to the planets they struck, in contrast with the distinct compositions of different planets existing in the same planetary system. This means that the planet—impactor pairs are robustly more similar in composition than are pairs of surviving planets in the same system. The impactor and planet having a similar composition is also applicable to the origin of the $\Delta^{17}O$ similarity between the Earth and the Moon.

This conclusion that the proto-Earth and the impactor have almost similar composition from the beginning solves a lot of problems in standard models.

3.4 Constraints from the stable isotope astrochemistry for the giant impact

3.4.1 Evaporation and condensation of elements after the giant impact

After the giant impact, large amounts of gas are formed by the heat of impact. As the gas cooled down, minerals formed and each element condensed with the minerals depending on their volatility. Finally, such minerals gathered to form the Earth and Moon. In order to define the volatility of each element, the temperature when the half-mass of each element is condensed (T_c or 50% T_c) is generally used. T_c is very useful to evaluate the stable isotope astrochemistry of one element, because we can know what temperature and which process are discussed by applying the stable isotope astrochemistry of the target element in the Moon origin.

Lodders (2003) determined the condensation order of each mineral based on the meteorite studies. At first, refractory metal alloys are formed with platinum group elements. Then lithophile trace elements (REE, Ba, Sr, Sc, Y, V, Nb, Ta, Th, and U) are condensed with calcium titanates (perovskite [$CaTiO_3$] and Ca aluminate hibonite [$CaAl_{12}O_{19}$]). Then, Ca-Al-rich inclusions (CAIs) made of gehlenite (the first Si-bearing mineral) and melilite (the first Mg-bearing mineral) formed. However, most Si and Mg form forsterite (Mg_2SiO_4) and enstatite ($MgSiO_3$). Metallic iron alloy also condensed at similar temperature of forming forsterite. The elements that condensed at higher temperature of Si, Mg, and Fe are called "refractory elements," and those condensed at lower temperature are called "volatile elements." Then the sulfur is removed as troilite. Elements that condense between Si-Mg-Fe and troilite are called "moderately volatile elements," and elements that condense below troilite are "volatile" or "highly volatile" elements. The T_c calculated by Lodders as well as host minerals are shown in Fig. 3.11.

Ideally, in the study of formation of the Moon, the stable isotope fractionation is only controlled by the condensation process, and little affected by common geological processes such as fractional crystallization, elemental diffusion in solids, etc. Or, the isotopic fractionation by these mechanisms should be perfectly understood quantitatively before being applied to Moon formation studies. However, the new isotopic fractionation studies for the condensation and the fractional crystallization are performed all at once, the new isotope studies sometimes proceed back and forth.

In this section, the stable isotopic composition of O, Zn, Rb, Ga, K, Cu, Li, Cr, Si, Fe, Mg, V, and Ti are reviewed according to the order of the T_c value.

3.4.2 Why has the stable isotope astrochemistry become so popular?

At the end of 20th century, the stable isotope chemistry is limited to volatile elements such as H, C, N, O and S. However, the stable isotope chemistry for solid elements have recently drastically accelerated by the development and improvement of MC-ICP-MS machines and measurement techniques. The sample are measured by (1) a standard-sample

(A)

50% Tc (K)	X	Host	Host	Host	Host
180	O	Rock + water ice			
252	Hg		Troilite		
532	Tl		Troilite		
535	I*	Cl apatite			
536	In		FeS		
546	Br	Cl apatite			
652	Cd		Troilite	Enstatite	
664	S		Troilite		
697	Se		Troilite		
704	Sn				Fe alloy
709	Te				Fe alloy
726	Zn			Forsterite+enstatite	
727	Pb				Fe alloy
734	F*	F apatite			
746	Bi*				Fe alloy
799	Cs*		Feldspar		
800	Rb		Feldspar		
883	Ge				Fe alloy
908	B		Feldspar		
948	Cl	Sodalite			
958	Na*		Feldspar		
968	Ga		Feldspar		Fe alloy
979	Sb				Fe alloy
996	Ag				Fe alloy
1006	K		Feldspar		
1037	Cu				Fe alloy
1060	Au*				Refractory metal alloy
1065	As*				Fe alloy
1142	Li			Forsterite+enstatite	
1158	Mn*			Forsterite+enstatite	
1229	P			Schreibersite	
1296	Cr				Fe alloy
1310	Si			Forsterite+enstatite	
1324	Pd				Fe alloy
1334	Fe				Fe alloy
1336	Mg			Forsterite	
1352	Co*				Fe alloy
1353	Ni				Fe alloy
1356	Eu		Feldspar		

*) Mono-isotopic elements

FIGURE 3.11 **Half-mass condensation temperature (50% Tc) of each element and major phases or host minerals of the element.** The table should be seen from the high Tc to low Tc. (A) Elements with 50% T_c between 180–1356 K. (B) Elements with 50% T_c between 1356–1821 K.

(B)

50% Tc (K)	X	Host	Host	Host
1356	Eu	Titanate	Hibonite	
1392	Rh*			Refractory metal alloy
1408	Pt			Refractory metal alloy
1429	V	Titanate		
1452	Be*	Melilite		
1455	Ba	Titanate		
1464	Sr	Titanate		
1478	Ce	Titanate	Hibonite	
1487	Yb	Titanate	Hibonite	
1517	Ca	Gehlenite	Hibonite	
1551	Ru			Refractory metal alloy
1559	Nb*	Titanate		
1573	Ta	Titanate	Hibonite	
1578	La	Titanate	Hibonite	
1582	Ti	Titanate		
1582	Pr*	Titanate	Hibonite	
1590	Mo			Refractory metal alloy
1590	Sm	Titanate	Hibonite	
1602	Nd		Hibonite	
1603	Ir			Refractory metal alloy
1610	U		Hibonite	
1653	Al*		Hibonite	
1659	Sc*		Hibonite	
1659	Y*		Hibonite	
1659	Gd		Hibonite	
1659	Tb*		Hibonite	
1659	Dy		Hibonite	
1659	Ho*		Hibonite	
1659	Er		Hibonite	
1659	Tm*		Hibonite	
1659	Lu		Hibonite	
1659	Th*		Hibonite	
1684	Hf			HfO_2
1741	Zr			ZrO_2
1789	W			Refractory metal alloy
1812	Os			Refractory metal alloy
1821	Re			Refractory metal alloy

*) Mono-isotopic elements

FIGURE 3.11 *(Continued).*

bracketing method (SSB method) and/or (2) internal standardization technique in which a purified target element solution is added with a near mass-number element solution (e.g., Fe with Ni; Cu with Zn) and the mass fractionation factor of the added element is used for the fractionation correction of the target element. In addition, hardware has improved, such as higher sensitivity by high vacuum, higher resolution ($M/\Delta M > 8000$), stabler ICP even at low power, collision cell technology, etc. Developments of peripherals, such as desolvator, microconcentric nebulizer, etc., also supported the precise measurement. (As a result, the price of the machine became higher and higher.)

Furthermore, the column chemistry is sophisticated employing the extraction resins developed by Eichrom as well as the conventional anion/cation exchange resins. In operation of MC-ICP-MS, the element and acid concentrations are precisely matched between sample and standard solutions, and the coexisting element effects in the sample solution are carefully evaluated.

3.4.3 Oxygen isotopic ratios

Oxygen has three isotopes, and one of isotopic mass fractionation using $^{18}O/^{16}O$ can be defined as:

$$\delta^{18}O = \left[\left(^{18}O/^{16}O\right)_{sample} / \left(^{18}O/^{16}O\right)_{VSMOW} - 1 \right] \times 1000 \qquad (3.1)$$

where VSMOW means the Standard Mean Ocean Water of Vienna.

Fig. 3.12 shows a histogram for the $\delta^{18}O$ values of lunar samples of low Ti basalts including soil, regolith, and apatite, and the Earth (an average and 2 SD are shown as a red-filled circle and a red bar, respectively). From Fig. 3.12, the $\delta^{18}O$ values of the Moon are identical to those of the terrestrial values. Both the oxygen isotope fractionations of the Moon and Earth are homogeneous. Combining $\delta^{18}O$ with another oxygen isotope ratio of $\delta^{17}O$, by which more detailed discussion can be done by the so-called three oxygen isotope ratios, is discussed in the next section (Section 3.4.4).

3.4.4 Constraints from three oxygen isotopes for the origin of the Moon

3.4.4.1 Three oxygen isotopes support the giant impact

Oxygen has three isotopes, ^{16}O, ^{17}O, and ^{18}O. The combination of two oxygen isotope ratios, $^{17}O/^{16}O$ and $^{18}O/^{16}O$ gives strong

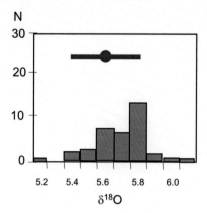

FIGURE 3.12   A histogram for the δ ^{18}O values of lunar samples. The lunar samples include low Ti basalts, soil, regolith, and apatites. The mean and 2 SD of terrestrial basalts are shown as a red circle and bar, respectively. *Source: The Moon data are from Wiechert et al. (2001), Spicuzza et al. (2007), Hallis et al. (2010) and Liu et al. (2010). The terrestrial data are from Eiler (2001). The figure is modified from Dauphas et al. (2014).*

constraints in cosmo-sciences. The oxygen isotope ratios are expressed by δ-notation:

$$\delta^{18}O = \left[\left(^{18}O/^{16}O\right)_{sample} / \left(^{18}O/^{16}O\right)_{VSMOW} - 1 \right] \times 1000 \qquad (3.2)$$

$$\delta^{17}O = \left[\left(^{17}O/^{16}O\right)_{sample} / \left(^{17}O/^{16}O\right)_{VSMOW} - 1 \right] \times 1000 \qquad (3.3)$$

The merit of the plot of two oxygen isotope ratios (i.d. the x and y axes are $\delta^{18}O$ and $\delta^{17}O$, respectively) is that if there are two components with different isotope ratios, the mixing line of the components becomes a linear plot. In addition, all terrestrial samples are plotted along one fractionation line of slope = ∼0.5. This line is called the Terrestrial Fractionation Line (TFL) which indicates that all terrestrial materials should be plotted on this line. This is because when mass-dependent fractionation of terrestrial samples occurs, any data always moves on TFL. Mass-independent fractionation rarely occurs, and such process for oxygen is only known as the ozone-related reaction at the stratospheric air samples on the Earth (Thiemens and Heidenreich III, 1983). Mass fractionation laws are explained in Section 1.3.14.

The two oxygen isotope ratios of the samples in our solar systems are shown in Fig. 3.13. The areas of the carbonaceous chondrites (CI, CM, CV, CO, CK, and CR), the ordinary chondrites (H, L, and LL), enstatite chondrites, and achondrites as well as those of the sun, Earth, Moon, and Mars are roughly shown in the figure. Enstatite chondrites (EH and EL) and

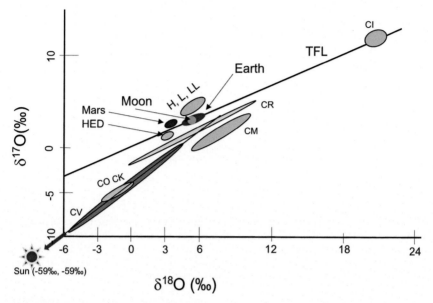

FIGURE 3.13 **A plot of δ ^{17}O against δ ^{18}O.** TFL with the slope of ~0.5 indicates a terrestrial fractionation line. The line with the slope of ~1 from Sun shows the line of CV chondrites. CI, CM, CV, CO, CK, and CR are the carbonaceous chondrites (Clayton and Mayeda, 1999). H, L, and LL indicate H, L, and LL ordinary chondrites, respectively (Clayton et al., 1991). HED indicates HED meteorites (howardites, eucrites, and diogenites). Mars data are from SNC meteorites (Clayton et al., 1991). The ranges of enstatite chondrites (EL and EH) and aubrites (achondrite) as well as the Moon data overlap with that of the Earth (Clayton et al., 1984). The sun data is from McKeegan et al. (2011). The δ^{18}O values of the Earth are 5–7‰. *Source: Data are from Clayton and Mayeda (1991, 1999), Clayton et al. (1984,1991), and McKeegan et al. (2011).*

aubrites are plotted at the same area of the Earth (Clayton et al., 1984). In this plot, the oxygen isotope ratios of the Moon and the Earth are overlapped. CI chondrites showed extremely heavier oxygen isotope ratios but on the TFL. The Moon, which overlaps in the small area of the Earth, is distinctively different from all other solar system bodies except enstatite chondrites and aubrites.

The data shows that the origin of the Moon does not contradict to the idea that the source material of the Moon came not from the chondritic asteroids but from the Earth, resulting in supporting the giant impact model. Furthermore, the oxygen isotope of the impactor (Theia) must be near Earth, otherwise the oxygen isotope ratios of the Moon will be shifted to Theia.

This observation has been robust over 30 years (Clayton and Mayeda, 1996; Wiechert et al., 2001; Spicuzza et al., 2007; Hallis et al., 2010). It is very accidental that the two bodies, the proto-Earth and

Theia, had the similar oxygen isotopic ratios on the TFL. It is required that two bodies mixed well and the evidences of two bodies were perfectly erased. This is consistent with other isotopes, such as Li, Mg, Si, and Ti which are discussed in the later sections (Sections 3.4.5–3.4.16).

3.4.4.2 $\Delta^{17}O$—a new oxygen isotope ratio presentation

Pack and Herwartz (2014) invented new data presentation for $\Delta^{17}O$ (read as "big-delta seventeen-oxygen"), and used it for the Earth mantle. They thought mass-independent fractionation should occur even in a small degree in rocks and minerals, which was anticipated by Matsuhisa et al. (1978).

Mass-dependent fractionation between reservoirs (A, B) is defined as:

$$\alpha_{A-B}^{2/1} = \left(\alpha_{A-B}^{3/1}\right)^{\theta_{A-B}} \tag{3.4}$$

The symbols of $\alpha_{A-B}^{2/1}$ and $\alpha_{A-B}^{3/1}$ are the fractionation factor in $^{17}O/^{16}O$ and $^{18}O/^{16}O$ between reservoirs A and B. In case of equilibrium, fractionation between A and B denotes 2 phases. θ_{A-B} is termed as the triple isotope fractionation exponent. For oxygen, θ_{A-B} can vary between 0.5 and 0.5305 (Matsuhisa et al., 1978; Young et al., 2002; Cao and Liu, 2011).

Small deviations of $^{17}O/^{16}O$ from a given reference line (RL) are expressed as:

$$\Delta^{17}O = \delta'^{17}O - \lambda \times \delta'^{18}O \tag{3.5}$$

where

$$\delta'^{17}O = 1000 \times \ln[\delta^{17}O/1000 + 1] \tag{3.6}$$

$$\delta'^{18}O = 1000 \times \ln[\delta^{18}O/1000 + 1] \tag{3.7}$$

and

$$\lambda = 0.5305. \tag{3.8}$$

Note that $\Delta^{17}O$ is not a measured value but an artificially calculated value, and a variable value depending on a slope of the reference line, λ.

3.4.4.3 Lunar oxygen isotopic ratios

Wiechert et al. (2001) measured oxygen isotopic ratios of $^{17}O/^{16}O$, $^{18}O/^{16}O$, and $\Delta^{17}O$ of 31 lunar samples collected by the 11, 12, 15, 16, and 17 Apollo missions. All oxygen isotopic ratios were within $\pm 0.016‰$ (2 SD) on a single mass-dependent fractionation line which is identical to the terrestrial fractionation line (TFL) within errors. This observation is consistent with the giant impact model. The proto-Earth and Theia, both of which had identical isotopic compositions, had mixed well. The similarity came from the identical heliocentric distances (distance from the sun) of both proto-Earth and Theia because the isotopic composition is determined by the heliocentric distance. They used

$$\Delta^{17}O = \delta^{17}O - 0.5245 \times \delta^{18}O \tag{3.9}$$

They attributed the depletion of volatiles and higher FeO content of the Moon than that of the Earth's mantle to secondary origin. Significant amounts of materials might be admixed to the Moon after the giant impact. However, they cannot change the oxygen isotopic ratios significantly. Their analytical precision can detect >3% and >5% of the primitive meteorites or differentiated planetesimals, respectively.

3.4.4.4 Development of accuracy of oxygen isotopic ratios

Herwartz et al. (2014) improved the accuracy of the oxygen isotope measurement. They carefully investigated the $\Delta^{17}O$ values of Wiechert et al. (2001) and concluded that the values of the lunar samples should be elevated, and not identical with that of the terrestrial values. They used $\Delta^{17}O$ to be

$$\Delta^{17}O = \delta'^{17}O - (0.532 \pm 0.006) \times \delta'^{18}O \tag{3.10}$$

which is different from Eq. (3.8). This slope is within the error of 0.5305, which is high-temperature oxygen isotope fractionation (Young et al., 2002). They defined the bulk silicate Earth (BSE) values to be $-0.101 \pm 0.002‰$ from their measurements of mantle minerals and mid-ocean ridge basalts (MORB). Three lunar basalt samples (high Ti basalt, low Ti gabbro, and low Ti basalt) gave oxygen isotopic ratios in a small range of $\Delta^{17}O = -0.089 \pm 0.002‰$ (1σ, $n = 20$), which is higher than that of BSE (see Fig. 3.14).

Theia was estimated to be made of not carbonaceous chondrites, but of non-carbonaceous chondrites such as Earth, Mars, ordinary chondrites, enstatite chondrites, or achondrites from Ti, Cr, and Ni isotopic composition (Warren, 2011). They reevaluated lunar oxygen isotope data from three previous studies (Wiechert et al., 2001; Spicuzza et al., 2007; Hallis et al., 2010), and the slightly higher $\Delta^{17}O$ was consistent with uncertainties.

It is suggested that the Moon received fractionally more impactor material than the Earth, and that the $\Delta^{17}O$ of Theia was most likely higher than that of the Earth and the Moon. Mixing 4% of material isotopically resembling Mars would be sufficient to explain the observed 12 ppm difference between the Earth and the Moon. It is against the general model (Cameron and Benz, 1991; Canup and Asphaug, 2001; Canup, 2012; Ćuk and Stewart, 2012; Reufer et al., 2012), therefore,

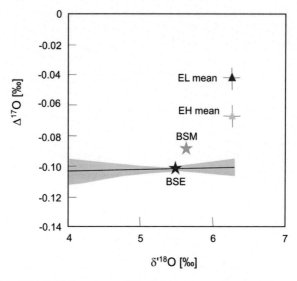

FIGURE 3.14 **The $\delta'\,^{18}O$ versus $\Delta\,^{17}O$ plots for BSE, BSM, and averages of EL and EH chondrites.** BSE is the bulk silicate Earth, and a horizontal red line and pink area show the average and 1 s area of mantle minerals. BSM means the bulk silicate Moon. EL and EH indicate the averages of EL and EH chondrites, respectively. Error bars are written as 1σ. *Source: Data are after Herwartz, D., Pack, A., Friedrichs, B., Bischoff, A., 2014. Identification of the giant impactor Theia in lunar rocks. Science 344, 1146–1150.*

larger fractions of Theia should be in the Moon. This implies that the $\Delta^{17}O$ composition of Theia was slightly higher than that of the Earth. Here the value of δf_T is introduced:

$$\delta f_T = \left(F_P/F_M - 1\right) \times 100 \tag{3.11}$$

where F is the mass fraction of proto-Earth material in the Moon (F_P) and in the final Moon (F_M), respectively. The δf_T quantifies the percent compositional deviation of the Moon (or disk) from the Earth. In most models the impactor material in the Moon is larger than that in the Earth, therefore, δf_T is usually negative. The model-independent δf_T is obtained only from the Nb/Ta ratios (see Section 3.1.5) to be 17.0 ± 0.8 and 14.0 ± 0.3 for the Moon's and the Earth's mantle, respectively. This translates δf_T to be $-21 \sim -44$. The model calculation indicates 70%–90% of the Moon is made of Theia.

They obtained new enstatite chondrite data, of which difference from the values of the Earth were 59 ± 8 ppm (1σ, n = 14) and 35 ± 10 ppm (1σ, n = 10) for those of EL (low iron enstatite chondrites) and EH (more metal-rich high iron enstatite chondrites), respectively. If the oxygen isotopic composition of Theia is similar to that of EL or EH chondrites, δf_T is $-21\% \pm 9\%$ and $-36\% \pm 15\%$ for EH, respectively. These estimates agree well with the more recent numerical models (Canup, 2012; Ćuk and Stewart, 2012; Reufer et al., 2012) and estimates from the Nb/Ta mass balance. If Theia resembled enstatite chondrites (ECs), the Moon will fall on a mixing line between ECs and the Earth. The high-precision Ti isotope results suggest that the Earth and ECs are rather similar (Zhang et al., 2012). The simple average of ε^{50}Ti of ECs are -0.12 (n = 5) and -0.29 (n = 2) for EH and EL, respectively, while ε^{50}Ti of the lunar and terrestrial averages were -0.03 ± 0.04 and 0.01 ± 0.01, respectively. Thus ECs, especially, EH fit as the composition of Theia. High FeO content in the lunar mantle does not fit low FeO content of ECs, however, ECs contents of Fe are high (EH and EL = 33 and 24 wt.%, respectively) (Javoy et al., 2010). Enstatite chondrites are sometimes considered to be the sole building blocks of the Earth (Javoy et al., 2010). However, there is no need to make the Earth from one known type of a meteorite.

In addition the late veneer (see Section 4.1) can change the isotope ratios of the Earth, and the present Moon shows the oxygen isotopes of the proto-Earth. From W, the late veneer is estimated to be 0.3%–0.8% of BSE (Walker, 2009). If 0.5% late veneer had low oxygen isotope composition of CV carbonaceous chondrites (Δ^{17}O = $-4‰$) (Clayton and Mayeda, 1999), Δ^{17}O of the bulk Earth would decrease ~ 20 ppm, which is consistent with the observed difference of Δ^{17}O between the Earth and the Moon. The water in the carbonaceous chondrites could be the source of the ocean water.

3.4.4.5 Other oxygen isotopic evidences

Young et al. (2016) proposed the newer model based on the most recent measurements of the oxygen isotopic ratios of the Moon and the Earth samples. They assumed that the San Carlos (SC) olivine is the standard of the Earth's mantle, because the oxygen isotope ratios of silicates are systematically different from those of the standard mean ocean water (SMOW). Thus, they obtained and expressed oxygen isotope ratios as δ'^{17}O, δ'^{18}O, and Δ'^{17}O based on the SC olivine. In this expression, Δ'^{17}O of the silicate Earth is $\sim 0.1‰$ lower than Δ^{17}O of that in SMOW. The following equation was the reference fractionation line for the SC olivine with the λ value of 0.528:

$$\Delta'^{17}O = \delta'^{17}O - 0.528 \times \delta'^{18}O \qquad (3.12)$$

Based on this criterion, Young et al. (2016) found no discernible difference between the Δ'^{17}O values of the terrestrial mantle (the SC olivine; -1 ± 2 ppm, 1 SE) and those of lunar basalts powders (0 ± 2 ppm, 1 SE) or lunar fused powder beads (0 ± 3 ppm, 1 SE). Adding all uncertainty, the Δ'^{17}O value of the terrestrial mantle is -1 ± 4.8 ppm (2 SE), which is indistinguishable from zero.

They compared the $\Delta'^{17}O$ value of the Earth (SC olivine) with that of the Moon in references. In five studies of Wiechert et al. (2001), Spicuzza et al. (2007), Hallis et al. (2010), Herwartz et al. (2014) and this study, two studies showed significant difference; one showed no difference; and the remaining two showed equivocal difference.

Especially, results of Young et al. (2016) do not agree with the conclusions of those of Herwartz et al. (2014), which is discussed in Section 3.4.4.4. However, the same lunar sample agreed with each other. Thus Young et al. (2016) explained that the discrepancies are caused by the unfortunate sample selection.

The lunar highland sample showed a significantly lower $\Delta'^{17}O$ value of -16 ± 3 ppm (1 SE), which is similar to the study of Wiechert et al. (2001). Considering the lower values of the terrestrial anorthosite sample into account, the lower values seem to be related to a mass fractionation process in forming this rock type (terrestrial anorthosite and lunar anorthositic troctolite).

Differences in the $\Delta'^{17}O$ value between Theia and the proto-Earth were assumed to be 0.15 or 0.05‰ by Kaib and Cowan (2015) and Mastrobuono-Battisti et al. (2015), respectively, based on the N-body simulations of standard terrestrial planet formation scenarios with assumed gradients in $\Delta'^{17}O$ across the inner solar system. Young et al. (2016) used a planetary accretion model (Rubie et al., 2015) which uses N-body accretion simulations based on the Grand Tack model (Walsh et al., 2011; see Section 2.3.5). They calculated to reproduce masses of the Earth and the Mars and the oxidation state of the Earth's mantle, using a multi-reservoir model (silicate, oxidized iron, and water) to describe the initial distribution of oxygen isotopes including the effects of mass accretion after the giant impact. When the late veneer (see Section 4.1) is considered to be <1% by mass into account, $\Delta'^{17}O_{Theia}$-$\Delta'^{17}O_{proto-Earth}$ became nearly zero from their measurement in all simulations. The $\Delta'^{17}O_{Moon}$-$\Delta'^{17}O_{Earth}$ values were $+20 \sim -60$‰ for the Mars-sized impactor scenario, and $+8 \sim -12$‰ for the proto-Earth-sized impactor scenarios.

As Young et al. (2016) observed that the Moon and Earth $\Delta'^{17}O$ values are similar, the oxygen isotopes of Theia and proto-Earth were thoroughly mixed by the Moon-forming impact for the difference to be less than 5 ppm levels. This also gives constraints on the late veneer, which is discussed in Section 4.1.

3.4.4.6 Recent oxygen isotopic studies

Cano et al. (2020) performed high-precision oxygen isotope analyses of various lunar lithologies and showed that the Earth and Moon had slightly different oxygen isotope compositions. Oxygen isotope values of lunar samples were dependent with lithology, and they proposed

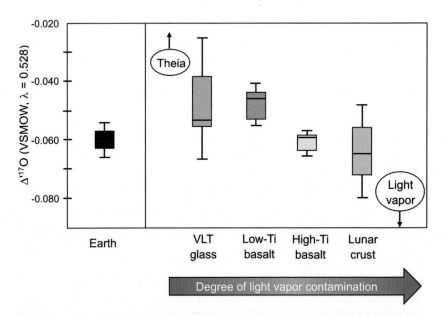

FIGURE 3.15   Box-and-whisker plots for the $\Delta'^{17}O$ values of the different lunar lithology and the Earth samples. The $\Delta'^{17}O$ values decrease with increasing degree of the light vapor contamination. The horizontal bar in each box indicates the median values. The whiskers show the upper and lower extremes, and the boxes indicate the upper and lower limits of data without extremes. *Source: Data from Cano, E.J., Sharp, Z.D., Shearer, C. K., 2020. Distinct oxygen isotope compositions of the Earth and Moon. Nat. Geosci. 13, 270–274.*

that the differences were caused by mixing between isotopically light vapor, generated by the impact, and the outermost portion of the early lunar magma ocean (Fig. 3.15). Their data suggested that samples derived from the deep lunar mantle, which had isotopically heavier than those of the Earth, had isotopic compositions of the protolunar impactor, Theia.

They further obtained the mineral separate data. Triple oxygen isotope equilibrium can be assessed from the θ values, where

$$\theta_{A-B} = (\delta'^{17}O_A - \delta'^{17}O_B)/(\delta'^{18}O_A - \delta'^{18}O_B). \tag{3.13}$$

At high temperatures, θ should be in the range of 0.528 to 0.530. Coexisting pyroxene—plagioclase pairs from terrestrial basalts were from 0.522 to 0.530 (± 0.004, 1σ). However, lunar samples have extremely varied θ values from 0.470 to 0.560 (figure is not shown), which can be easily explained by disequilibrium. Pyroxene from low-Ti basalts had $\Delta'^{17}O$ values that were close to the pristine lunar mantle (-0.044‰ and -0.041‰), whereas coexisting plagioclase had low $\Delta'^{17}O$ of -0.066‰ and -0.064‰. This can be explained by secondary contamination by light crustal material to the more easily exchangeable plagioclase during passage of the magmas through the crust. Incorporation of the light vapor should raise the $\delta^{18}O$ value of the most affected samples by $\sim 0.6‰$, which is consistent with the $\delta^{18}O$ shifts observed. These findings showed that the distinct oxygen isotope compositions of Theia and the Earth were not completely mixed by the Moon-forming impact, thus providing the evidence that Theia could have formed farther from the Sun than did the Earth.

Bindeman et al. (2022) found a slight but significant $\sim 0.2‰$ decrease in $\delta^{18}O$ in the shallowest continental lithospheric mantle (CLM) since the Archean without change in $\Delta'^{17}O$. Younger samples showed a decrease and greater heterogeneity of $\delta^{18}O$ due to the progress of plate tectonics and subduction. If $\delta^{18}O$ in the oldest Archean samples was the best $\delta^{18}O$ estimate for the Earth of 5.37‰ for olivine and 5.57‰ for bulk peridotite, these values are comparable to lunar rocks because of no plate tectonics in the Moon. As the continental lithospheric mantle is a large volume and even if its $\delta^{18}O$ are lower values, it can explain the increasing $\delta^{18}O$ of the continental crust since oxygen is progressively redistributed by fluids between these reservoirs via high-$\delta^{18}O$ sediment accretion and low-$\delta^{18}O$ mantle in subduction zones.

3.4.5 Zn isotopic composition of lunar rocks

Zn has Tc of 726K (Lodders, 2003), which is volatile element. Zn isotopic composition of lunar rocks is mainly measured by Lyon and IPGP groups in France. Zn isotopic ratios were determined by Moynier et al. (2006), Herzog et al. (2009), and Paniello et al. (2012) using MC-ICP-MS.

The Zn isotopic composition is defined as

$$\delta^{66}Zn(‰) = \left[\left({}^{66}Zn/{}^{64}Zn \right)_{sample} / \left({}^{66}Zn/{}^{64}Zn \right)_{standard} - 1 \right] \times 1000 \quad (3.14)$$

where the standard of Zn is "JMC Lyon standard JMC 400882B." The typical error of $\delta^{66}Zn$ was $\pm 0.05‰$ (2 SD). The $\delta^{66}Zn$ of the bulk silicate Earth (BSE) was estimated to be $0.27 \pm 0.10‰$ by Paniello et al. (2012). The averages of $\delta^{66}Zn$ of low-Ti and high-Ti basalts of Paniello et al. (2012) were $1.31 \pm 0.13‰$ (2 SD, n = 11) and $1.39 \pm 0.13‰$ (2 SD, n = 8), respectively (see Fig. 3.16). Thus, two types of the basalts are identical within the uncertainties. Their study also showed that the lunar regolith and soil samples reflect both sputtering and impact effects, therefore, these samples cannot represent primary magmatic composition of the Moon. Herzog et al. (2009) also measured high-Ti basalts to be the average of $1.47 \pm 0.54‰$ (2 SD, n = 5), which is also similar to that of

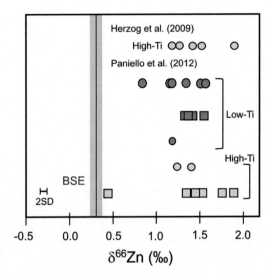

FIGURE 3.16 **The Zn isotopic compositions of lunar samples.** The vertical line and pink area indicate the 2 SD range of the bulk silicate Earth (BSE) estimated from terrestrial basalts. The data of high- and low-Ti basalts are from Herzog et al. (2009) and Paniello et al. (2012). The bulk silicate Earth (BSE) composition with 2 SD is shown as a dotted red line and pink box. The typical 2 SD for each $\delta^{66}Zn$ measurement is shown in the figure. *Source: Data are from Paniello, R.C., Day, J.M.D., Moynier, F., 2012. Zinc isotopic evidence for the origin of the Moon. Nature 490, 376—379.*

Paniello et al. (2012) within the error. However, Moynier et al. (2006) only measured lunar soil samples, so their data are not used here.

From Fig. 3.16, the high-Ti and low-Ti basalts are obviously heavier than that of BSE. The low-Ti mare basalts are predominant on the lunar surface at ~90% (Spicuzza et al., 2007), while the high-Ti mare basalts are formed from ilmenite cumulates formed after 95% crystallized magma ocean, in which the lunar mantle of >78% is formed of olivine and orthopyroxene. This clearly indicates that planetary volatile depletion processes removing light isotopes occurred, which is consistent with the giant impact origin of the Moon.

3.4.6 Rb isotopic composition of lunar rocks

Rubidium has two isotopes, ^{85}Rb and ^{87}Rb. Rb is a moderately volatile lithophile element with a T_c of 800K (Lodders, 2003). Rb is a heavy element compared to elements which have been used in traditionally stable isotope astrochemistry such as Zn, Cu, and Fe, therefore, the relative mass difference is small. Thus, the isotopic fractionation may be also small. However, Pringle and Moynier (2017) challenged the

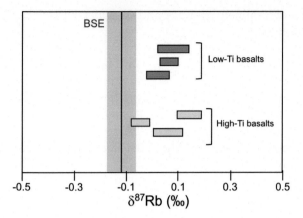

FIGURE 3.17   The Rb isotopic compositions of lunar samples. The red line and pink area indicate the average and 2 SD range of δ^{87}Rb of the bulk silicate Earth (BSE). Each box indicates the δ^{87}Rb value with 2SE range. *Source: Data from Pringle, E.A., Moynier, F., 2017. Rubidium isotopic composition of the Earth, meteorites, and the Moon: evidence for the origin of volatile loss during planetary accretion. Earth Planet. Sci. Lett. 473, 62–70.*

determination of Rb isotopic fractionation measurement by MC-ICP-MS. The Rb isotopic composition is expressed as δ^{87}Rb where:

$$\delta^{87}\text{Rb}(\text{‰}) = \left[\left(^{87}\text{Rb}/^{85}\text{Rb}\right)_{\text{sample}} / \left(^{87}\text{Rb}/^{85}\text{Rb}\right)_{\text{standard}} - 1\right] \times 1000 \quad (3.15)$$

The standard of Rb is the NIST SRM 984 RbCl standard. The typical error of δ^{87}Rb was ± 0.05‰ (2 SD) in Pringle and Moynier (2017).

The Rb isotopic composition of the lunar samples is shown in Fig. 3.17. They estimated the bulk silicate Earth (BSE) value to be -0.12 ± 0.05‰ (2 SD). Both low-Ti and high-Ti basalts showed heavier Rb isotopic compositions of ~ 0.2‰ than that of BSE. As Rb is a moderately volatile element, the heavy Rb isotope enrichment in lunar basalts should be due to volatile loss during Moon-forming giant impact. Similar behavior is observed in Zn, Cu, and K, which have similar T_c to Rb. This may be attributed to (1) volatile loss during the giant impact, (2) the magma ocean degassing, or (3) the incomplete condensation from the protolunar disk as K (Lock et al., 2019; Wang and Jacobsen, 2016a). In each model, volatile loss in the giant impact Moon-forming scenario is associated with the heavy isotopic characters of volatile elements.

3.4.7 Ga isotopic composition of lunar rocks

Gallium has two isotopes ^{69}Ga and ^{71}Ga, with T_c (half-mass condensation temperature) of 968K (Lodders, 2003). Ga isotopic ratios of lunar rocks were determined by Kato and Moynier (2017). The Ga isotopic ratios are defined as

$$\delta^{71}Ga(\text{‰}) = \left[\left({}^{71}Ga/{}^{69}Ga \right)_{sample} / \left({}^{71}Ga/{}^{69}Ga \right)_{standard} - 1 \right] \times 1000 \quad (3.16)$$

The standard of Ga is "IPGP standard." The typical error of $\delta^{71}Ga$ was ± 0.05‰ (2SE). The bulk silicate Earth (BSE) value was estimated to be 0.00 ± 0.06‰ (2 SD) (Kato et al., 2017).

The Ga isotopic ratios of Mg suits, anorthosites, low- and high-Ti basalts are shown in Fig. 3.18A. Three low-Ti basalts (0.09−0.32‰) and four high-Ti basalts (0.15−0.57‰) showed heavier isotopes of Ga compared to BSE. Two of nine samples showed lighter $\delta^{71}Ga$, one of which also showed the lightest $\delta^{66}Zn$ (Fig. 3.18B). These samples might be contaminated by the isotopically light vapor at the surface. Ferroan anorthosite samples (FANs) showed wide $\delta^{71}Ga$ variation.

The source of the mare basalts is enriched in heavier isotopes of Ga, compared to BSE. This confirms the heavier isotopic character of the mare basalts of Ga as well as Zn and K. However, this character would not have been caused by the magma ocean, but derived from accretion of planetesimals with highly depleted in volatiles.

3.4.8 K isotopic composition of lunar rocks

Potassium has three isotopes, ^{39}K, ^{40}K, and ^{41}K with T_c (half-mass condensation temperature) of 1006K (Lodders, 2003). Potassium is one of the moderately volatile elements. Therefore, it can give a clue for the depletion of volatile elements of the Moon compared to those of the Earth. Thus the Moon is thought to be the result of Moon-forming giant impact. Volatile-element-depleted bodies were expected to be enriched in heavy potassium isotopes during the loss of volatiles, however, Humayun and Clayton (1995) found no such enrichment. They separated chemically, and $^{41}K/^{39}K$ was determined using SIMS with repeatability of 0.5‰ (2SE).

The MC-ICP-MS had not been used because of both the peak tailing of the large $^{40}Ar^+$ generated by Ar plasma and the peak overlapping of $^{40}Ar^1H^+$ on $^{41}K^+$. To overcome the former problem, MC-ICP-MS with high-resolution (M/ΔM) of over 10,000 appeared; and/or the plasma of MC-ICP-MS was operated at low power (<800 W) to reduce formation of Ar^+. Furthermore, to lower the formation of $^{40}Ar^1H^+$, the desolvator that removes water and acids in the sample solution was used because they are the sources of H, and MC-ICP-MS equipped with the He collision cell in which ArH^+ is broken into Ar and H^+ by hitting with He appeared.

Employing one or combination of these techniques has made it possible to use MC-ICP-MS for the determination of $\delta^{41}K$. For example,

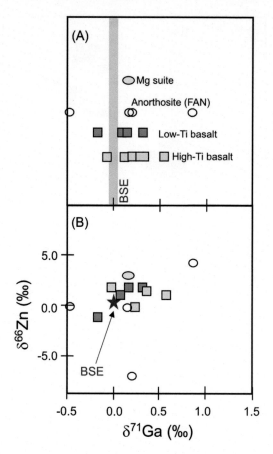

FIGURE 3.18 **The Ga isotopic compositions of lunar samples.** (A) The Ga isotopic composition of lunar samples. The vertical pink area indicates the bulk silicate Earth (BSE) composition. (B) Ga isotopic composition versus Zn isotopic composition of lunar samples. The BSE composition is shown in red star. The captions are the same with (A). *Source: Data from Kato, C., Moynier, F., 2017. Gallium isotopic evidence for extensive volatile loss from the Moon during its formation. Sci. Adv. 3, e1700571.*

Wang and Jacobsen (2016a) succeeded to determine potassium isotopic fractionation with the repeatability of ~0.03‰ (2SE) using MC-ICP-MS. The $\delta^{41}K$ values are defined as:

$$\delta^{41}K(‰) = \left[\left(^{41}K/^{39}K\right)_{sample} / \left(^{41}K/^{39}K\right)_{standard} - 1 \right] \times 1000 \qquad (3.17)$$

where the standard is NIST SRM 3141a. Wang and Jacobsen (2016a) presented $\delta^{41}K_{NIST}$ values of the Earth, the Moon, and chondritic meteorites with high precision (see Fig. 3.19). The lunar rocks were

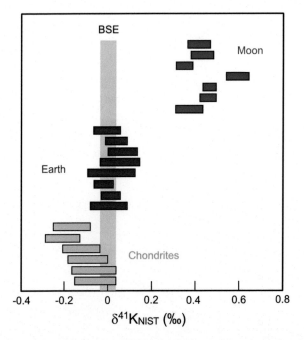

FIGURE 3.19 **The K isotopic compositions of the Earth, Moon, and chondrite samples.** Each data with 2SE is shown as a box. Vertical pink area represents the bulk silicate Earth (BSE) of $\delta^{41}K$ value ($-0.48 \pm 0.03‰$, 2 SD; Wang et al., 2015). The K isotopes of bulk chondrites and terrestrial igneous rocks are difficult to separate, but the lunar rocks are significantly enriched in heavy isotopes compared to the Earth and chondrite samples. *Source: Data from Wang, K., Jacobsen, S.B., Sedaghatpour, F., Chen, H., Korotev, R.L., 2015. The earliest Lunar Magma Ocean differentiation recorded in Fe isotopes. Earth Planet. Sci. Lett. 430, 202–208.*

significantly ($>2\sigma$) enriched in the heavy isotopes of potassium compared to the Earth (by $\sim 0.4‰$) and chondrites.

Tian et al. (2020) in the IPGP group reported the $\delta^{41}K_{NIST}$ for 19 Apollo lunar rocks and 22 lunar meteorites, spanning all major geochemical and petrologic types of lunar materials. Results for high- and low-Ti basalts are shown in Fig. 3.20. The K isotopic compositions of low-Ti and high-Ti basalts were indistinguishable, giving a lunar basalt average $\delta^{41}K_{NIST}$ of $-0.07 \pm 0.09‰$ (2 SD), which were considered to be the best estimate of the lunar mantle and the bulk silicate Moon (BSM). The significant enrichment of K in its heavier isotopes in the BSM, compared with the bulk silicate Earth (BSE; $\delta^{41}K_{NIST} = -0.48 \pm 0.03‰$), was consistent with previous analyses of K isotopes and other moderately volatile elements (e.g., Cu, Zn, Ga, and Rb).

The reason for the difference of BSE between Wang and Jacobsen (2016a) and Tian et al. (2020) is not clear. Although they used the same

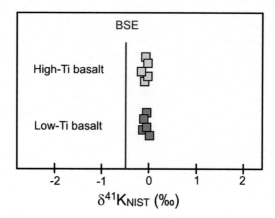

FIGURE 3.20   **The K isotopic compositions of lunar samples.** The red line indicates the bulk silicate Earth. The typical repeatability was 0.05‰, which is within the symbol. The average of K isotopic composition of low-Ti and high-Ti basalts gave the lunar basalt average of $\delta^{41}K$ of -0.07 ± 0.09‰ (2 SD), which is the bulk moon obtained by Tian et al. (2020). *Source: Data from Tian, Z., Jolliff, B.L., Korotev, R.L., Fegley, Jr, B., Lodders, K., Day, J. M.D., et al., 2020. Potassium isotopic composition of the Moon. Geochim. Cosmochim. Acta 280, 263−280.*

standard solution and measured several standard silicate materials, the author could not find why such discrepancies exist.

The enrichment of the heavy isotope of potassium in lunar rocks could be best explained as the result of the incomplete condensation of a bulk silicate Earth vapor at an ambient pressure that is higher than 10 bar. The K isotope result is inconsistent with the low-energy disk equilibration model (Fig. 3.21A; Pahlevan and Stevenson, 2007), but supports the high-energy, high-angular-momentum giant impact model (Lock et al., 2019) for the origin of the Moon (Wang and Jacobsen, 2016b). Briefly, such giant impact left the proto-Earth system spinning so fast that its mantle, atmosphere, and disk formed a well-mixed continuous BSE material that extended the Roche limit (see Fig. 3.21B), from which the Moon condensed and grew. Its pressure and temperature were ~20 bar and ~3700K, respectively. The material in the Roche limit went back to the Earth, and a tiny fraction formed the Moon. According to Rayleigh distillation, K isotopic fractionation is affected by the pressure effect at ~10 bar, resulting in enrichment of ^{41}K isotopes.

3.4.9 Cu isotopic composition of lunar rocks

Copper has two isotopes, ^{63}Cu and ^{65}Cu with Tc of 1037K, which is a moderately volatile element. Cu isotopic composition of lunar rocks is

(A)

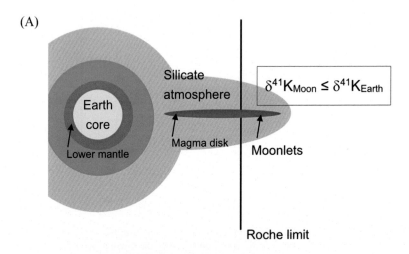

(B)

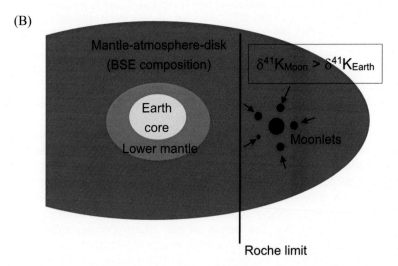

FIGURE 3.21 **Two models for the origin of the Moon.** (A) The Earth-Moon exchange model through a silicate vapor (Pahlevan and Stevenson, 2007). This model can explain the similarity of the oxygen isotope ratios between the Earth and Moon. (B) The Moon condensed model from a bulk silicate Earth (BSE) vapor. This new model of the Moon formation was proposed by Wang et al. (2015). Moon was condensed from a bulk silicate Earth vapor. The Moon was formed from the vapor out of the Roche limit (Lock et al., 2019). *Source: Data from Pahlevan, K., Stevenson, D.J., 2007. Equilibration in the aftermath of the lunar-forming giant impact. Earth Planet. Sci. Lett. 262, 438−449 and Wang, K., Jacobsen, S.B., Sedaghatpour, F., Chen, H., Korotev, R.L., 2015. The earliest Lunar Magma Ocean differentiation recorded in Fe isotopes. Earth Planet. Sci. Lett. 430, 202−208.*

mainly measured by Lyon and IPGP group in France. Cu isotopic ratios of lunar samples were determined by Herzog et al. (2009). Moynier et al. (2006) also presented copper isotopes of lunar soils, however, these data are not used here because of the same reason as discussed in the Zn isotopic composition.

The Cu isotopic composition is expressed as $\delta^{65}Cu$ where:

$$\delta^{65}Cu(\text{‰}) = \left[\left(^{65}Cu/^{63}Cu \right)_{sample} / \left(^{65}Cu/^{63}Cu \right)_{standard} - 1 \right] \times 1000 \quad (3.18)$$

The standard of Cu is "NIST SRM 976." The typical error of $\delta^{65}Cu$ was ± 0.05‰ (2 SD). Cu and Zn isotopic compositions of high Ti lunar basalts measured by Herzog et al. (2009) are shown in Fig. 3.22. The $\delta^{65}Cu$ of bulk silicate Earth (BSE) was estimated to be 0.07 ± 0.10‰ (2 SD) by Savage et al. (2015). Although the $\delta^{65}Cu$ variation $(0.1 \sim 0.5\text{‰})$ becomes smaller than that of $\delta^{66}Zn$ $(1.2 \sim 1.9\text{‰})$, both values are higher than that of BSE, and form positive slope correlation. Even copper in lunar basalts shows heavier isotopic character.

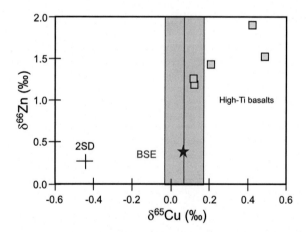

FIGURE 3.22 **The Cu isotopic compositions versus Zn isotopic composition of high-Ti basalts.** The vertical line and pink square indicate the 2 SD range of $\delta^{65}Cu$ of the bulk silicate Earth (BSE) composition estimated by Savage et al. (2015). The uncertainty (2 SD) of $\delta^{65}Cu$ and $\delta^{66}Zn$ of each sample is shown as the cross in the figure. Cu and Zn isotopic compositions of high-Ti basalts are from Herzog et al. (2009). *Source: Data are from Herzog, G.F., Moynier, F., Albarede, F., Berezhnoy, A.A. 2009. Isotopic and elemental abundances of copper and zinc in lunar samples, Zagami, Pele's hairs, and a terrestrial basalt. Geochim. Cosmochim. Acta 73, 5884–5904.*

3.4.10 Li isotopic composition of lunar rocks

Lithium has only two isotopes, ^6Li and ^7Li with T_c (half-mass condensation temperature) of 1142K (Lodders, 2003). Precise isotope ratio determination for the two isotope elements has been very difficult by thermal ionization mass spectrometry (TIMS; see Section 1.3.5.2). The Li isotope analysis could be available by MC-ICP-MS (see Section 1.3.5.2) with careful column chemistry, where the Li isotopic fractionation easily occurs.

The Li isotope ratios are expressed as follows:

$$\delta^7Li(‰) = \left[\left(^7Li/^6Li \right)_{sample} / \left(^7Li/^6Li \right)_{standard} - 1 \right] \times 1000 \qquad (3.19)$$

The NIST L-SVEC Li is used as the standard material.

In Fig. 3.23, a histogram for the δ^7Li values of low-Ti basalts including soil, regolith, and apatite. As shown in Fig. 3.23, Li isotopic signatures of lunar basalts are similar to those of the Earth's mantle (MORB and OIB) or the bulk silicate Earth (BSE). This indicates core formation, volatile loss, and the presence of the crust and hydrosphere have not significantly influenced differentiated planetary bodies of the Earth and Moon. In addition, 1142K of the T_c temperature of Li is too high to cause the fractionation of Li fractionation. Thus, it is suggested that even if the giant impact had occurred for the formation of the Moon,

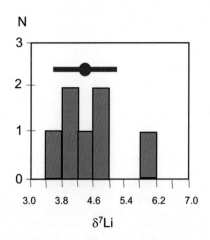

FIGURE 3.23 **A histogram for the δ ^7Li values of lunar samples.** A red-filled circle with a bar is the Earth average with 2 SD. Lunar samples include low-Ti basalts, soil, regolith, and apatite. The lunar data are from Magna et al. (2006) and Seitz et al. (2004). The Earth values are from Seitz et al. (2004) and Jeffcoate et al. (2007). *Source: Modified from Dauphas, N., Burkhardt, C., Warren, P.H., Teng, F.Z., 2014. Geochemical arguments for and Earth-like Moon-forming impactor. Phil. Trans. R. Soc. A 372, 20130244.*

moderately volatile elements such as Li could have accreted again without loss and formed the Moon. As there are few recent studies of the Li isotopic ratios of lunar samples, further constraints have not been obtained.

3.4.11 Cr isotopic composition of lunar rocks

Chromium has four isotopes with T_c (half-mass condensation temperature) of 1296K, and generally the $\delta^{53}Cr$ value is used, which is defined as:

$$\delta^{53}Cr(‰) = \left[\left(^{53}Cr/^{52}Cr\right)_{sample}/\left(^{53}Cr/^{52}Cr\right)_{standard} - 1\right] \times 1000 \quad (3.20)$$

where the standard is NBS979. Bonnard et al. (2016) presented $\delta^{53}Cr$ versus TiO_2 (wt.%) of high-Ti and low-Ti basalts by MC-ICP-MS with the double spike method (see Fig. 3.24). The typical error of each measurement was 0.022‰ (2SE).

The chromium isotope ratios showed no significant difference between neither $\delta^{53}Cr$ of high-Ti and low-Ti basalts nor $\delta^{53}Cr$ and TiO_2 (wt.%) (see Fig. 3.24). Spinel and pyroxene are the main phases controlling the Cr isotopic composition during fractional crystallization. Thus,

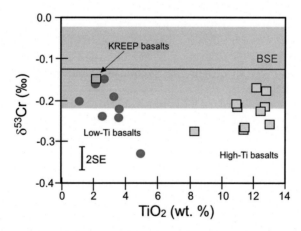

FIGURE 3.24   **The Cr isotopic composition versus TiO$_2$ (wt.%) of lunar basalts.** The red horizontal line and pink area indicate the average of BSE (the bulk silicate Earth) and $\pm 2\,SD$ area of $-0.124 \pm 0.101‰$, respectively (Schoenberg et al., 2008). Green and light blue squares and blue solid circles indicate high-Ti, KREEP, and low-Ti basalts, respectively. The typical 2SE range ($\pm 0.22‰$) of each point is shown in the figure. *Source: Based on Bonnard, P., Parkinson, I.J., Anand, M., 2016. Mass dependent fractionation of stable chromium isotopes in mare basalts: implications for the formation and the differentiation of the Moon. Geochim. Cosmochim. Acta 175, 208–221.*

the most evolved sample (the KREEP basalt) should have the lightest isotopic composition, however, there is no difference of $\delta^{53}Cr$ of the KREEP basalt (Fig. 3.24). Two hypotheses were proposed that (1) equilibrium fractionation where heavy isotopes were preferentially incorporated into the spinel lattice, and (2) Cr^{3+} existed in the melt at a temperature below 1200°C even at low oxygen fugacities. The least differentiated sample of the low-Ti basalts had the lowest Cr isotopic composition of $-0.222 \pm 0.025‰$, which is within the error of the BSE range. Thus, the similarity between mantles of the Earth and Moon is consistent with terrestrial origin of the Moon's mantle.

3.4.12 Si isotopic composition of lunar rocks

Silicon has three isotopes, ^{28}Si, ^{29}Si, and ^{30}Si with T_c (half-mass condensation temperature) of 1310K. The Si isotope ratios are expressed as follows:

$$\delta^{29}Si(‰) = \left[\left(^{29}Si/^{28}Si \right)_{sample} / \left(^{29}Si/^{28}Si \right)_{Standard} - 1 \right] \times 1000 \qquad (3.21)$$

$$\delta^{30}Si(‰) = \left[\left(^{30}Si/^{28}Si \right)_{sample} / \left(^{30}Si/^{28}Si \right)_{Standard} - 1 \right] \times 1000 \qquad (3.22)$$

The standard is the NBS-28 standard.

Fig. 3.25 shows a histogram for the $\delta^{30}Si$ values of low Ti basalts including soil, regolith, apatite, and the Earth (a red-filled circle with 2 SD as a bar). The Si isotope data are obtained by MC-ICP-MS. From Fig. 3.25, the $\delta^{30}Si$ values of the Moon samples are indistinguishable from those of the Earth.

Fig. 3.26 is a plot of $\delta^{30}Si$ versus Mg/Si of chondrites, BSE (Bulk Silicate Earth), and the Moon. The Si isotope ratios of all types of chondrites are systematically lower than those of the Earth (and the Moon). The difference between BSE and the chondrites is the equilibrium fractionation in silicon in metal and silicate during the core formation (Georg et al., 2007; Fitoussi et al., 2009; Armytage et al., 2011; Zambardi et al., 2013). In this model, the heavier silicon is in the mantle, and the lighter silicon is in the core. In addition, 3–16 wt.% of silicon must be in the core. This value is consistent with the density deficit of the core (Hirose et al., 2013).

In Fig. 3.26, the line and the dotted curves show the regression line and 95% confidence range of the chondrites. From this figure, the BSE is on the regression line of the chondrites. The Si isotopes of the Moon were not in the range of the chondrites, indicating the Moon cannot be made by mixing between neither the enstatite chondrites nor other

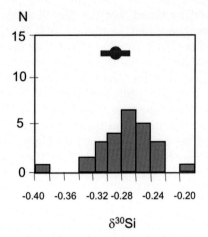

FIGURE 3.25   **A histogram for the δ ^{30} Si** values of low Ti basalts including soil, regolith, and apatite A red-filled circle with a bar is the Earth average with 2 SD. The Earth
data are from Fitoussi and Bourdon (2012), Zambardi et al. (2013), Fitoussi et al. (2009),
Armytage et al. (2011), and Savage et al. (2010, 2011). The Moon data are from Armytage
et al. (2012), Fitoussi and Bourdon (2012), and Zambardi et al. (2013). *Source: Modified from
Dauphas, N., Burkhardt, C., Warren, P.H., Teng, F.Z., 2014. Geochemical arguments for and
Earth-like Moon-forming impactor. Phil. Trans. R. Soc. A 372, 20130244.*

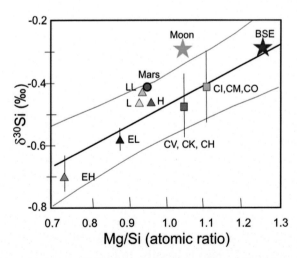

FIGURE 3.26   **A plot of δ ^{30}Si versus Mg/Si of chondrites, BSE (Bulk Silicate Earth),
and BSM (Bulk Silicate Moon).** The line is a regression line of all chondrites. The curves
indicate 95% confidence range. *Source: Modified from Dauphas, N., Burkhardt, C., Warren, P.
H., Teng, F.Z., 2014. Geochemical arguments for and Earth-like Moon-forming impactor. Phil.
Trans. R. Soc. A 372, 20130244.*

chondrites. It should be noted that the Mg/Si ratios of enstatite chondrites are very low compared to that of BSE (Dauphas et al., 2014). Fitoussi and Bourdon (2012) concluded that the Earth cannot be made of the enstatite chondrites. This does not contradict against generally accepted observations that the ordinary and carbonaceous chondrites do not match as the building block of the BSE, but the enstatite chondrites fit most isotope budgets (e.g., oxygen).

In order to make the Moon from the proto-Earth mantle, which is the same with BSE, and the enstatite chondritic Theia, decrease of Mg/Si is required. The mixture of BSE and EH in Fig. 3.26 is around the point of the carbonaceous chondrite CV, CK, and CH. So, a simple increase of $\delta^{30}Si$ is required to be the Moon point. Such Si isotopic fractionation requires the counterpart with very light Si isotope ratios. However, such materials are not found so far. The only possible process to remove the light isotopic Si materials is that the light-isotopic Si gases made by the giant impact were removed into the Sun as deleting the super-Earths after Batygin and Laughlin (2015) (see Section 2.3.7).

3.4.13 Fe isotopic composition of lunar rocks

Iron has three isotopes, ^{54}Fe, ^{56}Fe, ^{57}Fe, and ^{48}Fe with T_c (half-mass condensation temperature) of 1328K (Lodders, 2003). The $\delta^{56}Fe$ and $\delta^{57}Fe$ values are defined as:

$$\delta^{56}Fe(‰) = \left[\left(^{56}Fe/^{54}Fe \right)_{sample} / \left(^{56}Fe/^{54}Fe \right)_{standard} - 1 \right] \times 1000 \quad (3.23)$$

$$\delta^{57}Fe(‰) = \left[\left(^{57}Fe/^{54}Fe \right)_{sample} / \left(^{57}Fe/^{54}Fe \right)_{standard} - 1 \right] \times 1000 \quad (3.24)$$

where the standard IRMM-014 is used. The Fe isotope ratios were determined by TIMS using the double spike method (Johnson and Beard, 1999). But nowadays, the most popular method is MC-ICP-MS. As $\delta^{57}Fe$ is one and half times as large as $\delta^{56}Fe$, however, the isotopic abundance of ^{56}Fe is 92% and the largest, $\delta^{56}Fe$ is also used. As the double spike, ^{57}Fe and ^{58}Fe are often chosen, only $\delta^{56}Fe$ is obtained. In MC-ICP-MS, the 2SE of $\delta^{57}Fe$ are generally $< \pm 0.05‰$.

Iron has attracted many cosmochemists because of its large abundance in silicates. Poitrasson et al. (2019) carefully determined the Fe isotopic ratios using 17 low-Ti basalts, 16 high-Ti basalts including published data (Wiesli et al., 2003; Poitrasson et al., 2004; Weyer et al., 2005; Liu et al., 2010; Wang et al., 2015; Sossi and Moynier, 2017) (see Fig. 3.27).

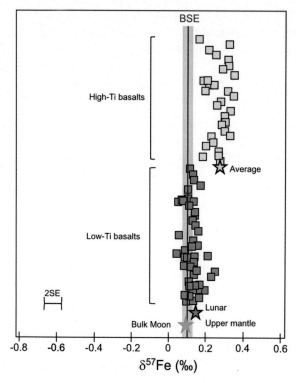

FIGURE 3.27 **The Fe isotopic compositions of lunar samples.** The $\delta^{57}Fe$ values are relative to IRMM-14. The typical $\pm 2SE$ of each point is shown in the figure. The red vertical line and pink area show the Earth reference value ($\delta^{57}Fe = 0.10 \pm 0.03‰$, 2SE). The bulk Moon value ($0.094 \pm 0.035‰$), the bulk lunar upper mantle (the average of low-Ti basalts; $0.142 \pm 0.026‰$), and the average of high-Ti basalts ($0.274 \pm 0.020‰$) are shown as the blue, black, and white stars by Poitrasson et al. (2019), respectively. *Source: Data from Wiesli et al. (2003), Poitrasson et al. (2004), Weyer et al. (2005), Liu et al. (2010), Wang et al. (2015), and Sossi and Moynier (2017).*

As shown in Fig. 3.27, Liu et al. (2010) first noticed that the Fe isotope compositions are different between low-Ti and high-Ti basalts. Craddock et al. (2013) pointed out that ilmenite and plagioclase may be in the ^{57}Fe-enriched phase. Based on the newly determined Fe isotope composition, Poitrasson et al. (2019) concluded that the bulk Moon was indistinguishable from that of the Earth, and heavier than other planetary bodies. This planetary isotope character was only observed for Fe and Si. Because significant amounts of both Fe and Si reside in the metallic core like the Earth's core, formation of metallic cores of the Earth and Moon had strong effects on the Fe and Si isotopic fractionation. The high-pressure metal—silicate fractionation at the core—mantle boundary would be more likely than partial vaporization of the liquid outer core after the giant impact.

3.4.14 Mg isotopic composition of lunar rocks

Magnesium has three isotopes, ^{24}Mg, ^{25}Mg, and ^{26}Mg with T_c (half-mass condensation temperature) of 1336K (Lodders, 2003), which makes Mg to be a relatively refractory element. The Mg isotope ratios are expressed as follows:

$$\delta^{25}Mg(‰) = \left[\left(^{25}Mg/^{24}Mg\right)_{sample}/\left(^{25}Mg/^{24}Mg\right)_{standard} - 1\right] \times 1000$$

$$(3.25)$$

$$\delta^{26}Mg(‰) = \left[\left(^{26}Mg/^{24}Mg\right)_{sample}/\left(^{26}Mg/^{24}Mg\right)_{standard} - 1\right] \times 1000$$

$$(3.26)$$

The DSM3 is the Mg standard solution made from pure Mg in Galy et al. (2003). As $\delta^{26}Mg$ (‰) is twice as large as $\delta^{25}Mg$, $\delta^{26}Mg$ is generally used to express the Mg isotopic composition.

Fig. 3.28 shows the $\delta^{26}Mg$ values of terrestrial and lunar samples analyzed and compiled by Sedaghatpour and Jacobsen (2019). All data are obtained by MC-ICP-MS. The repeatability and intermediate precision were $\pm 0.010‰$ (2SE) and $\pm 0.08‰$ (2 SD) over 18 months, respectively. When $\delta^{25}Mg$ is plotted against $\delta^{26}Mg$, all data are plotted on the line of the slope of ~ 2 (not shown), indicating the mass discrimination during measurement is appropriately corrected against the standard (Sedaghatpour et al., 2013).

From Fig. 3.28, the Mg isotopic ratios of the Moon by Sedaghatpour et al. (2013) were indistinguishable from those of the Earth obtained by Teng et al. (2010). As Mg is a relatively refractory element, the Mg isotopic results are consistent with the giant impact model.

In addition, as the data of the Earth and the Moon were also plotted at the similar place on the same line, showing the mass-dependent fractionation occurred at similar condition along this line. In addition, the data of the chondrites are on the line, indicating the same origin of these three components.

Sedaghatpour and Jacobsen (2019) presented high-precision Mg isotopic analyses of different types of lunar samples. As the Mg isotopic composition of the dunite 72415 ($\delta^{26}Mg = -0.291 \pm 0.018‰$) was the most Mg-rich and possibly the oldest lunar sample, they considered that the dunite 72415 might be the best estimate of the Mg isotopic

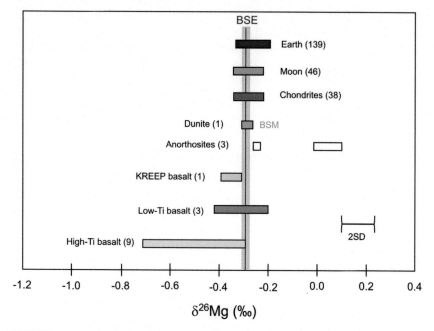

FIGURE 3.28 **The Mg isotopic compositions of lunar samples.** The red vertical lines with pink area are Mg isotopic composition of the bulk silicate Earth (BSE) with 2SE ($\delta^{26}Mg = -0.291 \pm 0.018‰$), respectively. The intermediate precision (2 SD) is shown in the figure. The data including 2SE of all samples are indicated by a box. The number in parentheses after the sample name indicates the sample number. The data of the Earth, Moon, and chondrites are from Teng et al. (2010), Sedaghatpour et al. (2013), and Magna et al. (2017), respectively. The bulk silicate Moon (BSM) value was estimated from the dunite (see text). *Source: Modified from Sedaghatpour, F., Jacobsen, S.B., 2019. Magnesium stable isotopes support the lunar magma ocean cumulate remelting model for mare basalts. PNAS 116, 73–78.*

composition of the bulk silicate Moon (BSM). The $\delta^{26}Mg$ value of BSM was similar to those of the Earth and chondrites, showing the relative homogeneity of Mg isotopes in the solar system and the lack of Mg isotope fractionation by the Moon-forming giant impact.

Three low-Ti basalts had the average of $-0.285 \pm 0.109‰$, which was similar to the BSM. In contrast, high-Ti basalts had the average of $-0.462 \pm 0.084‰$. Stable isotope studies of other elements such as Ti and Fe had similar dichotomy between low- and high-Ti basalts. Sedaghatpour et al. (2013) predicted the isotopically lighter ilmenite produced the isotopically light Mg shows in high-Ti basalts. It could also happen that Mg-Fe internal diffusion between minerals and melt causes isotopically lighter minerals.

In contrast to the homogeneity of the Mg isotopes in terrestrial basalts and mantle rocks, Mg isotopes on lunar samples showed large variations among the basalts and pristine anorthositic rocks. Sedaghatpour and Jacobsen (2019) considered that such large isotopic fractionation was caused by the early lunar magma ocean (LMO) differentiation. Calculation of evolutions of δ^{26}Mg values during the LMO differentiation was consistent with the observed δ^{26}Mg variations in lunar samples. Thus, it was concluded that the remelting of the LMO cumulates was occurred and caused the large Mg isotope variations in lunar basalts.

3.4.15 V isotopic composition of lunar rocks

Vanadium has two isotopes, ^{50}V and ^{51}V, with T_c of 1429K (Lodders, 2003) which makes V as a refractory element. V isotopic composition are measured by MC-ICP-MS, and expressed as δ^{51}V:

$$\delta^{51}V(‰) = \left[\left(^{51}V/^{50}V\right)_{sample} / \left(^{51}V/^{50}V\right)_{standard} - 1 \right] \times 1000 \qquad (3.27)$$

where the standard is "AA" (Alpha Aecer) standard. The repeatability of δ^{51}V was $\pm 0.04‰$ (2SE). Variation of V isotopic composition of carbonaceous chondrites was found, which was attributed to nucleosynthetic origin (Nielsen et al., 2019). However, all V isotopic variation of the bulk carbonaceous chondrites can be accounted for recent production of ^{50}V by galactic cosmic ray (GCR) spallation processes. Based on this fact, Nielsen et al. (2021) obtained corrected δ^{51}V values for GCR effects of the KREEP-rich sample, the low-Ti and high-Ti basalts, and found they are identical to each other. They also recalculated δ^{51}V values obtained by Hopkins et al. (2019) (see Fig. 3.29), which were also consistent with the results of Nielsen et al. (2021). Thus, they determined the Moon δ^{51}V to be $-1.037 \pm 0.031‰$ (2SE) and $\Delta^{51}V_{BSE-Moon} = 0.233 \pm 0.037‰$.

Nielsen et al. (2020) found the BSE value ($\delta^{51}V_{BSE} = -0.856 \pm 0.020‰$; 2SE, n = 76) was heavier than those of the chondrites ($\delta^{51}V_{chondrites} = -1.089 \pm 0.029‰$, 2SE, n = 14), which gave $\Delta^{51}V_{BSE-chondrites} = 0.233 \pm 0.037‰$. They proposed the model that the core formation processes before the giant impact had already caused the V isotope fractionation to make the bulk silicate of the proto-Earth heavier δ^{51}V than that of the chondrite. The Theia, which had the chondritic V isotopic composition hit the heavier-δ^{51}V proto-Earth mantle, and mixed with it. Thus, the Moon became a bit higher δ^{51}V than that of chondrite, while the BSE remained to be heavier-δ^{51}V than that of the chondrites.

3.4.16 Ti isotopic composition of lunar rocks

Titanium is a refractory element with T_c of 1582K (Lodders, 2003). Millet et al. (2016) presented high-precision Ti stable isotope

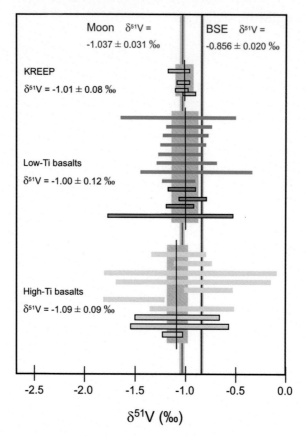

FIGURE 3.29 **The V isotopic compositions of lunar samples.** The red vertical line indicates the average and 2SE range of $\delta^{51}V$ of the bulk silicate Earth (BSE). The purple solid line with the purple area is the average and 2SE of $\delta^{51}V$ of the Moon. Each $\delta^{51}V$ value of a lunar sample with a 2SE range is shown by a horizontally long rectangle. The rectangles with and without black border lines indicate the corrected data for GCR effects of Hopkins et al. (2019) and Nielsen et al. (2021), respectively. The average and 2SE by both studies for each rock type, which are indicated by the black vertical line and gray area, respectively, are also shown in the figure. *Source: Data from Nielsen, S.G., Bekaert, D.V., Magna, T., Mezger, K., Auro, M., 2020. The vanadium isotope composition of mars: implications for planetary differentiation in the early solar system. Geochem. Perspect. Lett. 15, 35–39; Nielsen, S.G., Bekaert, D.V., Auro, M., 2021. Isotopic evidence for the formation of the Moon in a canonical giant impact. Nat. Commun. 12, 1817.*

measurements ($^{49}Ti/^{47}Ti$) of terrestrial materials together with lunar samples by MC-ICP-MS. Their objectives were to (1) evaluate isotopic effect of fractional crystallization and magma differentiation; (2) determine the bulk silicate Earth (BSE) value and assess the homogeneity; and (3) use the Ti isotope composition to lunar magmatic evolution.

Ti isotope ratios are shown as:

$$\delta^{49}Ti(‰) = \left[\left(^{49}Ti/^{47}Ti \right)_{sample} / \left(^{49}Ti/^{47}Ti \right)_{standard} - 1 \right] \times 1000 \qquad (3.28)$$

Millet et al. (2016) used OL-Ti as the standard. The intermediate precision of their method was $\pm 0.020‰$ (2 SD). They determined the bulk silicate Earth to be $0.005 \pm 0.005‰$ relative to OL-Ti. They found positive correlation between SiO_2 (wt.%) and $\delta^{49}Ti$, which indicated that $\delta^{49}Ti$ becomes higher as the fractional crystallization (magma differentiation) proceeds. This meant lighter Ti prefers crystals (Fe-Ti oxides such as ilmenite) to silicate melts. They also measured $\delta^{49}Ti$ of five high-Ti basalts and three low-Ti basalts (see Fig. 3.30).

High-Ti lunar basalts possessed Ti stable isotope compositions ranging from terrestrial values to slightly enriched in heavy isotopes ($+0.011$ to $+0.033‰$). Low-Ti lunar basalts had $\delta^{49}Ti$ values of -0.008 to $+0.011‰$, which were within the error of the BSE value. The heavy $\delta^{49}Ti$ values recorded in high-Ti basalts indicate that their source regions may have been enriched either by ilmenite cumulates formed in the latest stages of the lunar magma ocean (LMO).

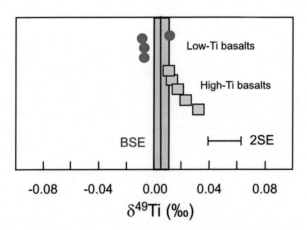

FIGURE 3.30 **The Ti isotope compositions of lunar mare basalts (high-Ti and low-Ti basalts).** The Ti isotope compositions of lunar mare basalts (high-Ti and low-Ti basalts). The red vertical line and pink area indicate the average of BSE (the bulk silicate Earth) and $\pm$ 2 SD area, respectively (Millet et al., 2016). Green and light blue squares and blue solid circles indicate high-Ti, KREEP and low-Ti basalts, respectively. The typical 2SE range ($\pm 0.22‰$) of each point is shown in the figure. *Source: Based on Bonnard, P., Parkinson, I.J., Anand, M., 2016. Mass dependent fractionation of stable chromium isotopes in mare basalts: implications for the formation and the differentiation of the Moon. Geochim. Cosmochim. Acta 175, 208–221.*

3.5 Summary

In Fig. 3.31, the summary of isotope compositions of lunar basalts (high-Ti and low-Ti basalts) are shown. The vertical axis is the difference of $\delta^N X$ of the average basalts of the Moon (BM) from $\delta^N X$ of the bulk silicate Earth (BSE). Please remember that the low-Ti basalts are less affected by the fractional crystallization than the high-Ti basalts. The $\Delta(\delta^N X_{BM}-\delta^N X_{BSE})$ per mass (denoted as DX; X is the element and N is its mass number) was plotted, because the $\Delta(\delta^N X_{BM}-\delta^N X_{BSE})$ value changes according to the difference of mass number between the numerator mass and the denominator mass. For example, $\Delta^{26}Mg$ and $\Delta^{58}Fe$ are twice as large as $\Delta^{25}Mg$ and $\Delta^{56}Fe$ (^{24}Mg and ^{54}Fe are denominator isotopes), respectively. There are no systematical differences between the ΔX values among the low- and high-Ti basalts. The relatively large differences exist in ΔGa and ΔV, respectively.

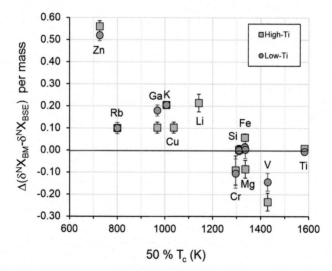

FIGURE 3.31 **Half-mass condensation temperature (50% T_c in K) versus $\Delta(\delta^N X_{BM}-\delta^N X_{BSE})$ per mass.** X is an element and N is the mass number of the element. Thus, $\delta^N X$ means $\delta^{66}Zn$, $\delta^{87}Rb$, etc., as shown in the figure. The $\Delta(\delta^N X_{BM}-\delta^N X_{BSE})$ value (denoted as ΔX) is the difference of $\delta^N X$ of the average basalts of the Moon (BM) from $\delta^N X$ of the bulk silicate Earth (BSE). The vertical axis is ΔX divided by the mass difference of the numerator isotope from the denominator isotope, namely, in case of Zn, Fe and V, ΔZn ΔFe and ΔV are divided by 2 (=66−64), 3 (=57−54) and 1 (=51−50), respectively. Both the averages of high-Ti and low-Ti basalts are chosen for BM, which are indicated as the green squares and the blue circles, respectively. The error bar is 2 SD per mass shown in each section. The horizontal red line indicates $\delta^N X_{BM} = \delta^N X_{BSE}$, namely the Moon values are same with the BSE values.

As the T_c increases from 750K, ΔZn ($\Delta = \sim 0.55$ at ~ 750K) to ΔV ($\Delta = \sim -0.15$ at ~ 1450K), systematically decreases. Then Δ increases to 0 again at Ti ($T_c = \sim 1600$K). At Tc = ~ 1350K, ΔSi, and ΔFe are almost $\Delta = 0$. ΔCr and ΔMg are slightly negative, but almost 0. ΔV is negative because the BSE value is larger than those of meteorites because of the core formation, in which liquid iron preferred lighter V isotopes. Similarly, Xia et al. (2019) performed partition experiments of Cu and Zn between silicate-metal-melt and sulfide-melt and found strong affinity of lighter isotopes into the sulfide-melt than than that of the silicate-metal-melt. If sulfides sink into the deep mantle or core, the Moon's mantle should become isotopically heavier for Cu isotope. In contrast, Zn is not so affected by the sulfides, therefore, the heavy Zn isotope character was formed by the volatility during the Moon formation and the light isotope loss by the large volcanic activity at the lunar magma ocean (LMO).

As mentioned above, it has been proved that many stable isotopes are affected by the formation of minerals or silicate/sulfide liquids. However, the effects on the isotopic signatures are relatively small, and the moderately volatile elements of the Moon have a bit heavier isotopic characters than those of the bulk silicate Earth (BSE). The stable isotopic signatures of the refractory elements of the Moon are almost the same as those of the BSE. These isotopic signatures cannot be explained by the simple accretion of the gas/liquid accretion made by the giant impact (Canup, 2004, 2012; Canup and Asphaug, 2001; Ćuk and Stewart, 2012).

Fifty years ago, Ringwood and his colleagues pointed out the depletion of abundances of moderately volatile elements (MVEs) such as Na and K (see Section 3.1.3). There have been few studies including dynamics, thermodynamics, and chemistry of the lunar origin. Canup et al. (2015) used the lunar disk models of Salmon and Canup (2012) and performed physicochemical calculations to link with the dynamics and thermodynamics of accretion and the canonical disk. Wang and Jacobsen (2016a) reported the K isotopes of the Moon, and presented a qualitative model (see Fig. 3.21B).

3.6 The "synestia" model

Lock et al. (2019) presented the new model for lunar origin within a terrestrial "synestia," which is an impact-generated structure with Earth mass and composition that exceeds the corotation limit (CoRoL). Synestias are evaporated rocks that rapidly rotate with a doughnut-like shape. Synestias are formed by a range of high-energy, high-angular momentum (AM) collisions. The synestia is a special dynamic structure compared to a planet with a condensate-dominated circumplanetary disk, and different processes dominate the early evolution of synestia.

There was no calculation that could fully cover the dynamics, thermodynamics, and chemistry of lunar accretion. Therefore, the approach of Lock et al. (2019) combined the physics with chemistry of satellite accretion through a terrestrial synestia. They showed the processes that control the pressure and temperature paths of the materials that formed a moon. They solved the problem in three steps: (1) first, they determined the pressure-temperature conditions of a moon that grows by accretion of condensing silicate vapor; then (2) they discussed the composition of a growing moon determined by equilibrium of materials of bulk silicate Earth (BSE) composition; finally (3) they showed that a variety of high-energy, high-AM giant impacts can generate initial conditions that can potentially lead to the formation of a lunar-mass moon with the observed geochemical characteristics of the Moon. Their model gave a promising pathway to explain all the key observations of the Moon discussed so far: the isotopic similarity between the Earth and the Moon; the magnitude and pattern of moderately volatile element depletion in the Moon; and the extraordinarily large mass of the Moon. If the Earth had a large obliquity after the giant impact, then a single event may also explain the inclination of the lunar orbit and the present-day AM of the Earth-Moon system.

The results of the calculation of Lock et al. (2019) are summarized as the illustration in Fig. 3.32. The Moon formation from the synestia could explain the main peculiarities of the present Moon. The cooling of the synestia mixed materials well (see Fig. 3.32A). The condensation from the surface made the moonlets which orbited in dense vapor of tens of bars of the bulk silicate Earth (BSE) composition, resulting in equilibrating with the BSE vapor (Fig. 3.32B). Thus, the moonlets and the growing Moon achieved the isotopic compositions and abundance patterns of the moderately volatile elements of the present Moon. Finally, the cooling synestia became smaller than the lunar orbit and the main stage of the Moon forming ended (Fig. 3.32C). The calculation can also explain the similarities between the Earth and the Moon compositions.[1]

3.7 C, N, and S were delivered by a giant impact

The timing and delivery mechanism of carbon (C), nitrogen (N), sulfur (S), and hydrogen (H) resulted in the Earth as the only life-sustaining planet. Their isotopic signatures of the terrestrial volatiles are derived from carbonaceous chondrites, while those of nonvolatile major

1   The recent result of a computer simulation movie can be seen in Kegerreis et al. (2020).

(A)

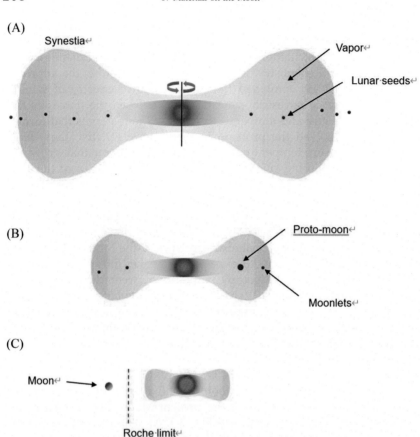

(B)

(C)

FIGURE 3.32   Evolution of the synestia and formation of the Moon. (A) A synestia was formed by a high-energy, high-angular momentum giant impact. Within the vapor structure, the synestia rapidly cooled by radiation forming condensates which rapidly accreted to be lunar seeds. In the synestia, turbulent convections were occurring. (B) Within a year, the synestia cooled and contracted forming moonlets from the lunar seeds. The moonlets chemically equilibrated with bulk silicate Earth vapor at the silicate vaporization temperature. The moonlets accreted to form proto-moon. (C) After a several tens of years, the Moon separated from the synestia and began to solidify out of the Roche limit. The synestia was within the Roche limit and fell below the CoRoL. *Source: Modified from Lock, S.J., Stewart, S.T., Petaev, M.I., Leinhardt, Z., Mace, M.T., Jacobsen, S.B., et al., 2019. The origin of the Moon within a terrestrial synestia. J. Geophys. Res. Planets 123, 910–951.*

and trace elements suggest that enstatite chondrite–like materials are the primary source of the Earth. However, the C/N ratio of the bulk silicate Earth (BSE) is superchondritic, which rules out volatile delivery by a chondritic late veneer. In addition, if delivered during the main phase of Earth's accretion, then, owing to the greater siderophile (metal loving) nature of C relative to N, core formation should have left behind a subchondritic C/N ratio in the BSE.

Grewal et al. (2019) challenged to constrain the fate of C-N-S volatiles during core-mantle segregation in the planetary embryo magma oceans by high pressure-temperature experiments and showed that C becomes much less siderophile in N-bearing and S-rich alloys, while the siderophile character of N remains largely unaffected in the presence of S. The new data and inverse Monte Carlo simulations suggested that the impact of a Mars-sized planet, having minimal contributions from carbonaceous chondrite-like material and coinciding with the Moon-forming event, can be the source of major volatiles in the BSE.

References

Agnor, C., Asphaug, E., 2004. Accretion efficiency during planetary collisions. Astrophys J 613, L157−L160.

Alibert, C., Norman, M.D., McCulloch, M.T., 1994. An ancient age for a ferroan anorthosite clast from lunar breccia 67016. Geochim. Cosmochim. Acta 58, 2921−2926.

Armytage, R., Georg, R., Savage, P., Williams, H., Halliday, A., 2011. Silicon isotopes in meteorites and planetary core formation. Geochim. Cosmochim. Acta 75, 3662−3676.

Armytage, R., Georg, R., Williams, H., Halliday, A., 2012. Silicon isotopes in lunar rocks: implications for the Moon's formation and the early history of the Earth. Geochim. Cosmochim. Acta 77, 504−514.

Asphaug, E., 2010. Similar-sized collisions and the diversity of planets. Chem Erde Geochem 70, 199−219.

Batygin, K., Laughlin, G., 2015. Jupiter's decisive role in the inner Solar System's early evolution. Proc. Natl. Acad. Sci. U. S. A. 112, 4214−4217.

Belbruno, E., Gott Jr, I.I.I., 2005. Where did the Moon come from? Astron J 129, 1724−1745.

Bindeman, I.N., Ionov, D.A., Tollan, P.M.E., Golovin, A.V., 2022. Oxygen isotope ($\delta^{18}O$, $\Delta'^{17}O$) insights into continental mantle evolution since the Archean. Nat. Comm. 13, 3779.

Bonnard, P., Parkinson, I.J., Anand, M., 2016. Mass dependent fractionation of stable chromium isotopes in mare basalts: implications for the formation and the differentiation of the Moon. Geochim. Cosmochim. Acta 175, 208−221.

Borg, L.E., Norman, M., Nyquist, L., et al., 1999. Isotopic studies of ferroan anorthosite 62236: a young lunar crustal rock from a light rare-earth element-depleted source. Geochim. Cosmochim. Acta 58, 2921−2926.

Borg, L.E., Connelly, J.N., Boyet, M., Carlson, R.W., 2011. Chronological evidence that the Moon is either young or did not have a global magma ocean. Nature 477, 70−72.

Bottke, W.F., Vokrouhlichy, D., Marchi, S., et al., 2015. Dating the Moon-forming impact event with asteroidal meteorites. Science 348, 321−323.

Boyet, M., Carlson, R.W., 2007. A highly depleted moon or a non-magma ocean origin for the lunar crust? Earth Planet. Sci. Lett. 262, 505−516.

Brandon, A.D., Lapen, T.J., Debaille, V., Beard, B.L., Rankenburg, K., Neal, C., 2009. Reevaluating $^{142}Nd/^{144}Nd$ in lunar mare basalts with implications for the early evolution and bulk Sm/Nd of the Moon. Geochim. Cosmochim. Acta 73, 6421−6445.

Cameron, A.G.W., 1985. Formation of the prelunar accretion disk. Icarus 62, 319−327.

Cameron, A.G.W., Benz, W., 1991. The origin of the Moon and the single impact hypothesis IV. Icarus 92, 204–216.

Cameron, A.G.W., Ward, W.R., 1976. The origin of the Moon., Lunar Planetary Institute Science Conference Aabstract, 7. LPI, pp. 120–122.

Cano, E.J., Sharp, Z.D., Shearer, C.K., 2020. Distinct oxygen isotope compositions of the Earth and Moon. Nat. Geosci. 13, 270–274.

Canup, R.M., 2012. Forming a Moon with an Earth-like composition via a giant impact. Science. 338, 1052–1055.

Canup, R.M., 2004. Simulations of a late lunar-forming impact. Icarus 168, 433–456.

Canup, R.M., Asphaug, E., 2001. Origin of the Moon in a giant impact near the end of the Earth's formation. Nature 412, 708–712.

Canup, R.M., Visscher, C., Salmon, J., Fegley, B., 2015. Jr: Lunar volatile depletion due to incomplete accretion within an impact-generated disk. Nat. Geosci. 8, 918–921.

Cao, X., Liu, Y., 2011. Equilibriumm mass-dependent fractionation relationships for triple oxygen isotopes. Geochim. Cosmochim. Acta 75, 7435–7445.

Carlson, R.W., Lugmair, G.W., 1988. The age of ferroan anorthosite 60025: oldest crust on a young Moon? Earth Planet. Sci. Lett. 90, 119–130.

Carlson, R.W., Borg, L.E., Gaffney, A.M., Boyet, M., 2014. Rb-Sr, Sm-Nd and Lu-Hf isotope systematics of the lunar Mg-suite: the age of the lunar crust and its relation to the time of Moon formation. Phil. Trans. R. Soc. A 372, 20130246.

Clayton, R.N., Mayeda, T.K., 1996. Oxygen isotope studies of achondrites. Geochim. Cosmochim. Acta 60, 1999–2017.

Clayton, R.N., Mayeda, T.K., 1999. Oxygen isotope studies of carbonaceous chondrites. Geochim. Cosmochim. Acta 63, 2089–2104.

Clayton, R.N., Mayeda, T.K., Rubin, A.E., 1984. Oxygen isotopic compositions of enstatite chondrites and aubrites. J. Geophys. Res. 89, C245–C249.

Clayton, R.N., Mayeda, T.K., Goswami, J.N., Olsen, E.J., 1991. Oxygen isotope studies of ordinary chondrites. Geochim. Cosmochim. Acta 55, 2317–2337.

Craddock, P.R., Warren, J.M., Dauphas, N., 2013. Abyssal peridotites reveal the near-chondritic Fe isotopic composition of the Earth. Earth Planet. Sci. Lett. 365, 63–76.

Ćuk, M., Stewart, S.T., 2012. Making the Moon from a fast-spinning Earth: a giant impact followed by resonant despinning. Science 338, 1047–1052.

Dahl, T.W., Stevenson, D.J., 2010. Turbulent mixing of metal and silicate during planet accretion-and interpretation of the Hf-W chronometer. Earth Planet Sci Lett 295, 177–186.

Daly, R.A., 1946. Origin of the Moon and its topography. Proc American Phil 90, 104–119.

Darwin, G.H., 1879. On the precession of a viscous spheroid, and on the remote history of the Earth. Phil Trans R Soc Lond 170, 447–538.

Dauphas, N., Burkhardt, C., Warren, P.H., Teng, F.Z., 2014. Geochemical arguments for and Earth-like Moon-forming impactor. Phil. Trans. R. Soc. A 372, 20130244.

Dowty, E., Prinz, M., Keil, K., 1974. Ferroan anorthosite: a widespread and distinctive lunar rock type. Earth Planet. Sci. Lett. 24, 15–25.

Drake, M.J., 1983. Geochemical constraints on the origin of the Moon. Geochim. Cosmochim. Acta 47, 1759–1767.

Drake, M.J., Newsom, C.J., Capobianco, C.J., 1989. V, Cr and Mn in the Earth, Moon, EPB and SPB and the origin of the Moon: experimental studies. Geochim. Cosmochim. Acta 53, 2101–2111.

Dreibus, G., Wänke, H., 1979. On the chemical composition of the moon and the eucrites parent body and comparison with composition of the Earth: the case of Mn. Cr V. Lunar Planet. Sci. Abst. 10, 315–317.

Eiler, J.M., 2001. Oxygen isotope variations of basaltic lavas and upper mantle rocks. Rev. Miner. Geochem. 43, 319–364.

Elkins-Tanton, L.T., Burgess, S., Yin, Q.-Z., 2011. The lunar magma ocean: reconciling the solidification process with lunar petrology and geochronology. Earth Planet. Sci. Lett. 304, 326−336.

Fitoussi, C., Bourdon, B., 2012. Silicon isotope evidence against an enstatite chondrite Earth. Science 335, 1477−1480.

Fitoussi, C., Bourdon, B., Kleine, T., Oberli, F., Reynolds, B.C., 2009. Si isotope systematics of meteorites and terrestrial peridotites: implications for Mg/Si fractionation in the solar nebula and for Si in the Earth's core. Earth Planet Sci Lett 287, 77−85.

Gaffney, A.M., Borg, L.E., 2013. A young age for KREEP formation determined from Lu-Hf isotope systematics of KREEP basalts and Mg-suite samples. Lunar Planet. Sci. 44, 1714.

Gaffney, A.M., Borg, L.E., 2014. A young solidification age for the lunar magma ocean. Geochim. Cosmochim. Acta 140, 227−240.

Galy, A., Yoffe, O., Janney, P.E., et al., 2003. Magnesium isotope heterogeneity of the isotopic standard SRM980 and new reference materials for magnesium-isotope-ratio measurements. J. Anal. At. Spectrom. 18, 1352−1356.

Georg, R.B., Halliday, A.N., Schauble, E.A., Reynolds, B.C., 2007. Silicon in the Earth's core. Nature 447, 1102−1106.

Grange, M.L., Nemchin, A.A., Pidgeon, R.T., Merle, R.E., Timms, N.E., 2013. What lunar zircon ages can tell? Lunar Planet Sci 44, 1844.

Gressmann, C.K., Rubie, D.C., 2000. The origin of the depletions of V, Cr and Mn in the mantles of the Earth and Moon. Earth Planet. Sci. Lett. 184, 95−107.

Grewal, D.S., Dasgupta, R., Sun, C., Tsuno, K., Costin, G., 2019. Delivery of carbon, nitrogen, and sulfur to the silicate Earth by a giant impact. Sci. Adv. 5, eaau3669.

Halliday, A.N., 2000. Terrestrial accretion rates and the origin of the Moon. Earth Planet Sci Lett 176, 17−30.

Hallis, L., Anand, M., Greenwood, R., Miller, M.F., Franchi, I., Russell, S., 2010. The oxygen isotope composition, petrology and geochemistry of mare basalts: evidence for Large-scale compositional variation in the lunar mantle. Geochim. Cosmochim. Acta 74, 6885−6899.

Hanan, B.B., Tilton, G.R., 1987. 60025-relict of primitive lunar crust. Earth Planet. Sci. Lett. 84, 15−21.

Hartmann, W.K., Davis, D., 1975. Satellite-sized planetesimals and lunar origin. Icarus 24, 504−515.

Herwartz, D., Pack, A., Friedrichs, B., Bischoff, A., 2014. Identification of the giant impactor Theia in lunar rocks. Science 344, 1146−1150.

Herzog, G.F., Moynier, F., Albarede, F., Berezhnoy, A.A., 2009. Isotopic and elemental abundances of copper and zinc in lunar samples, Zagami, Pele's hairs, and a terrestrial basalt. Geochim. Cosmochim. Acta 73, 5884−5904.

Hess, P.C., 1994. Petrogenesis of lunar troctolites. J. Geophys. Res. 99, 19083−19093.

Hirose, K., Labrosse, S., Hernlund, J., 2013. Composition and state of the core. Annu. Rev. Earth Planet. Sci. 41, 657−691.

Honniball, C.I., Lucey, P.G., Li, S., Shenoy, S., Orland, T.M., et al., 2021. Molecular water detected on the sunlit Moon by SOFIA. Nat. Astron. 5, 121−127.

Hopkins, S.S., Prytulak, J., Barling, J., Russell, S.S., Coles, B.J., Halliday, A.N., 2019. The vanadium isotopic composition of lunar basalts. Earth Planet. Sci. Lett. 511, 12−24.

Humayun, M., Clayton, R.N., 1995. Potassium isotope geochemistry: Genetic implications of volatile element depletion. Geochim. Cosmochim. Acta 59, 2131−2148.

Ida, S., Canup, R.M., Stewart, G.R., 1997. Lunar accretion from an impact-generated disk. Nature 389, 353−357.

Jacobsen, S.A., Morbidelli, A., Raymond, S.N., O'Brien, D.P., Walsh, K.J., Rubie, D.C., 2014. Highly siderophile elements in Earth's mantle as a clock for the Moon-forming impact. Nature 508, 84−87.

Javoy, M., Kaminski, E., Guyot, F., et al., 2010. The chemical composition of the Earth: enstatite chondrite models. Earth Planet. Sci. Lett. 293, 259–268.

Jeffcoate, A., Elliott, T., Kasemann, S., Ionov, D., Cooper, K., Brooker, R., 2007. Li isotope fractionation in peridotites and mafic melts. Geochim. Cosmochim. Acta 71, 202–218.

Johnson, C.M., Beard, B.L., 1999. Correction of instrumentally produced mass fractionation during isotopic analysis of Fe by thermal ionization mass spectrometry. Int. J. Mass. Spectrom. 193, 87–99.

Kaib, N.A., Cowan, N.B., 2015. The feeding zones of terrestrial planets and insights into Moon formation. Icarus 252, 161–174.

Kato, C., Moynier, F., 2017. Gallium isotopic evidence for extensive volatile loss from the Moon during its formation. Sci. Adv. 3, e1700571.

Kato, C., Moynier, F., Foriel, J., Teng, F.-Z., Puchtel, I.S., 2017. The gallium isotopic composition of the bulk silicate Earth. Chem. Geol. 448, 164–172.

Kegerreis, J.A., Eke, V.R., Catling, D.C., Massey, R.J., Teodoro, L.F.A., Zahnle, K.J., 2020. Atmospheric Erosion by Giant Impacts onto Terrestrial Planets: A Scaling Law for any Speed, Angle, Mass, and Density. Astrophys J 901 (L31). 10.3847/2041-8213/abb5fb.

Li, S., Lucey, P.G., Fraeman, A.A., Poppe, A.R., Sun, V.Z., et al., 2020. Widespread hematite at high latitudes of the Moon. Sci. Adv. 6, eaba1940.

Li, Q.L., Zhou, Q., Liu, Y., Xiao, Z., Lin, Y., et al., 2021. Two-billion-year-old volcanism on the Moon from Chang'er-5 basalts. Nature 600, 55–58.

Liu, Y., Spicuzza, M.J., Craddock, P.R., Day, J., Valley, J.W., Dauphas, N., et al., 2010. Oxygen and iron isotope constraints on near-surface fractionation effects and the composition of lunar mare basalt source regions. Geochim. Cosmochim. Acta 74, 6259–6262.

Lock, S.J., Stewart, S.T., Petaev, M.I., Leinhardt, Z., Mace, M.T., Jacobsen, S.B., et al., 2019. The origin of the Moon within a terrestrial synestia. J. Geophys. Res. Planets 123, 910–951.

Lodders, K., 2003. Solar system abundances and condensation temperatures of the elements. Astrophys. J. 591, 1220–1247.

Lugmair, G.W., Carlson, R.W., 1978. Sm-Nd constraints on early lunar differentiation and the evolution of KREEP. Lunar Planet. Sci. Conf. 9, 689–704.

Magna, T., Wiechert, U., Halliday, A.N., 2006. New constraints on the lithium isotope compositions of the Moon and terrestrial planets. Earth Planet. Sci. Lett. 243, 3336–3353.

Magna, T., Hu, Y., Teng, F., Mezger, K., 2017. Magnesium isotope systematics in Martian meteorites. Earth Planet. Sci. Lett. 474, 419–426.

Makishima, A., 2016. Thermal ionization mass spectrometry (TIMS). Silicate Digestion, Separation, Measurement. Wiley-VCH, Weinheim.

Makishima, A., Nakamura, E., Nakano, T., 1999. Determination of zirconium, niobium, hafnium and tantalum at ng g^{-1} levels in geological materials by direct nebulization of sample HF solutions into FI-ICP-MS. Geostand. Newslett. 23, 7–20.

Makishima, A., Tanaka, R., Nakamura, E., 2009. Precise elemental and isotopic analyses in silicate samples employing ICP-MS: application of HF solution and analytical techniques. Anal. Sci. 25, 1181–1187.

Mastrobuono-Battisti, A., Perets, H.B., Raymond, S.N., 2015. A primordial origin for the compositional similarity between the Earth and the Moon. Nature 520, 212–215.

Matsuhisa, Y., Goldsmith, J.R., Clayton, R.N., 1978. Mechanisms of hydrothermal crystallization of quartz at 250 °C and 15 kbar. Geochim Cosmochim Acta 42, 173–182.

McKeegan, K.D., Kallio, A.P.A., Heber, V.S., Jarzebinski, G., Mao, P.H., et al., 2011. The oxygen isotopic composition of the sun inferred from captured solar wind. Science 332, 1528–1532.

Millet, M.A., Dauphas, N., Greber, N.D., Burton, K.W., Dale, C.W., Debret, B., et al., 2016. Titanium stable isotope investigation of magmatic processes on the Earth and Moon. Earth Planet. Sci. Lett. 449, 197–205.

Moynier, F., Albarede, F., Herzog, G.F., 2006. Isotopic composition of zinc, copper, and iron in lunar samples. Geochim. Cosmochim. Acta 70, 6103–6117.

Münker, C., 2010. A high field strength element perspective on early lunar differentiation. Geochim. Cosmochim. Acta 74, 7340–7361.

Neal, C.R., Taylor, L.A., 1992. Petrogenesis of mare basalts – a record of lunar volcanism. Geochim. Cosmochim. Acta 56, 2177–2211.

Nebel, O., Van Westrenen, W., Vroon, P., Wille, M., Raith, M., 2010. Deep mantle storage of the Earth's missing niobium in late-stage residual melts from a magma ocean. Geochim. Cosmochim. Acta 74, 4392–4404.

Nemchin, A., Timms, N., Pidgeon, R., Geisler, T., Reddy, S., Meyer, C., 2009. Timing of crystallization of the lunar magma ocean continued by the oldest zircon. Nat. Geosci. 2, 133–136.

Nielsen, S.G., Auro, M., Righter, K., Davis, D., Prytulak, J., Wu, F., et al., 2019. Nucleosynthetic vanadium isotope heterogeneity of the early solar system recorded in chondritic meteorites. Earth Planet. Sci. Lett. 505, 131–140.

Nielsen, S.G., Bekaert, D.V., Magna, T., Mezger, K., Auro, M., 2020. The vanadium isotope composition of mars: implications for planetary differentiation in the early solar system. Geochem. Perspect. Lett. 15, 35–39.

Nielsen, S.G., Bekaert, D.V., Auro, M., 2021. Isotopic evidence for the formation of the Moon in a canonical giant impact. Nat. Commun. 12, 1817.

Nimmo, F., Agnor, C.B., 2006. Isotopic outcomes of N-body accretion simulations: Constraints on equilibration processes during large impacts from Hf/W observations. Earth Planet Sci Lett 243, 26–43.

Norman, M.D., Borg, L.E., Nyquist, L.E., Bogard, D.D., 2003. Chronology, geochemistry and petrology of a ferroan noritic anorthosite clast from Descartes Breccia 67215: clues to the age, origin, structure, and history of the lunar crust. Meteorit. Planet. Sci. 38, 645–661.

Nyquist, L.E., Wiesmann, H., Bansal, B., Shih, C.-Y., Keith, J.E., Harper, C.L., 1995. ^{146}Sm-^{142}Nd formation interval f or the lunar mantle. Geochim. Cosmochim. Acta 59, 2817–2837.

Nyquist, L.E., Shih, C.-Y., Reese, Y.D., et al., 2010. Lunar crustal history recorded in lunar anorthosites. Lunar Planet. Sci. 41, 1383.

Pack, A., Herwartz, D., 2014. The triple oxygen isotope composition of the Earth mantle and understanding Δ^{17}O variations in terrestrial rock and minerals. Earth Planet. Sci. Lett. 390, 138–145.

Pahlevan, K., Stevenson, D.J., 2007. Equilibration in the aftermath of the lunar-forming giant impact. Earth Planet. Sci. Lett. 262, 438–449.

Paniello, R.C., Day, J.M.D., Moynier, F., 2012. Zinc isotopic evidence for the origin of the Moon. Nature 490, 376–379.

Poitrasson, F., Halliday, A.N., Lee, D.C., Levasseur, S., Teutsch, N., 2004. Iron isotope differences between Earth, Moon, Mars and Vesta as possible records of contrasted accretion mechanisms. Earth Planet. Sci. Lett. 223, 253–266.

Poitrasson, F., Zambardi, T., Magna, T., Neal, C.R., 2019. A reassessment of the iron isotope composition of the Moon and its implications for the accretion and differentiation of terrestrial planets. Geochim. Cosmochim. Acta 267, 257–274.

Pringle, E.A., Moynier, F., 2017. Rubidium isotopic composition of the Earth, meteorites, and the Moon: evidence for the origin of volatile loss during planetary accretion. Earth Planet. Sci. Lett. 473, 62–70.

Rankenburg, K., Brandon, A.D., Neal, C.R., 2006. Neodymium isotope evidence for a chondritic composition of the Moon. Science 312, 1359–1372.

Reufer, A., Meier, M.M.M., Bentz, W., Wieler, R., 2012. A hit-and-run giant impact scenario. Icarus. 221, 296–299.

Ringwood, A.E., 1959. On the chemical evolution and densities of the planets. Geochim. Cosmochim. Acta 15, 257–283.

Ringwood, A.E., 1979. Origin of the Earth and Moon. Springer-Verlag, New York.

Ringwood, A.E., 1986. Terrestrial origin of the Moon. Nature 322, 323−328.

Ringwood, A.E., 1966. The chemical composition and origin of the Earth. In: Hurley, P.M. (Ed.), Advances in Earth Sciences. MIT press, Cambridge, pp. 287−356.

Ringwood, A.E., Kesson, S.E., 1977. Basaltic magmatism and the bulk composition of the Moon, II. Siderophile and volatile elements in Moon, Earth and chondrites: implications for lunar origin. Moon 16, 425−464.

Ringwood, A.E., Kato, T., Hibberson, W., Ware, N., 1990. High pressure geochemistry of Cr, V and Mn and implications for the origin of the Moon. Nature 347, 174−176.

Ringwood, A.E., Kato, T., Hibberson, W., Ware, N., 1991. Partitioning of Cr, V, and Mn between mantles and cores of differentiated planetesimals: implications for giant impact bypotheses of lunar origin. Icarus 89, 122−128.

Rubie, D.C., Jacobson, S.A., Morbidelli, A., et al., 2015. Accretion and differentiation of the terrestrial planets with implications for the compositions of early-formed Solar System bodies and accretion of water. Icarus 248, 89−108.

Salmon, J., Canup, R.M., 2012. Lunar accretion from a Roche-interior fluid disk. Astrophys. J. 760, 83.

Savage, P., Georg, R., Armytage, R., Williams, H., Halliday, A., 2010. Silicon isotope homogeneity in the mantle. Earth Planet. Sci. Lett. 295, 139−146.

Savage, P.S., Georg, R.B., Williams, H.M., Burton, K.W., Halliday, A.N., 2011. Silicon isotope fractionation during magmatic differentiation. Geochim. Cosmochim. Acta 75, 6124−6139.

Savage, P.S., Moynier, F., Chen, H., Shofner, G., Siebert, J., Badro, J., et al., 2015. Copper isotope evidence for large-scale sulphide fractionation during Earth's differentiation. Geochem. Persp. Lett. 1, 53−64.

Schoenberg, R., Zink, S., Staubwasser, M., von Blanckenburg, F., 2008. The stable Cr isotope inventory of solid Earth reservoirs determined by double spike MC-ICP-MS. Chem. Geol. 249, 294−306.

Sedaghatpour, F., Jacobsen, S.B., 2019. Magnesium stable isotopes support the lunar magma ocean cumulate remelting model for mare basalts. PNAS 116, 73−78.

Sedaghatpour, F., Teng, F.-Z., Liu, Y., Sears, D.W., Taylor, L.A., 2013. Magnesium isotopic composition of the Moon. Geochim. Cosmochim. Acta 120, 1−16.

Seitz, H.-M., Brey, G.P., Lahaye, T., Durali, S., Weyer, S., 2004. Lithium isotopic signatures of peridotite xenoliths and isotopic fractionation at high temperature between olivine and pyroxenes. Chem. Geol. 212, 163−177.

Shearer, C.K., Hess, P.C., Wieczorek, M.A., et al., 2006. Thermal and magmatic evolution of the Moon. Rev Mineral Geochem 60, 365−518.

Shearer, C.K., Papike, J.J., 2005. Early crustal building processes on the Moon: models for the petrogenesis of the magnesian suite. Geochim. Cosmochim. Acta 69, 3445−3461.

Sossi, P.A., Moynier, F., 2017. Chemical and isotopic kinship of iron in the Earth and Moon deduced from the lunar Mg-Suite. Earth Planet. Sci. Lett. 471, 125−135.

Spicuzza, M.J., Day, J., Taylor, L.A., Valley, J.W., 2007. Oxygen isotope constraints on the origin and differentiation of the Moon. Earth Planet. Sci. Lett. 253, 254−265.

Sprung, P., Kleine, T., Scherer, E.E., 2013. Isotopic evidence for chondritic Lu/Hf and Sm/Nd of the Moon. Earth Planet. Sci. Lett. 380, 77−87.

Stevenson, D.J., Hallliday, A.N., 2014. The origin of the Moon. Phil Trans Royal Soc A 372, 2024. 20140289.

Taylor, D.J., McKeegan, K.D., Harrison, T.M., 2009. Lu-Hf zircon evidence for rapid lunar differentiation. Earth Planet. Sci. Lett. 279, 157−164.

Teng, F.-Z., Li, W.-Y., Ke, S., et al., 2010. Magnesium isotopic composition of the Earth and chondrites. Geochim. Cosmochim. Acta 74, 4150−4166.

Terra, F., Wasserburg, G.J., 1974. U-Th-Pb systematics on lunar rocks and inferences about lunar evolution and the age of the moon. Proceedings of the 5th Lunar Science Conference. Pergamon Press, NewYork, pp. 1500–1571.

Thiemens, M.H., Heidenreich III, J.H., 1983. The mass-independent fractionation of oxygen: a novel isotope effect and its possible cosmochemical implications. Science 219, 1073–1075.

Tian, Z., Jolliff, B.L., Korotev, R.L., Fegley Jr, B., Lodders, K., Day, J.M.D., et al., 2020. Potassium isotopic composition of the Moon. Geochim. Cosmochim. Acta 280, 263–280.

Touboul, M., Kleine, T., Bourdon, R., Palme, H., Wider, R., 2007. Late formation and prolonged differentiation of the Moon inferred from W isotopes in lunar metals. Nature 450, 1206–1209.

Wade, J., Wood, B., 2001. The Earth's 'missing' niobium may be in the core. Nature 409, 75–78.

Walker, R.J., 2009. Highly siderophile elements in the Earth, Moon and Mars: update and implications for planetary accretion and differentiation. Chem. der Erde 69, 101–125.

Walker, D., Longhi, J., Hays, J.F., 1975. Differentiation of a very thick magma body and implications for the source regions of mare basalts. Proc. 6th Lunar Sci. Conf. Pergamon Press, New York, pp. 1103–1120.

Walsh, K.J., Morbidelli, A., Raymond, S.N., O'Brien, D.P., Mandell, A.M., 2011. A low mass for Mars from Jupiter's early gas-driven migration. Nature 475, 206–209.

Wang, K., Jacobsen, S.B., 2016a. Potassium isotopic evidence for a high-energy giant impact origin of the Moon. Nature 538, 487–490.

Wang, K., Jacobsen, S.B., 2016b. An estimate of the Bulk Silicate Earth potassium isotopic composition based on MC-ICPMS measurements of basalts. Geochim. Cosmochim. Acta 178, 223–232.

Wang, K., Jacobsen, S.B., Sedaghatpour, F., Chen, H., Korotev, R.L., 2015. The earliest Lunar Magma Ocean differentiation recorded in Fe isotopes. Earth Planet. Sci. Lett. 430, 202–208.

Wänke, H., Baddenhausen, H., Blum, K., et al., 1977. On the chemistry of lunar samples and achondrites. Primary matter in the lunar highlands: a re-evaluation. Proc. Lunar Sci. Conf. 8th 2191–2213.

Warren, P.H., 1985. The magma ocean concept and lunar evolution. Annu. Rev. Earth Planet. Sci. 13, 201–240.

Warren, P.H., 1993. A concise compilation of petrologic information on possibly pristine nonmare Moon rocks. Am. Mineral. 78, 360–376.

Warren, P.H., 2011. Stable-isotopic anomalies and the accretionary assemblage of the Earth and Mars: a subordinate role for carbonaceous chondrites. Earth Planet. Sci. Lett. 311, 93–100.

Warren, P.H., Wasson, J.T., 1979. The origin of KREEP. Rev. Geophys. Space Phys. 17, 73–88.

Weyer, S., Anbar, A.D., Brey, G.P., Münker, C., Mezger, K., Woodland, A.B., 2005. Iron isotope fractionation during planetary differentiation. Earth Planet. Sci. Lett. 240, 251–264.

Wiechert, U., Halliday, A., Lee, D.-C., Snyder, G., Taylor, L., Rumble, D., 2001. Oxygen isotopes and the Moon-forming giant impact. Science 294, 345–348.

Wiesli, R.A., Beard, B.L., Taylor, L.A., Johnson, C.M., 2003. Space weathering processes on airless bodies: Fe isotope fractionation in the lunar regolilth. Earth Planet. Sci. Lett. 216, 457–465.

Xia, Y., Kiseeva, E.S., Wade, J., Huang, F., 2019. The effect of core segregation on the Cu and Zn isotope composition of the silicate Moon. Geochem. Persp Lett. 12, 12–17.

Young, E.D., Galy, A., Nagahara, H., 2002. Kinetic and equilibrium mass-dependent isotope fractionation laws in nature and their geochemical and cosmochemical significance. Geochim. Cosmochim. Acta 66, 1095−1104.

Young, E.D., Kohl, I.E., Warren, P.H., Rubie, D.C., Jacobsen, S.A., Morbidelli, A., 2016. Oxygen isotopic evidence for vigorous mixing during the Moon-forming giant impact. Science 351, 493−496.

Zambardi, T., Poitrasson, F., Corgne, A., Meheut, M., Quitte, G., Anand, M., 2013. Silicon isotope variations in the inner Solar System: implications for planetary formation, differentiation, and composition. Geochim. Cosmochim. Acta 121, 67−83.

Zhang, J., Dauphas, N., Davis, A., Leya, I., Dedkin, A., 2012. The proto-Earth as a significant source of lunar material. Nat. Geosci. 5, 251−255.

Materials on the Earth

4.1 The late veneer

4.1.1 Introduction

The "late veneer" sometimes appears in astrochemistry. The late veneer means the addition of chondritic materials by small planetesimals after the giant impact and the core formation. Recently, the "late accretion" is used instead of the late veneer, which has the same meaning indicating the chondritic material addition to the Earth's mantle after the core formation.

The late veneer is required to reconcile the highly siderophile element (HSE) abundances, or platinum group elements (PGE) in the Earth's mantle. The W isotopic data also showed the necessity of the late veneer. Again, see Fig. 3.9, which is a cartoon showing the relation among the proto-Earth, Theia (the impactor), the giant impact, and the late veneer in order of the time to help you understand various events. The late veneer is required not only in astrochemistry, but also in astrophysics (e.g., Schlichting et al., 2012). In this Section 4.1, the late veneer is discussed in detail.

4.1.2 Constraints from platinum group elements

In Section 4.1.2, discussion around the HSE is given. HSE contains PGEs, Re, and Au. In this section, PGEs are used to indicate platinum group elements as well as Re and Au. Generally PGEs have very high evaporation/condensation temperature, however, when S, Se, and Te are included, PGEs become very low in evaporation temperature. When oxygen fugacity (fO_2) is high, Re and Os are oxidized into Re_2O_7 and OsO_4, which have low boiling temperatures and are very volatile.

Introductory Astrochemistry
DOI: https://doi.org/10.1016/B978-0-443-23938-0.00004-2

4.1.2.1 *Primitive upper mantle*

For the representative of the Earth's mantle, the primitive upper mantle (PUM) is used (see Table 4.1). Because, we have no idea for the HSE composition of the lower mantle. However, at least, the upper mantle concentration of HSE can be estimated from mantle peridotites. In Fig. 4.1, the Ir concentration versus Al_2O_3 (wt. %) for the mantle peridotites is plotted. The Ir concentrations of peridotites are rather in concentrated range of 3.5 ± 0.4 ng/g (2 SD). McDonough and Sun (1995) estimated the most fertile (least depleted) mantle, which is the PUM, has Al_2O_3 (wt.%) = 4.5%.

TABLE 4.1 The highly siderophile element (HSE) concentrations of primitive upper mantle (PUM), the averages of carbonaceous, ordinary, and enstatite chondrites.

Element	PUM (ng/g)	2 SD (ng/g)	Carbonaceous (ng/g)	Ordinary (ng/g)	Enstatite (ng/g)
Re	0.35	0.06	51.55	53.38	55.94
Os	3.9	0.5	642.6	678.6	637.4
Ir	3.5	0.4	608.5	584.6	582.7
Ru	7.0	0.9	862.2	880.4	873.3
Pt	7.6	1.3	1150	1185	1186
Pd	7.1	1.3	688.5	657.9	850.9

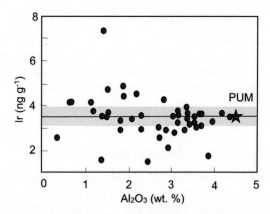

FIGURE 4.1 A plot of Ir content (ng/g) versus Al_2O_3 (wt.%) in peridotites. PUM as the red box means PUM. The red circles indicate the data from peridotites of all over the world. The horizontal red line indicates the average of Ir, and the pink area indicates the range of 2 SD of Ir concentration. The alumina data are from McDonough and Sun (1995). *Source: Modified from Becker et al. (2006) and Walker et al. (2009).*

TABLE 4.2 Elemental ratios of platinum group elements normalized to the carbonaceous chondrite.

Elemental ratio	PUM	Carbonaceous	Ordinary	Enstatite
Re/Ir	1.18	1	1.08	1.13
Os/Ir	1.06	1	1.10	1.04
Ru/Ir	1.41	1	1.06	1.06
Pt/Ir	1.15	1	1.07	1.08
Pd/Ir	1.79	1	0.99	1.29

Palladium, when plotted Pd/Ir ratios versus Al_2O_3 (wt.%), showed positive correlation from the origin (Pattou et al., 1996; Becker et al., 2006; not shown). When Al_2O_3 (wt.%) is 4.5%, Pd concentration was obtained to be 7.1 ± 1.3 ng/g (2 SD) by Becker et al. (2006).

Becker et al. (2006) plotted Re/Ir, Os/Ir, and Pt/Ir and obtained the PUM values similar to the chondritic values (see Table 4.2). However, Ru/Ir and Pd/Ir (already discussed above), were superchondritic values. In conclusion, Re, Os, Ru, and Pt concentrations of PUM were estimated to be 0.35 ± 0.06, 3.9 ± 0.5, 7.0 ± 0.9, and 7.6 ± 1.3 ng/g, respectively.

These PUM values as well as the chondritic vales (Horan et al., 2003; Walker et al., 2002) are summarized in Table 4.1. In addition, $(X/Ir)_A/(X/Ir)_{Carbonaceous}$, where X is HSE and A is the target material (e.g., PUM), is also shown. From the Re and Pd enrichments to Ir normalized to those of the carbonaceous chondrite, the PUM ratios resemble to the enstatite chondrite.

4.1.2.2 Comparison of primitive upper mantle with the Moon

As the Moon could be stratified, and our sample collection of the Moon is very limited, it is very difficult to estimate the HSE abundances of the Moon. One of the solutions is to compare the PGE abundances of similar major element abundances between PUM and lunar basalts, especially with similar MgO contents (Warren et al., 1999). In Fig. 4.2, a plot of Os contents versus MgO in terrestrial and lunar volcanic rocks (both low-Ti and high-Ti basalts) is shown. Although the lunar samples are biased, the Os concentration of the Moon is about 1/100 lower from the terrestrial basalt trend.

4.1.2.3 Platinum group element depletion of primitive upper mantle

For PUM, PGE is very depleted, about 1/100 of that of the CI chondrite. It is easy to explain by PGE partitioning into the metallic melts,

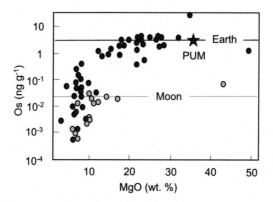

FIGURE 4.2   **A plot of Os (ng/g) content versus MgO (wt.%).** The horizontal axis is the concentration of MgO (wt.%). The vertical axis is logarithmic of Os concentration. Red and blue circles indicate terrestrial and lunar volcanic rocks, respectively. The solid red and blue lines are the estimated values of the PUM and the Moon, respectively. A star shows the PUM. The terrestrial data are mostly from the Caribbean Large Igneous Province (Walker et al., 1999). *Source: Data are from Walker et al. (2004) and Day et al. (2007).*

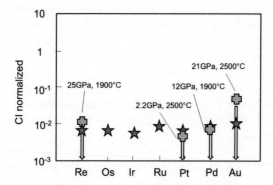

FIGURE 4.3   **Calculated concentration of PGEs in the upper mantle (PUM) normalized to those of CI chondrites.** The green cross is the concentrations of PGEs in PUM calculated using experimentally determined distribution coefficients (D values). The concentrations in PUM should be lower than the green cross as indicated by the arrows because of experimentally difficulty. The D values of Re, Pt, Pd, and Au of Righter and Drake (1997), Cottrell and Walker (2006), Righter et al. (2008), and Danielson et al. (2005) were used, respectively. The red star indicates the PGE concentrations of PUM. Note that the vertical axis is the logarithmic scale, normalized to CI chondrites determined by Anders and Grevisse (1989). *Source: Modified from Walker (2009).*

which formed the metallic core, from silicate mantle (PUM). In Fig. 4.3, using metal-silicate equilibrium constants at high pressure-temperature condition, the equilibrated mantle concentration for each PGE is shown as green crosses. It is very difficult to obtain accurate equilibrium constants because precise analysis of trace PGE in silicates of run products is required. The PGE in the run products is analyzed by the spot analytical methods (see Section 1.3.6.3). Electron probes (EPMA) have too poor detection limits ($\sim 200\,\mu g/g$) and LA-ICP-MS is generally used. Although the silicate spots ($\sim 50\,\mu m$ Φ) free from metals are chosen, PGEs often form micro particles (nuggets), which hinder precise analysis of silicates. As results, the partitioning constants can be lower than the true values. If the measurements or experiments are free from these effects (nugget effects), PGE concentration of the calculated mantle could be lower, which is shown as arrows in Fig. 4.3. As results, the PGE abundances of PUM seems to be overabundant than expected.

To explain high abundance composition of PUM, terms and models of "late accretion," "heterogeneous accretion," or "late veneer" appeared. These terms are generally used to indicate small mass addition ($\sim 0.1\%$) after almost all Earth is formed, to add desirable amounts of HSE or volatiles (water or carbons), etc. The late veneer often appears in later discussion, because it is easy to add or change elemental abundances or isotopic ratios. As conclusion, without the late veneer, the abundances of HSE cannot be explained.

4.1.3 Constraints for the late veneer from W isotopes

4.1.3.1 Problems in early works, and the solutions

Although the ^{182}Hf–^{182}W decay system is an ideal isotopic system for discussion of the giant impact and the Moon formation in the early solar system (Jacobsen, 2005; Halliday, 2008), early researches for the Moon were interfered with transforming effects of ^{181}Ta into ^{182}W by neutron capture of cosmic rays. Because the Moon samples were collected from the surface. In addition, the burn out process of ^{182}W could also occur (Leya et al., 2000).

In order to avoid such effects, using Ta-free metals separated from the KREEP basalts or melts by the impacts was one of the strategies (Kleine et al., 2009; Touboul et al., 2007). These studies using crystallization products of the lunar magma ocean revealed that the W isotopic composition was uniform and similar to the terrestrial mantle. These results suggested that the formation of the Moon and crystallization of the lunar magma ocean occurred after most ^{182}Hf decayed out, which is after 60 Myrs from the formation of the solar system (Touboul et al., 2007, 2009).

4.1.3.2 Lunar W isotopic ratios and percentage of the late veneer

Touboul et al. (2015) determined the Os isotope ratios, HSE (highly siderophile elements) concentration, and W isotope ratios of these metals from the impact melt rocks, to examine the early history of the

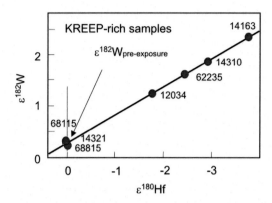

FIGURE 4.4 **A plot of ε^{182}W versus ε^{180}Hf determined for KREEP-rich samples.** The intercept at ε^{180}Hf = 0 indicates the pre-exposure of ε^{182}W (= +0.27 ± 0.04 at the 95% confidence level). *Source: Modified from Kruijer (2015).*

Moon. The HSE in the metals was thought to be equilibrated with the melt or vapor during impact (Tera et al., 1974). They discussed the incorporation of both stony or iron meteorites, and from two values of 68115, 114 metals and 68815, 394 and 68815, 396 metals, and obtained the average lunar W isotopic ratio to be ε^{182}W = +0.206 ± 0.051 (2 SD) relative to the present Earth mantle by negative thermal ionization mass spectrometry (N-TIMS) (Touboul and Walker, 2012).

Kruijer et al. (2015) developed the new strategy. They measured HSE contents and investigated the contribution of the meteorites. In addition, they measured ε^{180}Hf and ε^{182}W of the KREEP samples. The KREEP samples were chosen because they are residual liquids of the lunar magma ocean, and it is considered that the initial values of ε^{182}W (before irradiation values) were similar and similarly irradiated for Hf-Ta-W isotopes. Actually, when ε^{182}W is plotted against ε^{180}Hf, the samples formed a clean correlation line (Slope = −0.549 ± 0.019, MSWD = 0.36) (see Fig. 4.4). Thus only the less irradiated samples were used in discussion. They obtained the lunar W isotopic ratio to be ε^{182}W = +0.27 ± 0.04 (2 SD) relative to the present Earth mantle.

In Fig. 4.5, the ε^{182}W values of the Moon of Touboul et al. (2015) and Kruijer et al. (2015) are plotted as white and black circles, respectively. It is amazing that two data overlap within small range of error bars. The grand average is ε^{182}W = 0.24.

A red triangle in Fig. 4.5 indicates the relation between the ε^{182}W values and the mass fraction of the late-accreted material (the late veneer) by Touboul et al. (2015). These values are estimated by mass balance assuming W contents in BSE and the impacted Theia (the chondrite) to be 13 and 200 ppb, respectively. The upper and lower lines

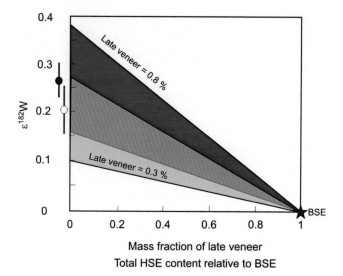

FIGURE 4.5   A plot of ε ^{182}W versus mass fraction of the late veneer to the present bulk silicate Earth (a red star). White and black circles are ε^{182}W of the Moon by Touboul et al. (2015) and Kruijer et al. (2015), respectively (the horizontal positions of these points have no meaning). The red and the green triangles are the estimated relation between the ε^{182}W values and the mass fractions of the late-accreted material by Touboul et al. (2015) or of total HSE content relative to BSE by Kruijer et al. (2015), respectively. *Source: Modified from Touboul et al. (2015) and Kruijer et al. (2015).*

from the bulk silicate Earth (BSE) indicate when the mass fraction of the late veneer was 0.8% and 0.3%, respectively.

Kruijer et al. (2015) estimated the late veneer component to be 80% of (CI + CV + CM) material and 20% of IVA-like iron meteorite material to explain HSE abundances including Se, Te, and S ratios in the Earth's PUM. The late veneer value of ε^{182}W of the mixture is −2.60, and its mass is 0.35% of the Earth's mass. The green triangle is the range of the 95% confidence level. The error mainly comes from the uncertainty of the W concentration of BSE of 13 ± 5 ppb.

As both estimations are highly dependent on many assumptions, the mass fraction of the late veneer cannot be defined with high accuracy. As far as the HSE concentrations of BSE cannot be determined to the definite values, and their compositions of added meteorites cannot be fixed, the precision of the estimation cannot be improved.

4.1.3.3 Discussion on formation age of the Moon

Thiemens et al. (2019) obtained high-precision trace element composition data for various types of lunar samples by ICP-MS, which showed that the Hf/W ratio of the silicate Moon (30−50) should be higher than that of the bulk silicate Earth (~25.6). By combining these data with experimentally derived partition coefficients, they proposed that the $+25 \ \mu^{182}W$ unit excess in lunar basalts (see Eq. (1.27) in Section 1.3.13 for $\mu^{182}W$ unit) can be explained by the decay of the now extinct ^{182}Hf to ^{182}W for the first 60 Myr after the Solar System formation. They concluded that the Moon formed earlier, about 50 Myr after the Solar System, and that the excess ^{182}W of the silicate Moon is unrelated to the late veneer.

However, Kruijer et al. (2021) argued that application of the Hf−W system to date Moon formation requires two conditions: (1) the BSE and Moon must have the same initial ^{182}W isotopic composition; and (2) the $\mu^{182}W$ of the lunar and terrestrial mantles must be closed in relation to the ^{182}Hf decay. However, mixing processes during the giant impact and subsequent late accretion to the lunar and terrestrial mantles must have modified the $\mu^{182}W$ compositions (Kruijer et al., 2015; Kruijer and Kleine, 2017). Therefore, even if the Hf/W ratios of the lunar and terrestrial mantles may be different, it should not be assumed that the present-day $\mu^{182}W$ difference is a purely radiogenic result. However, Thiemens et al. (2021) used the present-day $\mu^{182}W$ difference between the BSE and the Moon to date the giant impact, and they did not consider whether the two previous requirements were satisfied or not.

The late veneer can explain the highly siderophile element (HSE) budget of the BSE. Late accretion is a natural process in the planetary formation where the remaining planetesimals continue to be accreted to planets. In the case of the Earth, the mass was estimated to be $(4.8 \pm 1.6) \times 10^{-3}$ Earth masses $(M_{\oplus})$, which should have affected: (1) the BSE's $\mu^{182}W$; (2) the BSE's W concentration; (3) the chemical and ^{182}W compositions of the late-accreted materials; and (4) the late-accreted mass. Each parameter has uncertainty. Therefore, even if the mass and composition of the late veneer materials were known precisely, the $\mu^{182}W$ value of BSE before the late accretion remains uncertain.

The pre-late-accreted $\mu^{182}W$ of BSE is entirely independent of $\mu^{182}W$ anomalies in some terrestrial rocks that reflect late accretion (Archer et al., 2019; Willbold et al., 2011) or early Earth differentiation (Touboul et al., 2012). The late-accreted mass for the Moon is about three orders of magnitude smaller than that for the Earth, so the effect on the lunar $\mu^{182}W$ was much smaller (~ 1 ppm) (Kruijer and Kleine, 2017; Kruijer et al., 2015). However, the Moon's late-accreted mass may be ~ 10 times larger than the HSE-derived estimate, because some late-accretion-derived HSEs were removed to the lunar core (Morbidelli et al., 2018). In addition, mega-regolith contamination of lunar samples should give unprecise estimate on the amounts of late-accreted materials on the Moon (Brenan et al., 2019). As results, the late accreted-lunar $\mu^{182}W$ should be independently estimated to the HSE measurements.

Thus, Kruijer et al. (2021) used the theoretical upper bound of the Moon's late-accreted mass from the Earth/Moon impact flux ratio of ~ 20 (Morbidelli et al., 2018). Using this ratio and Earth's late-accreted mass of $\sim 4.8 \times 10^{-3}$ $M_{\oplus}$, an expected impact flux to the Moon was obtained to be $\sim 2.4 \times 10^{-4}$ $M_{\oplus}$. However, the fraction of impactor material that remains on the target (the impactor retention ratio) is only ~ 0.2 for the Moon (Zhu et al., 2019). Thus, only $\sim 4.8 \times 10^{-5}$ $M_{\oplus}$ would be accreted out of the impacted value of $\sim 2.4 \times 10^{-4}$ $M_{\oplus}$. Moreover, the Earth/Moon impact flux ratio was probably larger than 20 (Morbidelli et al., 2018). For example, the Earth/Moon impact flux ratio could be ~ 100 (Morbidelli et al., 2018), which results in a lower estimated late-accreted mass for the Moon to be $\sim 1 \times 10^{-5}$ $M_{\oplus}$. Thus, it is concluded that, 4.8×10^{-5} $M_{\oplus}$ is an upper limit for the Moon's late-accreted mass. Using these estimates, Kruijer et al. (2021) calculated a $\mu^{182}W$ difference between the prelate accretion BSE and the Moon, however, they could not solve it because the uncertainty was too large to constrain the timing of the Moon formation.

4.1.4 Constraints on the late veneer from the oxygen isotope ratios

By the oxygen isotope ratios, Young et al. (2016) tried to constrain the composition and size of the late veneer of primitive bodies that impacted on the silicate Earth. A large flux of the late veneer planetesimals is implied by high average tungsten isotope ratios for the Moon than for the Earth and by differences in highly siderophile element (HSE) concentrations between the mantles of the Earth and the Moon (Walker et al., 2015). The interpretation of these data is that the Earth and the Moon began with the same W isotopic ratios, but that the Earth inherited a greater fraction of low tungsten isotopic materials as chondritic planetesimals after the Moon forming giant impact (Kruijer et al., 2015; Touboul et al., 2015).

If the Earth-Moon system was mixed well by the giant impact, which is a constraint of the W isotope ratios, then the nearly identical $\Delta'^{17}O$ values of the Earth and the Moon can constrain the $\Delta'^{17}O$ values of the late veneer impactor. The mass flux ratio of the Earth/Moon is estimated in the range from 200 to 1200 (Walker, 2015; Schlichting et al., 2012; Bottke et al., 2010). Assuming a mass of the late veneer to be 2×10^{22} kg (Walker, 2015), and a maximum Earth/Moon flux ratio of 1200 (Bottke et al., 2010), the late veneer fraction is 0.00447.

Combining this value with the measured value for $\Delta'^{17}O_{Moon} - \Delta'^{17}O_{Earth}$ of zero (Young et al., 2016) requires that the late veneer impactor had average $\Delta'^{17}O$ values within $\sim 0.2‰$ or less of Earth, which is similar to enstatite chondrites (Newton et al., 2000). Alternatively, with the maximum value of $\Delta'^{17}O_{Moon} - \Delta'^{17}O_{Earth}$ of $\pm \sim 5$ ppm (Young et al., 2016; when the largest analytical uncertainty is assumed), the $\Delta'^{17}O$ value of the late veneer is $\pm 1.1‰$, which includes aqueously altered carbonaceous chondrites or some ordinary chondrites.

If the value of $\Delta'^{17}O_{Moon} - \Delta'^{17}O_{Earth}$ is 12 ppm (Herwartz et al., 2014 which is concluded to be impossible by Young et al., 2016), the late veneer of $\Delta'^{17}O$ becomes $-2.7‰$, suggesting that the late veneer impactor was mainly composed of relatively unaltered and dry carbonaceous chondrites (Clayton et al., 1976).

Thus, the result of Young et al. (2016) suggests that if the late veneer impactor was mainly composed of carbonaceous chondrites, the parent bodies must have included substantial fractions of high-$\Delta'^{17}O$ water either in the form of aqueous alteration minerals or as water ice.

4.1.5 Ejecta of the giant impact

Bottke et al. (2015) paid attention to the ejecta of the giant impact. They estimated numerous kilometer-sized ejecta fragments by the giant impact should have struck main-belt asteroids at >10 km/s velocities, which caused heating and degassing of the target rocks. The impacts produced >1000 times larger heat than usual impact at 5 km/s. By fitting the heating model of stony meteorites, the Moon was estimated to be formed at ~ 4.47 Ga.

4.2 The oldest geological records on the Earth

4.2.1 Introduction

After the late veneer, the Earth seems to be in a stabilized condition. In the Earth, the core was formed. On the surface, the atmosphere was formed, cooled, and the sea should have been appeared. Therefore, during 4.6 Ga, most evidences of large geological events on the surface have been washed away. In Table 4.3, early important Earth's differential events which can be discussed even now are shown (Carlson et al., 2014). In this section, each events summarized in Fig. 4.6 is explained. The pink boxes are based on observation of the chondrites. The orange boxes indicate events related to the Moon. The yellow boxes are related to topics on the early Earth. The green and blue boxes show constraints from extinct nuclei and astrophysics, respectively.

TABLE 4.3   Age estimates for early Earth's differentiation events.

Event	Age (Ga)	References
Hf−W age of core formation	4526	Kleine et al. (2009)
	4538−4468	Rudge et al. (2010)
U−Pb age of the Earth	4.55	Patterson (1956)
	~4450	Allegre et al. (2008)
I-Pu-Xe age of the Earth's atmosphere	~4450	Staudacher and Allegre (1982)
		Pepin and Porcelli (2006)
		Mukhopadhyay (2012)
$^{146}Sm-^{142}Nd$ model age for Isua	4350−4470	Caro et al. (2006)
		Rizo et al. (2011)
Peak in oldest Hadean zircons	4350	Holden et al. (2009)
$^{146}Sm-^{142}Nd$ model age for Nuvvuagittuq crust	4340−4400	O'Neil et al. (2012)

Data from Carlson et al. (2014).

4.2.2   Core formation age from Hf−W systematics

From the combination of $^{182}W-^{142}Nd$ evidence, Kleine et al. (2009) proposed that the bulk Earth may have superchondritic Sm/Nd and Hf/W ratios, and that the formation of the core must have terminated $>$ ~42 My (4526 Ma) after formation of CAIs (4568.3 ± 0.7 Ma; Burkhardt et al., 2008), which is consistent with the Hf−W age for the formation of the Moon.

Using the Hf−W and U−Pb isotopic systems, Rudge et al. (2010) suggested the rapid accretion of Earth's main mass within ~10 My from the formation of the solar system. The Earth's accretion terminated 30−100 My after formation of the solar system. They also proposed the disequilibrium model in which some fractions of the embryos' metallic cores were allowed to directly enter the Earth's core, without equilibrating with the Earth's mantle. Their results indicate that only 36% of the Earth's core must have formed in equilibrium with Earth's mantle.

Unfortunately, this model does not include the discussion of the highly siderophile elements (HSEs), because there are no appropriate radiogenic isotope systems for HSEs. However, if there were appropriate isotope systems, for example, the Re-Os isotope system could be

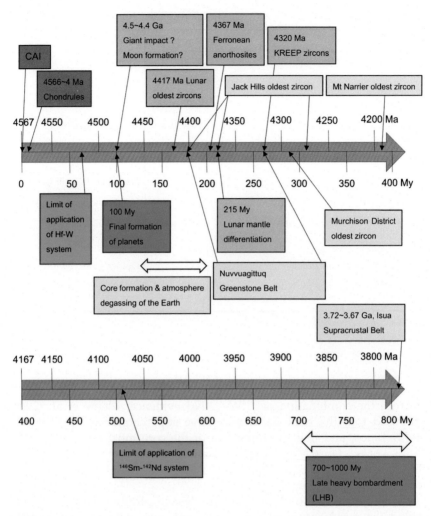

FIGURE 4.6   Summary of important events including hypothetical ones in the early Earth's history. The pink boxes are related to the chondrites. The orange boxes indicate events related to the Moon. The yellow boxes are related to the early Earth. The green and blue boxes show the limits of extinct nuclei and geophysical requirements.

applicable, the model could be extended to the HSEs. In addition, if the calculation results could explain the present concentration of the HSEs in the mantle, the late veneer would not be required. This is a delusion of the author, and of course, future studies are required!

4.2.3 The perspective of atmospheric evolution from the Hadean to the Archaean Earth

A Hadean atmosphere should contain N_2 and CO_2 and a Hadean ocean contains H_2O as a natural consequence of planetary accretion in the terrestrial planet region. The weakly reducing atmosphere with relatively high partial pressure of CO_2 should have formed (Holland, 1984; Walker, 1985; Kasting, 1993; Ferus et al., 2015; Furukawa, 2015). In this condition the important biological precursor compounds for the life were synthesized. It should be noted that high partial pressure of CO_2 in the early Hadean atmosphere can be presumed.

Atmospheric O_2 levels rose naturally and gradually, but not immediately, which occurred by photosynthesis and organic carbon burial. At the same time, the concentrations of CO_2 and other greenhouse gases declined to compensate for the brightening sun. The Earth's relatively stable climate was a result of the negative feedback between atmospheric CO_2, surface temperature, and the weathering rate of silicate rocks (Kasting, 1993).

In such atmospheric evolution, it is implied that the Earth is not a unique planet. If planets exist around other stars, some of them could reside in orbits where the illumination is similar to that received by the Earth. Planetary climates are buffered by the carbonate-silicate cycle. Therefore, the habitable zone around late F to mid K stars (see the top scale of the Hertzsprung-Russel diagram; see Section 1.2.2.1) may be wider, and more habitable planets may exist. If the origin of the life was not a fortuitous event, many of these planets could be inhabited and some may have even evolved into intelligent life. Both of these speculations can be tested: the first by spectroscopic investigations from large, space-based telescopes, and the second by monitoring microwave and radio emissions from space (Kasting, 1993).

Morbidelli et al. (2000) suggested that the most plausible sources of the water accreted by the Earth were in the outer asteroid belt, in the giant planet regions, and in the Kuiper Belt. It is plausible that the Earth accreted water, from the early phases when the solar nebula was still present to the late stages of gas-free scattered planetesimals. Asteroids and the comets from the Jupiter-Saturn region were the first water deliverers, when the Earth was less than half its present mass. The bulk water presently on the Earth was carried by a few planetary embryos, originally formed in the outer asteroid belt and accreted by the Earth at the final stage of its formation. (see Fig. 2.17 where water rich planetesimals are shown in blue, which could be sources of water of the Earth).

Finally, a late veneer (this could be the same or different from the late veneer for the HSEs discussed in Section 4.1), accounting for at most 10% of the present water mass, occurred due to comets from the Uranus-Neptune region and from the Kuiper Belt (see Section 1.2.10). The net result of accretion from these several reservoirs is that the D/H ratio of the water on the Earth is essentially the typical water condensed in the outer asteroid belt. This agrees with the observation that the D/H ratio in the oceans is very close to the mean value of the D/H ratio of the water inclusions in carbonaceous chondrites.

4.2.4 Transportation of materials on the Archaean Earth by Late Heavy Bombardment (LHB)

Following the solidification of the Moon ~ 4.5 Ga, the initially heavy impactor flux declined (Koeberl et al., 2000) and increased again during the Late Heavy Bombardment (LHB). As discussed in Section 2.3, a lot of materials fell on the Moon and of course on the Earth as well. The LHB occurred in $4 \sim 3.85$ Ga, in later than Hadean and in Archean, which is recorded in the lunar craters (Culler et al., 2000; Hartmann et al., 2000; Gomes et al., 2005; De Niem et al., 2012). This event should have delivered large amounts of materials including biologically important molecules, such as water or organic materials to the early Earth. The cause of LHB links to a dynamic instability of the outer solar system in the context of the so-called "Nice model" (see Section 2.3.3), when Jupiter's orbit changed as a result of resonances with Saturn and small cometary bodies (Tsiganis et al., 2005; Nesvorny and Morbidelli, 2012). These changes released the impactors from the previously stable asteroidal and cometary reservoirs on to the planets.

The impactor flux on the Earth should be ~ 10 times higher at LHB than in the period immediately preceding LHB, and slowly decayed afterward (Koeberl, 2006; Morbidelli et al., 2012; Geiss and Rossi, 2013).

4.2.5 U−Pb age of the Earth

By the U−Pb method, Patterson (1956) is a pioneer who obtained the age of the stony meteorites as the age of the Earth to be 4.55 ± 0.07 Ga.

Allegre et al. (2008) are against the rapid accretion and early differentiation of the Earth (30−40 Myr) after the birth of the solar system at 4.567 Ga (Amelin et al., 2002, 2006). They used Hf-W, U−Pb, and I−Xe systematics on the Earth. The W isotopic composition of the bulk silicate Earth can be explained by an incomplete isotopic re-equilibration between primitive metal and silicate components during the segregation of the Earth's core (e.g., Rudge et al., 2010).

The non-equilibrated fraction of primitive silicate material is estimated to be small, between 6% and 14%, enough to apply the $^{182}Hf-^{182}W$ chronometer. This incomplete metal/silicate re-equilibration only slightly affects the U$-$Pb chronometer. The mean age of the Earth's core's segregation is between 4.46 and 4.38 Ga. This evaluation overlaps the time of outgassing of the atmosphere based on the $^{129}I-^{129}Xe$ systematics, 4.46$-$4.43 Ga. Thus, they concluded that the period of $\sim$4.45 Ga relates to the major primitive differentiation of the Earth. This scenario comprehensively and quantitatively explains the $^{182}Hf-^{182}W$, $^{235,238}U-^{207,206}Pb$, $^{129}I-^{129}Xe$, and $^{146}Sm-^{142}Nd$ terrestrial records. In addition, it is compatible with the formation of the Moon and coherent with the $\sim$102 Ma time scale for the accretion of the Earth.

4.2.6 Application of ^{142}Nd isotope systematics to the oldest crust on the Earth

4.2.6.1 Application of $^{146,147}Sm-^{142,143}Nd$ systematics to West Greenland samples

Caro et al. (2006) developed a new ultrahigh-precision $^{142}Nd/^{144}Nd$ measurement method and applied to early Archaean rocks. The $^{142}Nd/^{144}Nd$ ratio of the Nd standard solution can be determined with intermediate precision of 2 ppm (2σ), allowing resolution of 5 ppm. The 3.6$-$3.8 Ga West Greenland metasediments, metabasalts, and orthogneisses, which are considered to be the oldest rocks on the Earth, displayed positive ^{142}Nd anomalies ranging from 8 to 15 ppm. Using a simple two-stage model with an initial $\varepsilon^{143}Nd$ value of 1.9 $\pm$ 0.6 ε units, both $^{147}Sm-^{143}Nd$ and $^{146}Sm-^{142}Nd$ systematics constrained mantle differentiation to 50$-$200 Ma after formation of the solar system. This chronological constraint fits the differentiation of the Earth's mantle in the late stage of crystallization of a magma ocean.

They developed a two-box model describing ^{142}Nd and ^{143}Nd isotopic evolution of the depleted mantle in the crust$-$mantle system. Early terrestrial proto-crust had a lifetime of c. 0.7$-$1 Ga to produce the observed Nd isotope signature of Archaean rocks. In this two-box mantle$-$crust system, the evolution of isotopic and chemical heterogeneity of depleted mantle was modeled as a function of the mantle stirring time. Using the dispersion of $^{142}Nd/^{144}Nd$ and $^{143}Nd/^{144}Nd$ ratios observed in the early Archaean rocks, the stirring time of the early Earth's mantle was constrained to be 100$-$250 Ma, a factor of 5 faster than that of the modern oceanic basalts.

4.2.6.2 Application of the $^{146,147}Sm-^{142,143}Nd$ and $^{176}Lu-^{176}Hf$ systematics to West Greenland samples

Rizo et al. (2011) first applied a combined $^{146,147}Sm-^{142,143}Nd$ and $^{176}Lu-^{176}Hf$ study to mafic rocks (amphibolites) from the western part of the Isua Supracrustal Belt (ISB, SW Greenland). The whole-rock isochrons of Sm−Nd and Lu−Hf gave identical ages within error, 3.72 ± 0.08 and 3.67 ± 0.07 Ga, respectively. The excess of ^{142}Nd in Isua samples was observed to be 7−16 ppm relative to the terrestrial Nd standard. This indicates that early-differentiated reservoirs escaped complete homogenization by mantle convection until the Archean. The intercept of the Sm−Nd whole-rock isochron is consistent with ^{142}Nd results and with a superchondritic initial $^{143}Nd/^{144}Nd$ ratio $(\varepsilon^{143}Nd_{3.7Ga} = +1.41 \pm 0.98)$. In contrast, the corresponding initial $\varepsilon^{176}Hf_{3.7Ga} = -1.41 \pm 0.57$ is subchondritic. Since Lu/Hf and Sm/Nd fractionate similarly during mantle processes, the Sm−Nd and Lu−Hf isotope systems display inconsistent parent−daughter behavior in the source of Isua amphibolites.

Based on high-pressure and -temperature phase partition coefficients, a model was proposed that satisfies $^{147}Sm-^{143}Nd$, $^{176}Lu-^{176}Hf$, ^{142}Nd results, and trace element characters. A deep-seated source composed largely of magnesium perovskite (98% MgPv) containing 2% calcium perovskite satisfactorily explains the Nd and Hf isotopic discordance observed for Isua amphibolites. The negative HFSE anomalies characterizing Isua basalts could have been inherited from such an early $(4.53 \sim 4.32$ Ga) deep mantle. A deep-seated source was involved in the formation of ISB lavas.

4.2.6.3 Application of the $^{146,147}Sm-^{142,143}Nd$ systematics to the Ujaraaluk unit in the Nuvvuagittuq Greenstone Belt (NGB), Canada

O'Neil et al. (2012) studied the Ujaraaluk unit in the Nuvvuagittuq Greenstone Belt (NGB) in Northern Quebec, Canada. NGB is dominated by mafic and ultramafic rocks metamorphosed to upper amphibolite facies. Rare felsic intrusive rocks provide zircon ages of up to ~ 3.8 Ga (David et al., 2009; Cates and Mojzsis, 2007) establishing the minimum formation age of the NGB as Eoarchean. Primary U-rich minerals that provide reliable formation ages for the dominant mafic lithology, called the Ujaraaluk unit, had found.

Metamorphic zircons, rutiles, and monazites were present in the unit and gave variably discordant results with $^{207}Pb/^{206}Pb$ ages ranging from 2.8 to 2.5 Ga. The younger ages overlaped 2686 ± 4 Ma zircon ages for intruding pegmatites (David et al., 2009) and Sm−Nd ages for garnet formation in the Ujaraaluk rocks suggesting this era as the time of peak metamorphism and metasomatism in the NGB, simultaneous to regional metamorphism of the Superior craton.

The slope of Sm–Nd data for Ujaraaluk was 3.6 ± 0.2 Ga with scattering on the isochron (MSWD = 134). This "isochron" seemed to consist of a series of younger ~ 3.2–2.5 Ga slopes for the different geochemical groups within the Ujaraaluk, possibly older than 4 Ga. The ^{146}Sm–^{142}Nd systematics is less affected by metamorphism at 2.7 Ga because of ^{146}Sm extinction prior to ~ 4 Ga. The ^{142}Nd dataset for the Ujaraaluk and associated ultramafic rocks showed a good correlation between Sm/Nd ratio and ^{142}Nd/^{144}Nd that corresponds to an age of 4388_{-17}^{+15} Ma. The dataset included samples with superchondritic Sm/Nd ratios.

The upper Sm/Nd ratio end of the Ujaraaluk correlation is defined by rocks that are interpreted to be cumulates to compositionally related extrusive rocks indicating that this crystal fractionation had to occur while ^{146}Sm decay was active, well before 4 Ga.

Intruding gabbros gave ^{143}Nd and ^{142}Nd isochron ages of 4115 ± 100 Ma and 4313_{-69}^{+41} Ma, respectively, also supporting a Hadean age for the gabbros and providing a minimum age for the intruded Ujaraaluk unit. 3.6 Ga tonalites surrounding the NGB, 3.8 Ga trondhjemitic intrusive veins, and a 2.7 Ga pegmatite showed a lack of ^{142}Nd compared to the terrestrial standard.

A subset of least disturbed Ujaraaluk samples had coherent isotopic compositions for both short-lived and long-lived Nd isotopic systems giving ^{143}Nd and ^{142}Nd isochron ages overlapping within an error of 4321 ± 160 Ma (MSWD = 6.3) and 4406_{-17}^{+14} Ma (MSWD = 1.0), respectively (see Fig. 4.7). The NGB thus preserved over 1.6 Gyr of early Earth history in mafic crust formed in the Hadean.

4.2.7 Oldest zircon on the Earth

High-resolution secondary ion mass spectrometry (HR-SIMS) made a great success in the zircon analysis. From the Mt. Narryer in Western Australia, using the HR-SIMS, SHRIMP, zircons having U–Pb ages of 3800–4500 My were discovered from Archaean metasediments. Four zircons had near-concordant U–Pb ages between ~ 4180 My. This result shows that pre-3800 My silica-saturated rocks were present on the Earth's crust.

Compston and Pidgeon (1986) found 4276 ± 6 Ma detrital zircon from Jack Hills area, also from Western Australia. The frequency of old zircons is 12%, 5 times higher than the Mt. Narryer area. The high U and low Th suggested that they came from the felsic igneous rocks.

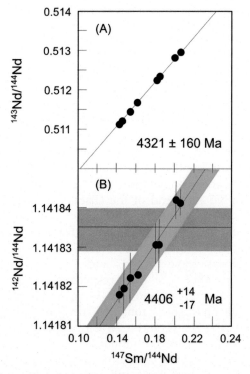

FIGURE 4.7 **Isochrons for the Ujaraaluk samples.** (A) $^{147}Sm/^{144}Nd$ versus $^{143}Nd/^{144}Nd$ isochron diagram. The slanted line indicates the isochron. The age was obtained to be 4321 ± 160 Ma. (B) $^{147}Sm/^{144}Nd$ versus $^{142}Nd/^{144}Nd$ fossil isochron diagram. The slanted line shows the best fit line. The green area shows the error in the regression (2 SD) and the pink area shows the terrestrial standard value and the 4.5ppm intermediate precision range. The number with the errors is the estimated age. *Source: Modified from O'Neil et al. (2012).*

Holden et al. (2009) developed the method using SHRIMP II to survey $^{207}Pb/^{206}Pb$ for 5 s, and candidates were analyzed in longer time. They analyzed 10,000 grains from Jack Hills, and found 7% were older than 3800 My, and the oldest grain was 4372 ± 6 Ma. The oldest population was 4350 Ga.

4.2.8 First water on the early Earth from oxygen isotopic data of zircon by HR-SIMS

The oldest zircons in Mt. Narryer and Jack Hills are felsic silicate origin, which is similar to the present continental crust. In addition, they are in metamorphosed sedimentary rocks, which indicates liquid water existed at ~ 4350 Ma. The oldest rock (note that "zircon" is only a mineral, but "rock" means the "whole rock," which is an assemblage of various minerals) is gneiss (metamorphosed sedimentary rocks) of Isua area in Greenland, which also indicates that water already existed at 3800 Ma.

High-resolution secondary ion mass spectrometry (HR-SIMS) also gives us oxygen isotopic data of diameter of less than 10 μm. Wilde et al. (2001), based on a detailed micro-analytical study of Jack Hills zircons, discovered a detrital zircon with an age as old as 4404 ± 8 My about 130 million years older than any previously identified zircons on the Earth. They found the zircon was zoned for REEs and oxygen isotopic ratios ($\delta^{18}O$ values of 7.4–5.0‰). These imply that it formed from an evolving magmatic source. The high $\delta^{18}O$ value and micro-inclusions of SiO_2 were consistent with growth from a granitic melt with a $\delta^{18}O$ value of 8.5–9.5‰. Magmatic oxygen isotopic ratios indicated the involvement of supracrustal material that underwent low temperature interaction with a liquid hydrosphere. Therefore, this zircon is the earliest evidence for continental crust and oceans on the Earth.

The similar study was presented by Mojzsis et al. (2001) in the following article of Nature. They found detrital zircons from quartzitic rocks in the Murchison District of Western Australia. They found that 3910 − 4280 My old zircons which had $\delta^{18}O$ values from 5.4 ± 0.6‰ to 15 ± 0.4‰. These data indicated that the zircons of ~4300 Ma formed from magmas containing a significant component of continental crust, which formed in the presence of water near the Earth's surface. These data are consistent with the presence of a hydrosphere interacting with the crust by 4300 My ago.

4.2.9 Enstatite chondrites can be the source materials of the Earth

The origin of Earth's water is still not fixed. As enstatite chondrites (ECs) have similar isotopic composition to terrestrial rocks, ECs may be representative of the material that formed Earth. ECs are assumed to be devoid of water, because they were formed in the inner solar system. Therefore, Earth's water is generally attributed to the late addition of a small fraction of hydrated materials, such as carbonaceous chondrite meteorites, which were formed in the outer solar system where water was more abundant.

Piani et al. (2020) found that ECs have bulk hydrogen contents (reported as water equivalents) ranging from 0.08 to 0.54 wt.% H_2O. The EH3 meteorites are generally richer in H (0.44 ± 0.04 wt.% H_2O; all uncertainties are 1 STD) than the more metamorphosed groups EH4 (0.2 ± 0.1 wt.% H_2O), EH5 (0.3 ± 0.1 wt.% H_2O), and EL6 (0.3 ± 0.3 wt.% H_2O). The Norton County aubrite is 0.3 ± 0.2 wt.% H_2O.

Thus, ECs contain sufficient hydrogen to fill the ocean at least three times the mass of water. EC hydrogen and nitrogen isotopic compositions are also similar to those of the Earth's mantle, so EC-like asteroids might have contributed these volatile elements to Earth's crust and mantle.

4.3 The oldest evidence of Life on the Earth

4.3.1 Introduction

Although a lot of researchers are trying to solve the origin of Life as summarized in Section 1.5.21, it is not solved at all. Here, not biological way, but the geologists are trying to approach to the origin of Life by studying the oldest records of Life on the Earth by space science methods. The famous Archean areas where the possibly first lives were recorded are the Pilbara area in Western Australia and the Isua Supracrustal Belt (ISB). Thus, in this section, topics of the oldest records of Life from these two areas are picked up.

Morphological evidences for microfossils were discovered from the Pilbara area in Australia. However, the morphology or the shape of microfossils cannot be any conclusive evidences for the Life, which will be discussed in Section 7.3. Instead, the chemo-fossils (graphite) in ISB with low carbon isotope ratios ($\delta^{13}C$) are searched for the oldest Life. Because long time from Archean to today have destroyed the microfossils by metamorphism, deformation, tectonic activity, etc. Nowadays, $\delta^{13}C$ of the chemo-fossils in the graphitic form can be measured by HR-SIMS. Such analytical development can give us definitive conclusions for the Life at ISB.

The stromatolites (see Section 1.5.18) are one of the critical signs for the Life. In 2008, Van Kranendonk et al. concluded that the 3.5 Ga Dresser Formation at the Pilbara area contained the oldest evidences for the existence of the stromatolites. Following this, in 2016, Nutman et al. reported the evidences of the oldest stromatolites in the ISB about 3700 Ma. However, other researchers insisted that the evidences are abiogenic, which caused furious discussion. In this section, this hot discussion for the oldest Life is also picked up.

4.3.2 The oldest microfossils (?) in the Apex chert, from the Pilbara area, Western Australia

Based on the environment with water and initial atmosphere, the Life SOMEHOW started. It is unknown when and how the first Life appeared on the Earth. In section here, it is tried to describe when the Life started from the geological records and ages. The earliest record of the Life may be microfossils, but they should have experienced intense metamorphism, which would have obliterated any fragile microfossils contained therein. Schopf (1993) thought the most promising area for searching the microfossils was the Pilbara Block of northwestern Australia. The region is underlain by a 30 km thick sequence of relatively well-preserved sedimentary and volcanic rocks of 3000 ~ 3500 Ma.

He detected a diverse assemblage of filamentous microbial fossils in the Early Archean (~3465 Ma by the zircon dating). He believed to have discovered eleven taxa (including eight undescribed species) of cellularly preserved filamentous microbes, in a bedded chert unit of the Early Archean Apex Basalt of northwestern Western Australia. This prokaryotic assemblage establishes that cyanobacterium-like microorganisms were extant and morphologically diverse at least as early as ~3465 Ma and suggests that O_2-producing photoautotrophs may have already evolved by this early stage in biotic history (see Fig. 4.8), which are some sketches of microfossils reported by Schopf (1993, 1994). However, the validity of these microfossils is challenged later.

4.3.3 Questioning the evidence for the Earth's oldest fossils in Apex cherts

Structures resembling remarkably preserved bacterial and cyanobacterial microfossils from 3465 Ma Apex cherts of the Warrawoona Group in Western Australia (Fig. 4.8) provided the oldest morphological evidence for the Life on the Earth and had been taken to support an early beginning for oxygen producing photosynthesis.

However, Brasier et al. (2002) newly researched re-collected materials, involving optical and electron microscopy, digital image analysis, micro-Raman spectroscopy, and other geochemical techniques. They reinterpreted the observed microfossil-like structure as secondary artifacts formed from amorphous graphite within multiple generations of metalliferous hydrothermal vein chert and volcanic glass. Although there is no support for primary biological morphology, a Fischer-Tropsch-type synthesis of carbon compounds and carbon isotopic fractionation (see Section 1.5.15) were inferred for one of the oldest known hydrothermal systems on the Earth.

4.3.4 Carbon isotopic fractionation

Carbon has two stable isotopes of ^{12}C and ^{13}C, and one radioactive isotope of ^{14}C (half-life is 5730 years). As the half-life of ^{14}C is

(A)

(B)

(C)

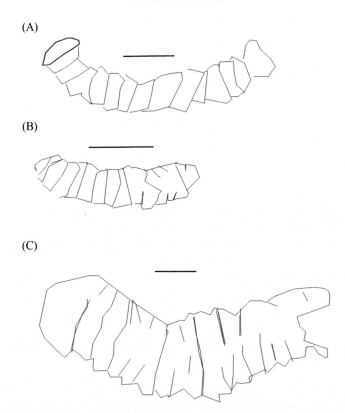

FIGURE 4.8   **A carbonaceous microfossils in thin section of the Early Archean Apex chert of Western Australia.** (A) *Primaevifilum delicatulum*, Schopf (1993). The most popular microbial taxum in the Apex chert. (B) *Primaevifilum conicoteminatum*, Schopf (1993). (C) *Archaeoscillatoriopsis maxima*, n.gen., n. sp. (holotype). Each bar is a scale bar of 10μm. *Source: Modified from the drawings of Schopf (1993).*

appropriately short, ^{14}C is used for dating up to $\sim$30,000 years. The carbon isotopes of ^{12}C and ^{13}C are used for the fractionation of carbon as:

$$\delta^{13}C[\permil] = \left[\left(^{13}C/^{12}C\right)_{sample} / \left(^{13}C/^{12}C\right)_{standard} - 1 \right] \times 1000 \qquad (4.1)$$

Usually, the Pee Dee Belemnite (PDB) is used as the carbon isotope standard, which is based on a Cretaceous marine fossil of the Pee Dee Formation in South Carolina. All $\delta^{13}C$ data without notices are $\delta^{13}C_{PDB}$ in this book.

The carbon isotope fractionation is useful because the carbon isotopic fractionation of the organic carbon and that of the inorganic carbon are significantly different. As shown in Fig. 4.9, the organic carbon is light ($\delta^{13}C_{PDB}$ is $-35 \sim -15\permil$; e.g., Schopf and Kudryavtsev, 2014), while the

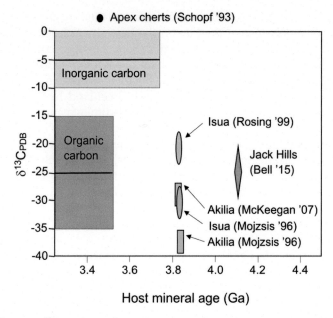

FIGURE 4.9   δ ^{13}C for carbon samples measured by SIMS vs. host mineral age compared with inorganic and organic carbon. Organic and inorganic carbon values from Schopf and Kudryavtsev (2014) and Hoefs (1973), respectively.

inorganic carbon is heavy ($\delta^{13}C_{PDB}$ is $-10 \sim 0‰$; e.g., Hoefs, 1973). Therefore, the carbon isotopic ratio is a good indicator to discriminate carbon in a fossil-like object to be derived from the organic or inorganic origins. Furthermore, carbon isotope ratios of mm-scale microfossils can be determined when high-resolution secondary ion mass spectrometry (HR-SIMS) is applied.

4.3.5   The evidences of the Life older than 3800 Ma at the Isua supracrustal belt and the Akilia island, West Greenland

Mojzsis et al. (1996) reported the carbon isotopic fractionation of carbonaceous inclusions within apatite (calcium phosphate) grains by HR-SIMS from the oldest known sediment sequences of ~ 3800 Ma banded iron formation from the Isua Supracrustal Belt (ISB), West Greenland, and from a similar formation from the nearby Akilia island that is possibly older than 3850 Ma.

The carbon in carbonaceous inclusions was isotopically lighter (see Fig. 4.9), indicative of biological activity, and no known abiotic process can explain the data. Unless some unknown abiotic process exists, which is able both to create such isotopically light carbon and then selectively incorporate it into apatite grains, their results provide the evidence for emergence of the Life on the Earth by at least 3800 Ma.

Ueno et al. (2002) also measured carbon isotopic compositions of graphite in seven metasediments, which are altered sediments such as mudstone, sandstone, and chert, etc., by metamorphism, and two carbonate rocks from the $\sim$3.8 Ga ISB, West Greenland by HR-SIMS. The $\delta^{13}C$ values of graphite globules in the metasediments and the carbonate rocks were $-18 \sim -2‰$ and $-7 \sim -3‰$, respectively. The highest $\delta^{13}C$ value of graphite globules in the metasediment rose from -14 to $-5‰$, as the metamorphic grade increased. Thus, they concluded that the existing light carbon is modified by metamorphism and decomposition of carbonate or exchanged with ^{13}C-enriched reservoir. Therefore, the light carbon found by Mojzsis et al. (1996) was probably pre-metamorphic origin. This supported the existence of the Life at Isua time of $\sim$3.8 Ga.

McKeegan et al. (2007) constructed three-dimensional molecular-structural images of apatite grains and associated minerals embedded in a banded quartz-pyroxene-magnetite from supracrustal rocks from Akilia, south western Greenland by using confocal Raman spectroscopy (see Section 1.5.16). The sample was the same rock used by Mojzsis et al. (1996), which contained isotopically light graphite inclusions in apatite. The graphite inclusions were perfectly contained within apatite without cracks. The carbon isotopic composition of such inclusion was isotopically light ($\delta^{13}C = -29 \pm 4‰$), which was determined by HR-SIMS, in agreement with earlier analyses. This result is consistent with the hypothesis that graphite contained in apatite grains of the >3830 Ma Akilia metasediments are chemo-fossils of the early Life. Although the original form of the Life is perfectly lost, the constituent carbon remained in an apatite grain, being transformed into graphite. Thus, it can be called as a chemo-fossil.

4.3.6 Objection to the earliest Life of the Akilia Island

Hayes (1996) critically commented the difficulty of the HR-SIMS analysis and why graphite inclusions in apatites were used in "News and Views" of the same issue of Nature. If the data are obtained as the whole-rock analysis, the data might be acceptable.

The $\delta^{13}C$ values of the Isua graphite have been interpreted by previous investigators (e.g., Schidlowski and Aharon, 1992; Mojzsis et al., 1996) as metamorphic modification of organic matter with initial $\delta^{13}C$ values of $\sim -30‰$. However, Naraoka et al. (1996) proposed a two-component mixing model in which a major component had $\delta^{13}C$ values of $\sim -12‰$ and a minor component of $\sim -25‰$. The minor component might be biogenic, but it was not certain whether they were Archean biological products or more recent ones.

Their thermodynamic consideration of the temperature and composition of metamorphic fluids and carbon isotope calculations suggested that the major component of graphite with $\delta^{13}C$ values around $-12\permil$ was form by an inorganic process. The graphite appears at temperatures between 700°C and 400°C by one or both of the following two processes:

$$CO_2 + CH_4 \rightarrow 2C + 2H_2O \qquad (4.2)$$

during cooling of fluids with a CO_2/CH_4 molar ratio of ~ 1.

$$4FeO + CO_2 \rightarrow 2Fe_2O_3 + C \qquad (4.3)$$

where FeO comes from olivine in ultramafic rocks.

The $\delta^{13}C_{Total\text{-}carbon}$ values of the fluids for both Eq. (4.2) and Eq. (4.3) were between -12 and $-5\permil$, suggesting that the carbon-bearing fluids could be derived from the mantle.

Similarly, earlier studies (e.g., Schidlowski and Aharon, 1992; Mojzsis et al., 1996) on isotopic characteristics of graphite occurring in rocks of the approximately 3.8 Ga ISB in southern West Greenland were criticized by van Zuilen et al. (2002). They showed that graphite occurred abundantly in secondary carbonate veins in the ISB that were formed at depth in the crust by injection of hot fluids reacting with older crustal rocks. During these reactions, graphite formed from Fe(II)-bearing carbonates at high temperature. These metasomatic rocks, which had no biological relevance, were earlier thought to be of sedimentary origin and this graphite gave the basis for the early Life.

Mojzsis et al. (1996) reported the carbon isotope ratios in grains of apatite. Sano et al. (1999) directly measured the age of the apatite using U–Pb and Pb-Pb dating by HR-SIMS. They obtained 1504 ± 336 and 1459 ± 160 Ma, respectively. This threw doubt on the conclusion of Mojzsis et al. (1996). Mojzsis et al. (1999) replied that by the metamorphic events the carbonaceous inclusions were not affected but the host apatites were.

Mojzsis et al. (1996) interpreted a quartz-pyroxene rock as a banded iron formation (BIF) from the island of Akilia, southwest Greenland, containing ^{13}C-depleted graphite that had been an evidence for the oldest (3850 My ago) Life on the Earth (Section 4.3.5). However, Fedo and Whitehouse (2002) observed that field relationships on Akilia recorded multiple intense deformation events that had resulted in parallel transposition of Early Archean rocks, the tails of which commonly formed the banding in the quartz-pyroxene rock. Geochemical data possessed distinct characteristics consistent with an ultramafic igneous, not BIF, protolith for this lithology and the adjacent schists. Later metasomatic silica and iron introduction have merely resulted in the rock that superficially resembled a BIF. Thus, Fedo and Whitehouse (2002) concluded that an ultramafic igneous origin did not support the claims of Mojzsis et al. (1996) that the carbon isotopic composition of graphite inclusions represented evidence for the Life at the time of crystallization.

4.3.7 ^{13}C-depleted carbon microparticles in >3700 Ma sea-floor sedimentary rocks from West Greenland

Rosing (1999) reported that turbiditic and pelagic sedimentary rocks (see Section 1.4.9) from ISB in western Greenland (3779 ± 81 Ma by the ^{147}Sm−^{143}Nd isochron of the sediment and amphibolite samples) might contain reduced carbon of biogenic origin. The carbon was in 2−5 mm graphite globules and had an isotopic composition of δ^{13}C about −19%. These data and the occurrence indicated that the graphite represented biogenic detritus, which was perhaps derived from planktonic organisms.

The largest difference from the study of Mojzsis et al. (1996) is that Rosing (1999) chose samples with well-preserved sedimentary structures, which showed clear depositional contact with the basaltic Garbenschiefer Formation (they became amphibolites by recent metamorphism), with occasionally pillow structures The sedimentary rocks can be traced ∼100 m. In the least deformed samples, mica and chlorite defined a schistosity parallel to the sedimentary bedding. The black slaty units with thicknesses of <10 cm containing graphite are finely laminated by dark layers (greywacke) also containing graphite.

In more deformed rocks, all sedimentary structures have been obliterated, and up to 1 cm of biotite and garnet porphyroblasts (round crystals of garnet) are present. In such recrystallized samples, graphite is absent. These relations indicate that the graphite globules formed before the earliest growth of garnet and biotite disappeared during progressive metamorphism.

Sedimentological and geochemical evidences indicate that the carbon forming the graphite globules is of a biogenic origin. In analogy to modern oceanic pelagic shales, the precursor organic detritus of the graphite globules could have been derived from planktonic organisms that sedimented from surface waters. Thus, the organism should be photoautotrophic.

4.3.8 Geological evidence of recycling of altered crust in Hadean

By SIMS, Cavosie et al. (2005) measured δ^{18}O in 4400−3900 Ma igneous zircons from the Jack Hills, Western Australia, which gave a record of the oxygen isotope composition of magmas in the earliest Archean. The main finding of Cavosie et al. (2005) is that evidence for the recycling of altered crust was preserved in igneous zircons from Jack Hills with magmatic $\delta^{18}O$ values as high as 6.5‰ by 4325 Ma, and up to 7.3‰ by 4200 Ma. The $\delta^{18}O$ values in these zircons were indistinguishable from the range of igneous zircons found throughout the Archean, and demonstrate from an oxygen isotope perspective that no fundamental, planetary-scale change in magmatic process was recorded in zircon from at least 4325 to 2500 Ma.

The detrital magmatic zircons possibly represent many igneous rocks, and are clearly not from the same magmatic event. The magmatic $\delta^{18}O$ values obtained from micro-volumes of zircon are interpreted to preserve igneous chemistry, including U–Pb concordance, Th/U chemistry, and internal zoning. Although some grains have internal zoning patterns that are ambiguous, both the lower (5.3‰) and upper (7.3‰) limits of magmatic $\delta^{18}O$ values for the zircons in the study are defined by multiple grains, which are average of many growth zoning patterns. The range of magmatic $\delta^{18}O$ of zircons observed is within the range previously reported for detrital zircons of similar age from Jack Hills.

Valley et al. (2002) proposed the hypothesis that the early Earth was cool. No known rocks have survived from the first 500 My of the Earth's history, but studies of single zircons suggest that some continental crust formed as early as 4.4 Ga, 160 My after accretion of the Earth, and that surface temperatures were low enough for liquid water. Surface temperatures are inferred from high $\delta^{18}O$ values of zircons. The range of $\delta^{18}O$ values is constant throughout the Archean (4.4–2.6 Ga), suggesting uniformity of processes and conditions. The hypothesis of a cool early Earth suggests long intervals of relatively temperate surface conditions from 4.4 to 4.0 Ga that were conducive to liquid water oceans and possibly Life.

4.3.9 Back to the Isua Supracrustal Belt, Western Greenland

Ohmoto et al. (2014) analyzed the geochemistry and structure of the ^{13}C-depleted graphite in ISB, Western Greenland. Raman spectroscopy and geochemical analyses indicated that the schists were formed from clastic marine sediments that contained ^{13}C-depleted carbon at the time of their deposition. Transmission electron microscope (TEM) observations showed that the graphite in the schist formed nanoscale polygonal and tube-like grains. In contrast, abiotic graphite in carbonate veins exhibits a flaky morphology. The graphite grains in the schist contained distorted crystal structures and disordered stacking of sheets of graphene. The observed morphologies were consistent with pyrolyzed and pressurized organic compounds during metamorphism. Therefore, they concluded that the graphite contained in the Isua metasediments represents traces of early Life that flourished in the oceans at least 3.7 Ga.

Fedo and Whitehouse (2002) wondered what evidence could be used. The Apex chert was denied by Brasier et al. (2002). The Akilia island chert was denied by themselves. The graphite particles in ISB in Western Greenland in 3700–3800 Ma of Rosing (1999) might be negatively the oldest record of the Life, because it was not denied yet.

4.3.10 Origin of Life back to 4.1 billion-year-old

Some of the detrital zircons from Jack Hills, Western Australia have ages up to nearly 4.4 Ga. Bell et al. (2015), from a population of over 10,000 Jack Hills zircons, identified one >3.8-Ga zircon that contains primary graphite inclusions. These inclusions were judged as primary, because (1) their enclosure in a crack-free host was shown by transmission X-ray microscopy using the Stanford Synchrotron Radiation Lightsource (SSRL); and (2) the graphite crystal habit. Then they measured carbon isotopic ratios in inclusions in a concordant, 4.10 ± 0.01-Ga zircon using HR-SIMS. They obtained $\delta^{13}C_{PDB}$ of $-24 \pm 5‰$, which is consistent with a biogenic carbon and may be evidence that the Life had appeared by 4.1 Ga, or ~300 My earlier than has been previously proposed.

4.3.11 Stromatolites: convincing evidence of the oldest Life at 3480 Ma in the Pilbara Craton

The most convincing oldest Life so far is the 3480 Ma Dresser Formation of the Pilbara Craton, by multidisciplinary methods in Australia (Walter et al., 1980; van Kranendonk et al., 2008). This area was studied based on stratigraphic and petrological data of surface outcrops and diamond drillcore through the lowermost, c. 3.5 Ga, chert–barite horizon. Dresser Formation was found to be deposition within a volcanic caldera of hydrothermal processes. Intense hydrothermal activity and felsic volcanism were found. Stromatolites at shallow water were found with carbonate deposition, followed by the main caldera formation and hydrothermal fluid circulation, marked by coarse conglomerates and soft-sediment deformation. The final caldera collapsed, and sandstones and carbonates deposited.

Signs of the Life in the Dresser Formation included stromatolites of diverse morphology (mat-like, columnar, domic, and coniform), clasts of laminated carbonaceous material in bedded micritic carbonate (eroded microbial mats), and diagenetic crystals of pyrite and carbonate in bedded micritic carbonate. Stromatolites consisted of pyrite of hydrothermal origin, which was considered to be a replacement of carbonate. Stromatolitic structures were not correlated with crystal growth forms, thus, became good morphological indicator of the early Life on the Earth.

4.3.12 Stromatolites in the Isua Supracrustal Belt are the oldest Life in 3700 Ma

Nutman et al. (2016) reported 3700 Ma rocks of the Isua supracrustal belt in Greenland with low deformation and a closed metamorphic system, where primary sedimentary features including putative conical and domical stromatolites were preserved. The morphology, layering, mineralogy, chemistry, and geological context of the structures were attributed to the formation of microbial mats in a shallow marine environment by 3700 Ma, at the start of Earth's rock record. They reported this is the evidence for ancient Life from a newly exposed outcrop of 3700 Ma metacarbonate rocks in the ISB, which contained $1 \sim 4$ cm high stromatolites.

The ISB stromatolites grew in a shallow marine environment, as indicated by seawater-like REE + Y patterns of the metacarbonates, and by interlayered detrital sedimentary rocks with cross-lamination and storm-wave generated breccias. The ISB stromatolites were younger by 220 Myr than the generally accepted oldest Life of Dresser Formation of the Pilbara Craton, Australia in 3480 Ma.

The presence of the ISB stromatolites demonstrated the shallow marine carbonate production with biotic CO_2 started at 3700 Ma, which was almost simultaneous with the start of Earth's sedimentary record.

4.3.13 Counterarguments to the Isua Supracrustal Belt structures as stromatolites

Allwood et al. (2018) argued that such sedimentary structures are of a non-biological and post-depositional origin. They three-dimensionally analyzed the morphology and orientation of the structures within the context of host rock fabrics, combining texture-specific major and trace element chemistry.

They used PIXL (planetary instrument for X-ray lithochemistry) micro-X-ray fluorescence maps of elemental composition and found that distribution maps of Ca, Fe, and Mn showing "stromatolitic" structures were actually a dolomitic alteration rim on a quartzose interior and that there was no other compositional relict of internal lamination in the structures. Furthermore, Ti and K were depleted in the "stromatolites" as well as quartzose layers because of less K- and Ti-bearing mica compared to the dark layers. In addition, from Fe and Si maps the Fe-rich/Si-poor layers were dark, while Si-rich/Fe-poor layers light, indicating that the rock may have originally consisted of intercalated cherty and iron-rich layers. Therefore, this observation shows carbonate-altered banded iron formation and cherty metasediments.

They also separated carbonate and silicate (quartz and mica) fractions and measured REE + Y patterns. Both fractions showed Archean ~ Paleoproterozoic seawater properties. Therefore, the carbonate REE + Y patterns were inherited from diagenetic and/or metasomatic fluids.

In summary, these results showed that the "stromatolites" are more plausibly an assemblage of deformation structures formed in carbonate-altered metasediments long after burial. The investigation of the structures of the Isua supracrustal belt highlights the importance of three-dimensional, integrated analysis of morphology, rock fabrics, and geochemistry at appropriate scales.

Zawaski et al. (2020) newly mapped the discovered sites by Nutman et al. (2016) at the appropriate scale and higher resolution. Their new map was used to investigate micro- and macro-structures and to collect comprehensive geochemical samples. They analyzed structures in detail to show linearly inverted ridges aligned with azimuths of local and regional fold axes and parallel to linear structures. Combined major element (e.g., Ca, Mg, Si) scanning μXRF maps, and electron backscatter diffraction (EBSD) patterns collected on fresh surfaces without any residual sedimentary laminae within these structures' cores. Internal layering inferred for the "stromatolites" arises from variable weathering of outcrop surfaces that conceals granoblastic quartz ± dolomite cored boudins that sit between semi-continuous competent layers of enveloping quartzite in a calc-silicate schist. The morphology of boudins reflects viscosity contrasts of the different ductile layers during deformation. Therefore, these features are not of sedimentary origin.

Collectively, Zawaski et al. (2020) showed that the ISB is the expected result of a tectonic fabric that preserves no fine-scale primary sedimentary structures and were probably never stromatolites.

4.3.14 Reply of Nutman et al. (2021) to the counterarguments

Nutman et al. (2021) claim to the counterarguments that if the stromatolites are formed by boudinage, the stromatolite structure should appear in two directions in 2D (e.g., up and down). However, the stromatolite shapes always appear like the chevron (shows only one direction). In addition, the lack of layering could occur because the terranes

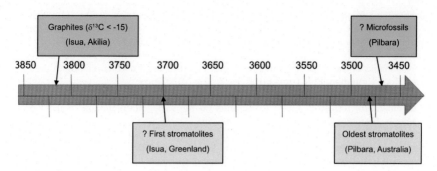

FIGURE 4.10 **The time scale of the early Lives on the Earth.** The time scale from 3850 to 3450 Ma, when the early Lives on the Earth appeared.

are highly deformed, and the recrystallization during diagenesis and metamorphism can easily erase fine-scale layering.

It is very difficult to judge which group is correct until new analytical methods are developed to find the biological evidences. In Fig. 4.10, the microfossils, graphites, the oldest stromatolites, and the possible first stromatolites are plotted on the timescale.

References

Allegre, C.J., Mahnes, G., Goppel, C., 2008. The major differentiation of the Earth at ~ 4.45Ga. Earth Planet Sci Lett 267, 386–398.

Allwood, A.C., Rosing, M.T., Flannery, D.T., Hurowitz, J.A., Heirwegh, C.M., 2018. Reassessing evidence of life in 3,700-million-year old rocks of Greenland. Nature 563, 241–244.

Amelin, Y., Krot, A.M., Hutcheon, I.D., Ulyanov, A.A., 2002. Lead isotopic ages of chondrules and calcium-aluminum-rich inclusions. Science 297, 1678–1683.

Amelin, Y., Wadhwa, M., Lugmair, G.W. Pb-isotopic dating of meteorites using 202Pb-205Pb double spike: comparison with other high-resolution chronometers. Lunar Planet Sci Conf XXXVII #1970, 2006.

Anders, E., Grevesse, N., 1989. Abundances of the elements meteoritic and solar. Geochim Cosmochim Acta 53, 197–214.

Archer, G.J., Brennecka, G.A., Gleissner, P., et al., 2019. Lack of late-accreted material as the origin of 182W excesses in the Archean mantle: Evidence from the Pilbara Craton, Western Australia. Earth Planet Sci Lett 528, 115841.

Becker, H., Horan, M.F., Walker, R.J., Gao, S., Lorand, J.-P., Rudnick, R.L., 2006. Highly siderophile element composition of the Earth's primitive upper mantle constraints from new data on peridotite massifs and xenoliths. Geochim Cosmochim Acta 70, 4528–4550.

Bell, E.A., Boehnke, P., Harrison, T.M., Mao, W.L., 2015. Potentially biogenic carbon preserved in a 4.1 billion year-old zircon. Proc Natl Acad Sci USA 112, 14518–14521.

Bottke, W.F., Vokrouhlichy, D., Marchi, S., et al., 2015. Dating the Moon-forming impact event with asteroidal meteorites. Science 348, 321–323.

Brasier, M.D., Green, O.R., Jephcoat, A.P., et al., 2002. Questioning the evidence for Earth's oldest fossils. Nature 416, 76–81.

Brenan, J.M., Mungall, J.E., Bennett, N.R., 2019. Abundance of highly siderophile elements in lunar basalts controlled by iron sulfide melt. Nature Geosci 12 (9), 701–706.

Burkhardt, C., Kleine, T., Palme, H., Bourdon, B., Zipfel, J., Friedrich, J., Ebel, D., 2008. Hf-W mineral isochron for Ca, Al-rich inclusions: age pf the solar system and the timing of core formation in planetesimals. Geochim Cosmochim Acta 72, 6177−6197.

Carlson, R.W., Borg, L.E., Gaffney, A.M., Boyet, M., 2014. Rb-Sr, Sm-Nd and Lu-Hf isotope systematics of the lunar Mg-suite: the age of the lunar crust and its relation to the time of Moon formation. Phil Trans Royal Soc A 372, 20130246.

Caro, G., Bourdon, B., Birck, J.-L., Moorbath, S., 2006. High-precision ^{142}Nd/^{144}Nd measurements in terrestrial rocks: Constraints on the early differentiation of the Earth's Mantle. Geochim Cosmochim Acta 70, 164−191.

Cates, N.L., Mojzsis, S.J., 2007. Pre-3750 Ma supracrustal rocks from the Nuvvuagittuq supracrustal belt, northern Quebec. Earth Planet Sci Lett 255, 9−21.

Cavosie, A.J., Valley, J.W., Wilde, S.A., 2005. Magmatic delta O-18 in 4400-3900 Ma detrital zircons: A record of the alteration and recycling of crust in the Early Archean. Earth Planet Sci Lett 235, 663−681.

Clayton, R.N., Onuma, N., Mayeda, T.K., 1976. A classification of meteorites based on oxygen isotopes. Earth Planet. Sci. Lett 30, 10−18.

Compston, W., Pidgeon, R.T., 1986. Jack Hills, evidence of more very old detrital zircons in Western Australia. Nature 321, 766−769.

Cottrell, E., Walker, R.J., 2006. Constraints of core formation from Pt partitioning in mafic silicate liquids at high temperature. Geochim Cosmochim Acta 70, 1565−1580.

Culler, T.S., Becker, T.A., Muller, R.A., Renne, P.R., 2000. Lunar impact history from 40Ar/39Ar dating of glass spherules. Science 287, 1785−1788.

Danielson, L.R., Sharp, T.G., Hervig, R.L. Implications for core formation of the Earth from high pressure temperatureAu partitioning experiments. Lunar and Planetary Science Conference XXXVI, Abstract #1955 (CD-ROM), Houston, 2005, Lunar and Planetary Institute.

David, J., Godin, L., Stevenson, R., O'Neil, J., Francis, D., 2009. U−Pb ages (3.8−2.7 Ga) and Nd isotope data from the newly identified Eoarchean Nuvvuagittuq supracrustal belt, Superior Craton, Canada. Geol Soc Amer Bull 121, 150−163.

Day, J.M.D., Pearson, D.G., Taylor, L.A., 2007. Highly siderophile element constraints on accretion and differentiation of the Earth-Moon system. Science 315, 217−219.

De Niem, D., Kuehrt, E., Morbidelli, A., Motschmann, U., 2012. Atmospheric erosion and replenishment induced by impacts upon the Earth and Mars during a heavy bombardment. Icarus 221, 495−507.

Fedo, C.M., Whitehouse, M.J., 2002. Metasomatic origin of quartz-pyroxene rock, Akilia, Greenland, and implications for Earth's earliest life. Science 296, 1448−1452.

Ferus, M., Nesvorný, D., Sponer, J., et al., 2015. High-energy chemistry of formamide: A unified mechanism of nucleobase formation. PNAS 112, 657−662.

Furukawa, Y., Nakazawa, H., Sekine, T., Kobayashi, T., Kakegawa, T., 2015. Nucleobase and amino acid formation through impacts of meteorites on the early ocean. Earth Planet Sci Lett 429, 216−222.

Geiss, J., Rossi, A.P., 2013. On the chronology of lunar origin and evolution: Implications for Earth, Mars and the Solar System as a whole. Astron Astrophys Rev 21 (1), 54.

Gomes, R., Levison, H.F., Tsiganis, K., Morbidelli, A., 2005. Origin of the cataclysmic Late Heavy Bombardment period of the terrestrial planets. Nature 435, 466−469.

Halliday, A.N., 2008. A young Moon-forming giant impact at 70-110 million years accompanied by late-stage mixing, core formation and degassing of the Earth. Philos Trans R Soc A 366, 4163−4181.

Hartmann, W.K., Ryder, G., Dones, L., 2000. Grinspoon: The time-dependent intense bombardment of the primordial Earth/Moon system. 493−512. In: Canup, R.M., Righter, K. (Eds.), Origin of the Earth and Moon. Tucson. The University of Arizona Press.

Hayes, J.M., 1996. The earlist memories of life on Earth. Nature 384, 21−22.

Herwartz, D., Pack, A., Friedrichs, B., Bischoff, A., 2014. Identification of the giant impactor Theia in lunar rocks. Science 344, 1146–1150.

Hoefs, J., 1973. Stable Isotope Geochemistry. Springer, Berlin.

Holden, P., Lanc, P., Ireland, T.R., Harrison, T.M., Foster, J.J., Bruce, Z., 2009. Mass-spectrometric mining of Hadean zircons by automated SHRIMP multi-collector and single-collector U/Pb zircon age dating: the first 100,000 grains. Int J Mass Spec 286, 53–63.

Holland, H.D., 1984. The Chemical Evolution of the Atmosphere and Oceans. Princeton. Princeton University Press.

Horan, M.F., Walker, R.J., Morgan, J.W., Grossman, J.N., Rubin, A., 2003. Highly siderophile elements in chondrites. Chem Geol 196, 5–20.

Jacobsen, S.B., 2005. The Hf-W isotopic system and the origin of the Earth and Moon. Annu Rev Earth Planet Sci 33, 531–570.

Kasting, J.F., 1993. Earth's early atmosphere. Science 259, 920–926.

Kleine, T., Palme, H., Metzger, K., Halliday, A.N., 2005. Hf-W chronometry of lunar metals and the age and early differentiation of the Moon. Science 320 1671–1674.

Kleine, T., Touboul, M., Bourdon, B., et al., 2009. Hf-W chronology of the accretion and early evolution of asteroids and terrestrial planets. Geochim Cosmochim Acta 73, 5150–5188.

Koeberl, C., Reimold, W.U., McDonald, I., Rosing, M., 2000. Impacts and the early Earth. In: Gilmour, I., Koeberl, C. (Eds.), Lecture Notes in Earth Sciences. Springer, Berlin, pp. 73–97.

Koeberl, C., 2006. Impact processes on the early Earth. Elements 2, 211–216.

Kruijer, T.S., Kleine, T., 2017. Tungsten isotopes and the origin of the Moon. Earth Planet Sci Lett 475, 15–24.

Kruijer, T.S., Kleine, T., Fischer-Gödde, M., Sprung, P., 2015. Lunar tungsten isotopic evidence for the late veneer. Nature 520, 534–537.

Kruijer, T.S., Archer, G.J., Klene, T., 2021. No [182]W evidence for early Moon formation. Nat Geosci 14, 714–715.

Leya, I., Wieler, R., Halliday, A.N., 2000. Cosmic-ray production at tungsten isotopes in lunar samples and meteorites and its implications for Hf-W astrochemistry. Earth Planet Sci Lett 175, 1–12.

McDonough, W.F., 1995. Sun S-s: The composition of the Earth. Chem Geol 120, 223–253.

McKeegan, K.D., Kudryavtsev, A.B., Schopf, J.W., 2007. Raman and ion microscopic imagery of graphitic inclusions in apatite from older than 3830 Ma Akilia supracrustal rocks, West Greenland. Geology 35, 591–594.

Mojzsis, S.J., Arrhenius, G., McKeegan, K.D., Harrison, T.M., Nutman, A.P., Friend, C.R.L., 1996. Evidence for life on Earth before 3,800 million years ago. Nature 384, 55–59.

Mojzsis, S.J., Harrison, T.M., Arrhenius, G., McKeegan, K.D., Grove, M., 1999. Reply-Origin of life from apatite dating? Nature 400, 127–128.

Mojzsis, S.J., Harrison, T.M., Pidgeon, R.T., 2001. Oxygen-isotope evidence from ancient zircons for liquid water at the Earth's surface 4300 Myr ago. Nature 409, 178–180.

Morbidelli, A., Chambers, J., Lunine, J.I., et al., 2000. Source regions and timescales for the delivery of water to the Earth. Meteorit Planet Sci 35, 1309–1320.

Morbidelli, A., Marchi, S., Bottke, W.F., Kring, D.A., 2012. A sawtooth-like timeline for the first billion years of lunar bombardment. Earth Planet Sci Lett 355, 144–151.

Morbidelli, A., Nesvorny, D., Laurenz, V., Marchi, S., et al., 2018. The timeline of the lunar bombardment: Revisited. Icarus 305, 262–276.

Mukhopadhyay, S., 2012. Early differentiation and volatile accretion in deep mantle neon and xenon. Nature 486, 101–110.

Naraoka, H., Ohtake, M., Maruyama, S., Ohmoto, H., 1996. Non-biogenic graphite in 3.8-Ga metamorphic rocks from the Isua district, Greenland. Chem Geol 133, 251–260.

Nesvorny, D., Morbidelli, A., 2012. Statistical study of the early solar system's instability with four, five, and six giant planets. Astron J 144, 20–68.

Newton, J., Franchi, I.A., Pillinger, C.T., 2000. The oxygen-isotopic record in enstatite chondrite. Meteorit Planet Sci 35, 689–698.

Nutman, A.P., Bennett, V.C., Friend, C.R.L., Van Kranendonk, M.J., Chivas, A.R., 2016. Rapid emergence of life shown by discovery of 3,700-million-year-old microbial structures. Nature 537, 535–538.

Nutman, A.P., Bennett, V.C., Friend, C.R.L., Van Kranendonk, M.J., 2021. In support of rare relict ~3700 Ma stromatolites from Isua (Greenland). Earth Planet Sci Lett 562, 116850.

O'Neil, J., Carlson, R.W., Paquette, J.-L., Francis, D., 2012. Formation age and metamorphic history of the Nuvvuagittuq greenstone belt. Precambrian Res 23–44, 220–221.

Ohmoto, H., Watanabe, Y., Lasaga, A.C., Naraoka, H., Johnson, I., et al., 2014. Oxygen, iron, and sulfur geochemical cycles on early Earth: Paradigms and contradictions. Earth's Early Atmosphere and Surface Environment 504, 55–95.

Patterson, C., 1956. Age of meteorites and the Earth. Geochim. Cosmochim. Acta 10, 230.

Pattou, L., Lorand, J.P., Gros, M., 1996. Non-chondritic platinum-group element ratios in the Earth's mantle. Nature 379, 712–715.

Pepin, R.O., Porcelli, D., 2006. Xenon isotope systematics, giant impacts, and mantle degassing on the early Earth. Earth Planet Sci Lett 250, 470–485.

Piani, L., Marrocchi, Y., Rigaudier, T., Vacher, L.G., Thomassin, D., et al., 2020. Earth's water may have been inherited from material similar to enstatite chondrite meteorites. Science 369, 1110–1113.

Righter, K., Drake, M.J., 1997. Metal-silicate equilibrium in a homogeneously accreting Earth: new results for Re. Earth Planet Sci Lett 146, 541–553.

Righter, K., Huayun, M., Danielson, L., 2008. Partitioning of palladium at high pressures and temperatures during core formation. Nat. Geosci 1, 321–323.

Rizo, H., Boyet, M., Blichert-Toft, J., Rosing, M., 2011. Combined Nd and Hf isotope evidence for deep-seated source of Isua lavas. Earth Planet Sci Lett 312, 267–279.

Rosing, M.T., 1999. ^{13}C-depleted carbon microparticles in > 3700-Ma sea-floor sedimentary rocks from west Greenland. Science 283, 674–676.

Rudge, J.F., Kleine, T., Bourdon, B., 2010. Broad bounds on Earth's accretion and core formation constrained by geochemical models. Nat Geosci 3, 439–443.

Sano, Y., Terada, K., Takahashi, Y., Nutman, A.P., 1999. Origin of life from apatite dating? Nature 400, 127.

Schidlowski, M., Aharon, P., 1992. Carbon cycle and carbon isotope record: geochemical impact of life over 3.8 Ga of earth history. In: Schidolowski, M., Kimberley, S., Golubic, M.M., McKirdy, D.M., Trudinger, P.A. (Eds.), Early Organic Evolution: Implications for Mineral and Energy Resources. Springer, Berlin, pp. 147–175.

Schlichting, H.E., Warren, P.H., Yin, Q.-Z., 2012. The last stages of terrestrial planet formation: dynamical friction and the late veneer. Astrophys J 752, 8.

Schopf, J.W., Kudryavtsev, A.B., 2014. Biogenicity of Earth's earliest fossils. Evolution of Archean Crust and Early Life. In: Dilek, Y., Fumes, H. (Eds.), Modern Approaches in Solid Earth Sciences. Dordrecht, Vol. 7. Springer, pp. 333–349.

Schopf, J.W., 1993. Microfossils of the Early Archean Apex Chert: New Evidence of the Antiquity of life. Science 260, 640–646.

Schopf, J.W., 1994. Disparate rates, differing fates: Tempo and mode of evolution changed from the Precambrian to the Phanerozoic. Proc Natl Acad Sci USA 91, 6735–6742.

Staudacher, T., Allegre, C.J., 1982. Terrestrial xenology. Earth Planet Sci Lett 60, 389–406.

Tera, F., Papanastassiou, D.A., Wasserburg, G.J., 1974. Isotopic evidence for a terminal lunar cataclysm. Earth Planet Sci Lett 22, 1–21.

Thiemens, M.M., Sprung, P., Fonseca, R.O.C., Leitzke, F.P., Munker, C., 2019. Early Moon formation inferred from hafnium–tungsten systematics. Nat Geosci 12, 696–700.

Touboul, M., Walker, R.J., 2012. High precision tungsten isotope measurement by thermal ionization mass spectrometry. Int J Mass Spectrom 309, 109–117.

Touboul, M., Kleine, T., Bourdon, B., Palme, H., Wieler, R., 2007. Late formation and pro-longed differentiation of the Moon inferred from W isotopes in Lunar metals. Nature 450, 1206−1209.

Touboul, M., Kleine, T., Bourdon, B., Palme, H., Wieler, R., 2009. Tungsten isotopes in ferroan anorthosites: implications for the age of the Moon and the lifetime of its magma ocean. Icarus 199, 245−249.

Touboul, M., Puchtel, I.S., Walker, R.J., 2015. Tungsten isotopic evidence for dispropor-tional late accretion to the Earth and Moon. Nature 520, 530−533.

Tsiganis, K., Gomes, R., Morbidell, I.A., Levison, H.F., 2005. Origin of the orbital architec-ture of the giant planets of the Solar System. Nature 435, 459−461.

Ueno, Y., Yurimoto, H., Yoshioka, H., Komiya, T., Maruyama, S., 2002. Ion microprobe analysis of graphite from ca. 3.8 Ga metasediments, Isua supracrustal belt, West Greenland: Relationship between metamorphism and carbon isotopic composition. Geochim Cosmochim Acta 66, 1257−1268.

Valley, J.W., Peck, W.H., King, E.M., Wilde, S.A., 2002. A cool early Earth. Geology 30, 351−354.

Van Kranendonk, M.J., Philippot, P., Lepot, K., Bodorkos, S., Pirajno, F., 2008. Precambrian Res 167, 93−124.

Van Zuilen, M., Lepland, A., Arrhenius, G., 2002. Reassessing the evidence for the earliest traces of life. Nature 418, 627−630.

Walker, R.J., Storey, M., Kerr, A., Tarne, J., Arndt, N.T., 1999. Implications of ^{187}Os hetero-geneities in mantle plume: evidence from Gorgona Island and Curacao. Geochim Cosmochim Acta 66 63, 713−728.

Walker, R.J., Horan, M.F., Morgan, J.W., Becker, H., Grossman, J.N., Rubin, A., 2002. Comparative ^{187}Re-^{187}Os systematics of chondrites: implications regarding early solar system processes. Geochim Cosmochim Acta 66, 4187−4201.

Walker, R.J., Horan, M.F., Shearer, C.K., Papike, J.J., 2004. Low abundances of highly side-rophile elements in the lunar mantle: evidence for prolonged late accretion. Earth Planet. Sci. Lett 224, 399−413.

Walker, R.J., Berminghan, K., Liu, J.G., Puchtel, I.S., Touboul, M., Worsham, E.A., 2015. In search of late-stage planetary building blocks. Chem Geol 411, 125−142.

Walker, J.C.G., 1985. Carbon-dioxide on the early earth. Orig Life Evol Biosph 16, 117−127.

Walker, R.J., 2009. Highly siderophile elements in the Earth, Moon and Mars: Update and implications for planetary accretion and differentiation. Chemie der Erde 69, 101−125.

Walter, M.R., Buick, R., Dunlop, S.R., 1980. Stromatolites 3,400−3,500 Myr old from the North Pole area. West Australia. Nature 284, 443−445.

Warren, P.H., Kallemeyn, G.W., Kyte, F.T., 1999. Origin of planetary cores: evidence from highly siderophile elements in martian meteorites. Geochim Cosmochim Acta 63, 2105−2122.

Wilde, S.A., Valley, J.W., Peck, W.H., Graham, C.M., 2001. Evidence from detrital zircons for the existence of continental crust and oceans on the Earth 4.4 Gyr ago. Nature 409, 175−178.

Willbold, M., Elliot, T., Moorbath, S., 2011. The tungsten isotopic composition of the Earth's mantle before the terminal bombardment. Nature 477, 195-U91.

Young, E.D., Kohl, I.E., Warren, P.H., Rubie, D.C., Jacobsen, S.A., Morbidelli, A., 2016. Oxygen isotopic evidence for vigorous mixing during the Moon-forming giant impact. Science 351, 493−496.

Zawaski, M.J., Kelly, N.M., Orlandini, O.F., Nichols, C.I.O., Allwood, A.C., Mojzsis, S.J., 2020. Reappraisal of purported ca. 3.7 Ga stromatolites from the Isua Supracrustal Belt (West Greenland) from detailed chemical and structural analysis. Earth Planet Sci Lett 545, 116409.

Zhu, M.H., Artemieva, N., Morbidelli, A., Yin, Q.Z., et al., 2019. Reconstructing the late-accretion history of the Moon. Nature 571, 226−229.

Origins of life-related molecules on Earth

5.1 Transportation of life-related molecules of extraterrestrial origin on Earth

5.1.1 Introduction

In this section, how to bring life-related molecules (LRMs) onto the Earth is discussed. LRMs are essential molecules for building life. The LRMs are divided into two groups: one group is called CHOs that are the molecules made of carbon (C), hydrogen (H), and oxygen (O), such as aldoses, ketoses, and sugars including ribose; another group is named as CHONs that are made of C, H, O, and N (nitrogen), such as formamide, nucleobases, and amino acids.

There are two main proposals on the origin of LRMs on the Earth. One is the synthesis of LRMs on the Earth, and the other is the extraterrestrial origin of LRMs. The latter is subdivided into: (1) large energy (e.g., ultraviolet light) or impact shocks activated organic syntheses of LRMs on asteroids and comets, and LRMs fell on the Earth without decomposition; and (2) LRMs were synthesized by impact shocks or other high energy such as lightning on the Earth. In this section, the possibility of transportation of LRMs synthesized in space onto the Earth's surface and the syntheses of LRMs on the Earth's surface are discussed. The syntheses of LRMs in the space are discussed in Section 8.6.

5.1.2 Transportation of life-related molecules of space origin onto the Earth

It is widely known that extraterrestrial bodies and materials such as asteroids, comets, carbonaceous chondrites, and interplanetary dust

Introductory Astrochemistry
DOI: https://doi.org/10.1016/B978-0-443-23938-0.00005-4

particles (IDPs) contain a lot of LRMs. Transportation of such LRMs onto the Earth's surface is one of the hypotheses of the origin of LRMs on the Earth.

Major counterarguments to this simple idea are that the organic matters cannot survive the extremely high temperature ($> 10^4$K) reached at impact, which will atomize the meteoritic bodies. Only small particles of $10^{-12} - 10^{-6}$ g, which are gently decelerated by the atmosphere, can deliver LRMs without decomposition onto the Earth. However, we can actually obtain hand specimens of carbonaceous chondrites containing LRMs.

In the comet or asteroid interactions with the atmosphere, surface impact, and resulting organic pyrolysis, LRMs will not survive impacts at velocities greater than ~ 10 km/s and that even comets and asteroids as small as 100 m in radius cannot survive. Anders (1989) estimated the amount of such soft-landed organic carbon to be about 20 g/cm^2 in the few hundred million years during the late heavy bombardment (LHB). It may have included LRMs that were not formed by the abiotic synthesis on the Earth.

However, for plausible dense early atmospheres (CO$_2$ pressure of 1×10^6 Pa), Chyba et al. (1990) calculated that the Earth at 4.5 Gyr ago can accrete intact cometary organics at a rate of at least approximately 10^6–10^7 kg/y, which declined with a half-life of 10^8 y. These results may increase the terrestrial oceanic and total biomasses, to be 3×10^{12} kg and 6×10^{14} kg, respectively.

Chyba and Sagan (1992) proposed that the sources of LRMs on the early Earth are divided into three categories: (1) delivery by extraterrestrial objects (Chapters 5 and 8), (2) organic synthesis by other energy (such as an electrical spark or ultraviolet light) on the Earth like the Miller experiment (Section 5.2.1), and (3) organic synthesis driven by impact shocks on the Earth afterward (Section 5.2.2). Estimates in the early terrestrial atmosphere suggested that quantities of organics either delivered by (1) or produced by (2) were comparable to those produced by (3).

5.1.3 Counterarguments to the space origin of life-related molecules

There are some counterarguments to the space origin of LRMs. First, the amount of LRMs in IDPs is unclear (Glavin et al., 2004; Maurette, 2006). Second, LRMs in carbonaceous chondrites and IDPs which contains LRMs (see Section 8.3) can be easily decomposed by heat in entering on the Earth.

The impact shock origin of LRMs became one of the strong counterarguments. This is the third proposal of Chyba and Sagan (1992) and Anders (1989) that the synthesizing energy is obtained by the impact shocks of meteorites to the Earth. The major difference from the space origin is that the carbons of LRMs were supplied from the terrestrial atmosphere because the partial pressure of CO_2 was high enough in Hadean and Archaean.

In the next section, experiments of synthesis of LRMs on the Earth's origin are presented.

5.2 Synthetic experiments of life-related molecules on the Earth

5.2.1 Classic experiments

5.2.1.1 *The primordial soup theory*

A. I. Opalin (1894−1980) was a notable biochemist in the Soviet Union. He was very much interested in how the life began. He thought the early Earth was reducing conditions containing CH_4, NH_3, H_2, and water vapor. He insisted on the hypothesis that life was developed by the gradual chemical evolution of carbon molecules in the primordial soup on the Earth.

J.B.S. Haldane (1892−1964) was a British-Indian biologist and mathematician who published an article explaining the primordial soup theory in "The Origin of Life" in 1929. He described the primitive ocean as a "vast chemical laboratory" containing inorganic compounds as a "hot dilute soup." The sunlight generated a variety of organic compounds from the oxygenless atmosphere with CO_2, NH_3, and water vapor. First, the molecules reacted with each other and produced more complex compounds of LRMs, and finally the cellular components are produced resulting in Life.

In 1924 Oparin presented a similar idea in Russia, and in 1936 he introduced it to the English-speaking people (Oparin, 1938).

5.2.1.2 *The Miller experiment (or the Miller−Urey experiment)*

To prove the Oparin-Haldane hypothesis, S. Miller conducted the famous Miller experiment (or Miller−Urey experiment) under the supervision of H. Urey in 1952 (Miller, 1953). In Fig. 5.1, the apparatus is indicated. After a week, the water in the flask became deep red.

The paper chromatogram showed glycine, α-alanine, β-alanine, aspartic acid, and unidentified organic compounds. Although synthesized amino acids were racemic, it was proved that some amino acids (LRMs) can be synthesized by such simple conditions.

5.2.2 The icy planet that fell on Earth's Ocean

5.2.2.1 *Introduction*

Amino acids and nucleobases are LRMs essential for life, which are composed of CHON. Large impacts can cause chemical reactions among

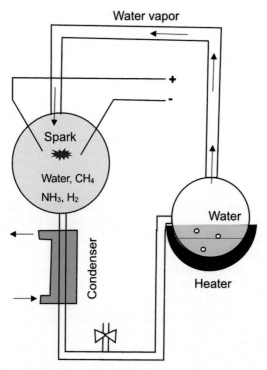

FIGURE 5.1 The Miller's experiment. Two flasks were connected with two pipes. One flask with pure water was heated to boil for making water vapor which was lead through one pipe to another reaction flask containing methane, ammonia, and hydrogen gases, in which electrical sparks were supplied. The water vapor was cooled by a condenser, and the liquid water was returned to the boiling flask through another pipe. *Source: Modified from Miller, S.L., 1953. A production of amino acids under possible primitive earth conditions. Science 117, 528–529.*

meteoritic materials, the ocean, and the atmosphere on the Earth. Formation of reduced volatiles from inorganic materials has been reported in simulations of post-impact reactions on the early Earth (Fegley et al., 1986; Mukhin et al., 1989; Gerasimov et al., 2002; Schaefer and Fegley, 2010; Kurosawa et al., 2013; Furukawa et al., 2014).

The Furukawa's group has investigated such post-impact reactions with experimental simulations and demonstrated the formation of glycine and aliphatic carboxylic acids from inorganic carbon in meteorites (Nakazawa, 2008; Furukawa et al., 2009). Amino acid formation in impacts involving simulated cometary ice composed of ammonia, methanol, and carbon dioxide has also been proposed (Goldman et al., 2010). These studies support the importance of impact-induced reactions as a mechanism for providing LRMs on the early Earth.

Furukawa's group synthesized glycine using solid amorphous carbon as the carbon source (Furukawa et al., 2009). They presumed high partial pressure of CO_2 in the early Hadean atmosphere (Holland, 1984; Walker, 1985; Kasting, 1993), and therefore, dissolution of large quantities of CO_3^{2-} in the early oceans should have occurred (Morse and Mackenzie, 1998). Hence, large amounts of carbon would have been available in the post-impact plumes. The huge carbon reservoir on the early Earth might have been used in impact-induced reactions to form various kinds of organic compounds important for life.

Ammonia can be synthesized through the reduction of terrestrial nitrogen species in the ocean, crust, and impact plumes (Summers and Chang, 1993; Brandes et al., 1998; Nakazawa et al., 2005; Smirnov et al., 2008; Schaefer and Fegley, 2010; Furukawa et al., 2014). Thus, both CO_3^{2-} and ammonia were easily available in the impact environments and could have been carbon and nitrogen sources for prebiotic synthesis on the early Earth.

The purpose of this section is to investigate what kind of LRMs are synthesized from CO_3^{2-} and ammonia in simulating the fall of meteorites, following the experiments of Furukawa et al. (2015).

5.2.2.2 Shock-recovery experiments

Furukawa et al. (2015) employed the shock experiments conducted using a single-stage propellant gun (Sekine, 1997). In the shock-recovery experiment, forsterite (Mg_2SiO_4), metallic iron, magnetite (Fe_3O_4) and metallic nickel, ^{13}C-labeled ammonium bicarbonate solution (representative of the ocean), and gaseous nitrogen (representative of the atmosphere) were mixed to be representative of simplified meteorite components. The mixtures of IMx, OCx, and CCx ($x = 1$ or 2) represent an iron meteorite, an ordinary chondrite, and a carbonaceous chondrite, respectively (see Table 5.1).

5.2.2.3 Run products of the impact experiments

Ultra-high performance liquid chromatography (UHPLC) coupled with tandem mass spectrometry (MS/MS) was used for the detection of ^{13}C-labeled nucleobases. The ^{13}C-labeling ensured the accurate identification of products by removing contamination. The combination of UHPLC and MS/MS facilitated the identification of the specific products. Run products are summarized in Table 5.2.

Various amino acids were detected including glycine (Gly), alanine (Ala), serine (Ser), aspartic acid (Asp), glutamic acid (Glu), valine (Val), leucine (Leu), isoleucine (Ile), proline (Pro), β-alanine (β-Ala), sarcosine (Sar), α-amino-n-butyric acid (α-ABA), and β-aminoisobutyric acid (β-AIBA), all of which were ^{13}C-labeled. The experimental results were

TABLE 5.1   Composition of starting materials and impact velocity.

Type		IM2	IM1	CC1	CC2	OC1	OC2
Starting materials (mg)	Fe	200	300	50	50	100	100
	Fe_3O_4	0	0	100	100	0	0
	Ni	20	30	15	15	30	30
	Mg_2SiO_4	0	0	200	300	200	200
	NH_4HCO_3	170	170	170	40	170	30
	H_2O	130	130	130	150	130	150
	N_2 (gas)	Filled	Filled	Filled	Filled	Filled	Filled
Impact velocity	km/s	0.82	0.86	0.89	0.86	0.86	0.87

Data from Furukawa, Y., Samejima, T., Nakazawa, H., Kakegawa, T., 2014. Experimental investigation of reduced volatile formation by high-temperature interactions among meteorite constituent materials, water, and nitrogen. Icarus 231, 77–82.

the first observation that the simultaneous formation of various amino acids and nucleobases was synthesized in the single shock experiment.

β-Ala, Sar, α-ABA, and β-AIBA are non-proteinogenic amino acids (see Fig. 5.2; "non-proteinogenic" means that the amino acid is NOT used in the formation of naturally existing proteins), therefore, this further excluded the contamination. Gly was produced in the largest amount. Asp, Glu, and Val were produced only in the IM1 experiment. The ^{13}C-labeled primary amines (methylamine, ethylamine, propylamine, and butylamine) decreased as the length of the alkyl chain increased.

Uracil was formed only in the IM compositions, while cytosine was formed in all experiments. When the yields of the amino acids were normalized to the initial amount of $NH_4H^{13}CO_3$, the yields of amino acids decreased in the order of IM1, IM2, OC1, and CC1, depending on the amounts of metallic iron and nickel in the starting materials. The yields also depended on the initial amounts of NH_4C. The conversion rates were 3.5×10^{-1}–1.8×10^{-2} mol%, which were far higher than those using solid amorphous carbon of 1.0×10^{-6} mol% (Furukawa et al., 2009). This meant that the molecular carbon ($H^{13}CO_3^{-}$) was more reactive than solid carbon.

5.2.2.4 Implication for the prebiotic Earth

In this experiment, the impact velocities were lower (~ 0.9 km/s) than typical impact velocities of large extraterrestrial materials (~ 20 km/s), which are limited simply by the resistance of the container

TABLE 5.2   Run products of the impact experiments.

Products			IM2	IM1	CC1	CC2	OC1	OC2
Nucleobase	(n mol)	Cytosine	8.8	5.3	tr.	tr.	0.11	0.16
		Uracil	0.096	0.023	BD	BD	BD	BD
Amino acids	(n mol)	Gly	520	2900	350	34	370	55
		Ala	48	210	12	0.21	13	0.59
		Ser	0.3	536	0.2	BD	0.51	BD
		Asp	0.9	1.9	BD	BD	BD	BD
		Glu	BD	0.9	BD	BD	BD	0.3
		Val	BD	0.9	BD	BD	BD	BD
		Ile	BD	tr.	BD	tr.	BD	BD
		Leu	BD	tr.	tr.	BD	BD	BD
		Pro	BD	tr.	BD	BD	BD	BD
		Sar	1.9	140	2	0.13	2.2	0.1
		β-Ala	19	86	13	0.11	11	tr.
		α-ABA	19	86	5.1	BD	4.5	BD
		β-AIBA	2.5	22	0.6	BD	0.6	BD
	Total		610	3500	380	35	400	56
	C conversion rate (%)		0.062	0.35	0.037	0.018	0.039	0.03
Amines	(n mol)	Methylamine	3900	20000	1700	2500	1000	4200
		Ethylamine	460	1400	100	40	52	81
		Propylamine	34	240	15	2.5	6.2	6.1
		Butylamine	0	37	5.1	BD	1.1	2
	Total		4400	22000	1800	2500	1100	4300

BD and tr. represent "below detection limit" and "detected in trace amounts," respectively. The C conversion rate is from NH_4HCO_3 into amino acids and nucleobases (atom%). *Data from Furukawa, Y., Samejima, T., Nakazawa, H., Kakegawa, T., 2014. Experimental investigation of reduced volatile formation by high-temperature interactions among meteorite constituent materials, water, and nitrogen. Icarus 231, 77–82.*

against the speed. The container burst at higher velocity impacts. It is estimated that the higher velocity would cause higher temperature, thus resulting in higher production of organic compounds by carbonate reduction reactions. It should be noted that the labile molecules, such as

β-alanine (β-Ala)   sarcosine (Sar)

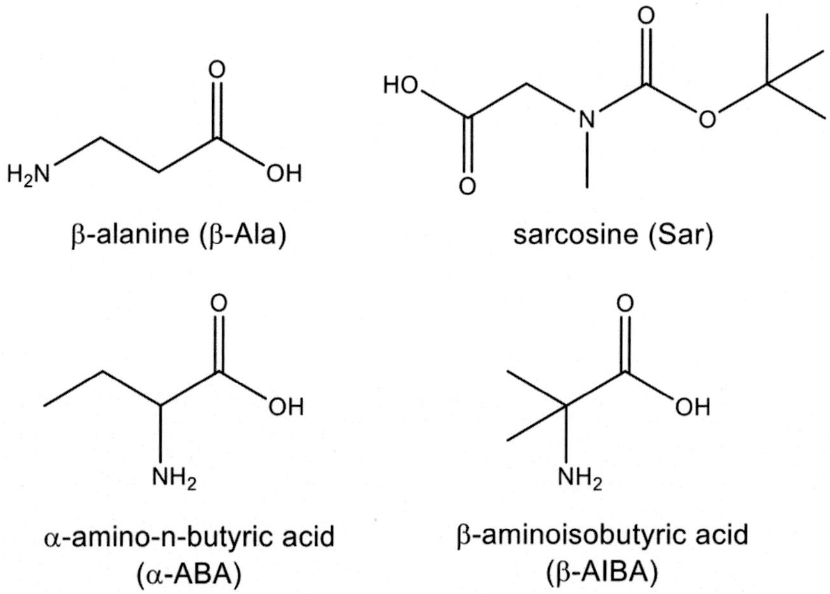

α-amino-n-butyric acid   β-aminoisobutyric acid
(α-ABA)   (β-AIBA)

FIGURE 5.2   **Non-proteinogenic amino acids synthesized by** Furukawa et al. (2009).

nucleobases and amino acids, were not formed at peak temperatures, instead, they were formed in post-impact conditions with lower temperature pressure as in this experiment.

As nucleobases and amino acids are generated from carbon reservoirs, such organics can be formed in a CO_2-rich atmosphere. Ammonia could be formed from nitrogen oxides in the N_2-atmosphere by lightning and meteorite impacts, and subsequent reduction by ferrous iron in the ocean or iron sulfides from hydrothermal vents (Summers and Chang, 1993; Brandes et al., 1998; Summers, 1999; Summers et al., 2012). Summers (1999) calculated the amount of ammonia in the prebiotic ocean formed by the reduction of nitrogen oxides by ferrous iron to be $7 \times 10^{-5} - 8 \times 10^{-7}$ mol/L. Brandes et al. (1998) also estimated the reduction rate of nitrogen oxides to ammonia was dependent on the iron sulfide catalysts. Nakazawa et al. (2005) proposed that meteorite impacts reduced atmospheric nitrogen to ammonia providing NH_3-rich areas in the ocean.

Hartmann et al. (2000) and Valley et al. (2002) estimated the impact rates during LHB of 4.0–3.8 Ga to be $10^3 - 10^5$ times higher than those of today. Thus, even if ammonia and bicarbonate concentrations in the Archean ocean were low, the accumulation of bombardment could make large amounts of nucleobases or amino acids. Anders (1989) estimated that hypervelocity projectiles should have delivered intact organic materials as 276 times greater than meteorites.

It is interesting that, in contrast to this study, carbonaceous chondrites contain more purine than pyrimidine (Shapiro, 1999; Callahan et al., 2011). In other words, pyrimidine bases were preferentially formed in this study.

5.2.2.5 Implication for life-related molecules in early Earth and origin of life

Ferus et al. (2015) experimentally synthesized five nucleobases by simulating icy comets hitting the icy comets or early Earth. Details are shown in Section 8.6. Based on this experiment, they proposed that the formation of the prebiotic nucleobases through pure formamide ($HCONH_2$) during impacts is the key reaction. However, Miyakawa et al. (2002) contradicted that formamide accumulation in oceans would not be high enough to facilitate nucleobase synthesis. The formamide-rich lakes are one of the solutions; however, the land areas were small from geological evidence (Armstrong and Harmon, 1981; McCulloch and Bennett, 1994).

The continuous formation of LRMs from atmospheric CO_2 and N_2 in the early ocean environment, and impact-induced reactions using terrestrial carbon sources seemed possible. However, Meinert et al. (2016) have shown the importance of LRMs syntheses on ice in space again. Thus, the most feasible origin of LRMs in the early Earth should be the hybrid of the space (ice−ice collision in space) and terrestrial (impact of ice onto the Earth, especially on the ocean) synthetic sources of LRMs.

5.2.3 Origin of life-related molecules by computer simulations

The prebiotic chemistry is challenging to follow the syntheses of LRMs starting from small kinds of primordial substrates (H_2O, N_2, HCN, NH_3, CH_4, and H_2S). Wołos et al. (2020) invented a forward-synthesis algorithm that described a full network of prebiotic chemical reactions of the substrates under generally accepted conditions. This network contained both reported and previously unidentified routes to biotic targets, and plausible synthesizing routes of abiotic molecules (see Fig. 5.3). The network also showed three types of nontrivial chemical emergence: (1) the molecules within the network behaved as catalysts of downstream reaction types; (2) forming functional chemical systems, including self-regenerating cycles; and (3) producing surfactants relevant to primitive forms of biological compartmentalization.

To support these claims, computer-predicted, prebiotic syntheses of several biotic molecules as well as a multistep, self-regenerative cycle of iminodiacetic acid were evaluated by experiments.

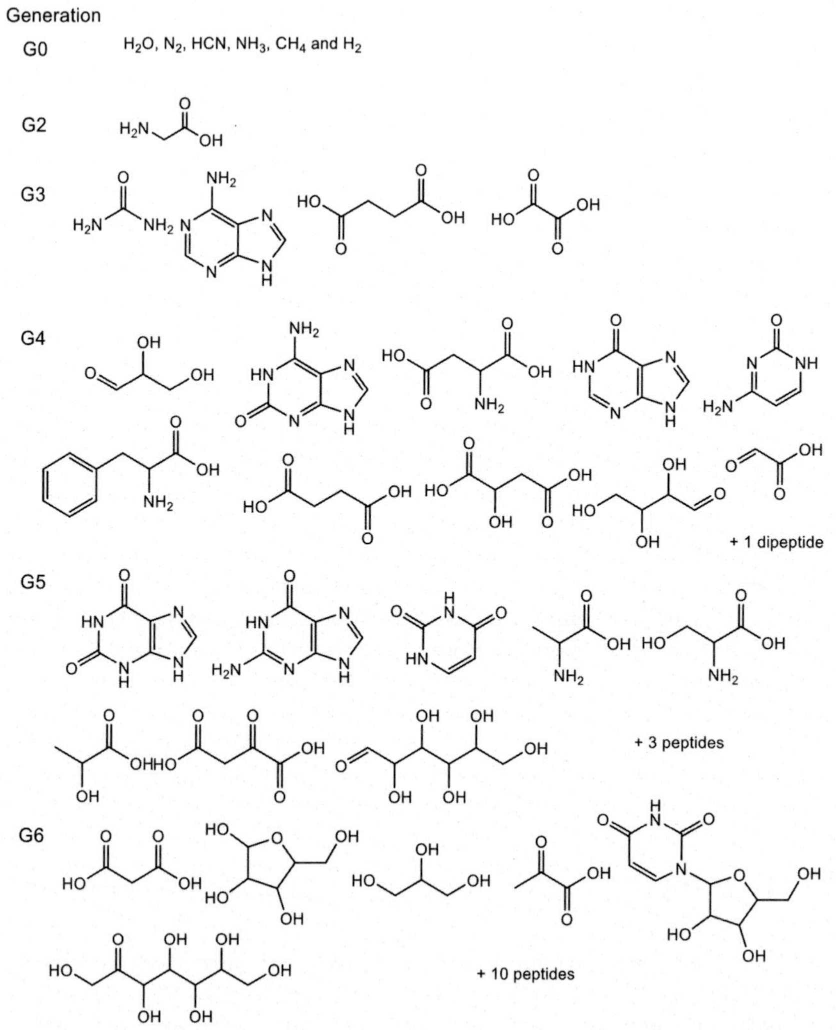

FIGURE 5.3   The six generations of LRMs in computer simulations. Life-related molecules glycine is in the second generation (G2); urea, adenine, butenedioic acid, and oxalic acid are in G3; glyceraldehyde, isoguanine, aspartic acid, hypoxanthine, cytosine, phenylalanine, succinic acid, malic acid, glyoxylic acid, and aldotetrose are in G4; xanthine, alanine, serine, guanine, uracil, lactic acid, oxaloacetic acid, and aldohexose are in G5; malonic acid, pentofuranose, glycerol, pyruvic acid, cytidine, and ketoheptose are in G6. Threonine, methionine, proline, glutamic acid, citric acid, acetic acid, thymine, adenosine, guanosine, uridine, and uric acid are in G7 (not shown). *Source: Data from Wołos, A., Roszak, R., Żądło-Dobrowolska, A., Beker, W., Mikulak-Klucznik, B., et al., 2020. Synthetic connectivity, emergence, and self-regeneration in the network of prebiotic chemistry. Science 369, eaaw1955.*

References

Anders, E., 1989. Pre-biotic organic-matter from comets and asteroids. Nature 342, 255–257.

Armstrong, R.L., Harmon, R.S., 1981. Radiogenic isotopes: the case for crustal recycling on a near-steady-state no-continental-growth earth and discussion. Philos. Trans. R. Soc. A301, 443–472.

Brandes, J.A., Boctor, N.Z., Cody, G.D., et al., 1998. Abiotic nitrogen reduction on the early Earth. Nature 395, 365–367.

Callahan, M.P., Smith, K.E., Cleaves, H.J., et al., 2011. Carbonaceous meteorites contain a wide range of ex-traterrestrial nucleobases. Proc. Natl. Acad. Sci. U. S. A. 108, 13995–13998.

Chyba, C., Sagan, C., 1992. Endogenous production, exogenous delivery and impact-shock synthesis of organic molecules: an inventory for the origins of life. Nature 355, 125–132.

Chyba, C.F., Thomas, P.J., Brookshaw, L., Sagan, C., 1990. Cometary delivery of organic molecules to the early. Earth Planet. Sci. Lett. 249, 366–373.

Fegley, B., Prinn, R.G., Hartman, H., Watkins, G.H., 1986. Chemical effects of large impacts on the Earth's primitive atmosphere. Nature 319, 305–307.

Ferus, M., Nesvorný, D., Sponer, J., et al., 2015. High-energy chemistry of formamide: a unified mechanism of nucleobase formation. Proc. Natl. Acad. Sci. U. S. A. 112, 657–662.

Furukawa, Y., Nakazawa, H., Sekine, T., Kobayashi, T., Kakegawa, T., 2015. Nucleobase and amino acid formation through impacts of meteorites on the early ocean. Earth Planet. Sci. Lett. 429, 216–222.

Furukawa, Y., Samejima, T., Nakazawa, H., Kakegawa, T., 2014. Experimental investigation of reduced volatile formation by high-temperature interactions among meteorite constituent materials, water, and nitrogen. Icarus 231, 77–82.

Furukawa, Y., Sekine, T., Oba, M., Kakegawa, T., Nakazawa, H., 2009. Biomolecule formation by oceanic impacts on early Earth. Nat. Geosci. 2, 62–66.

Gerasimov, M.V., Dikov, Y.P., Yakovlev, O.I., Wlotzka, F., 2002. Experimental investigation of the role of water in impact vaporization chemistry. Deep-Sea Res. Part 2 Top. Stud. Oceanogr. 49, 995–1009.

Glavin, D.P., Matrajt, G., Bada, J.L., 2004. Re-examination of amino acids in Antarctic micrometeorites. Adv. Space Res. 33, 106–113.

Goldman, N., Reed, E.J., Fried, L.E., Kuo, I.F.W., Maiti, A., 2010. Synthesis of glycine containing complexes in impacts of comets on early Earth. Nat. Chem. 2, 949–954.

Hartmann, W.K., Ryder, G., Dones, L., 2000. Grinspoon: the time-dependent intense bombardment of the primordial Earth/Moon system. In: Canup, R.M., Righter, K. (Eds.), Origin of the Earth and Moon. The University of Arizona Press, Tucson, pp. 493–512.

Holland, H.D., 1984. The Chemical Evolution of the Atmosphere and Oceans. Princeton University Press, Princeton.

Kasting, J.F., 1993. Earth's early atmosphere. Science 259, 920–926.

Kurosawa, K., Sugita, S., Ishibashi, K., et al., 2013. Hydrogen cyanide production due to mid-size impacts in a redox-neutral N_2-rich atmosphere. Orig. Life Evol. Biosph. 43, 221–245.

Maurette, M., 2006. Micrometeorites and the Mysteries of Our Origins. Springer, New York.

McCulloch, M.T., Bennett, V.C., 1994. Progressive growth of the Earth's continental crust and depleted mantle: geochemical constraints. Geochim. Cosmochim. Acta 58, 4717–4738.

Meinert, C., Myrgorodska, I., de Marcellus, P., et al., 2016. Ribose and related sugars from ultraviolet irradiation of interstellar ice analogs. Science 352, 208–212.

Miller, S.L., 1953. A production of amino acids under possible primitive earth conditions. Science 117, 528–529.

Miyakawa, S., Cleaves, H.J., Miller, S.L., 2002. The cold origin of life: A. Implications based on the hydrolytic stabilities of hydrogen cyanide and formamide. Orig. Life Evol. Biosph. 32, 195–208.

Morse, J.W., Mackenzie, F.T., 1998. Hadean ocean carbonate geochemistry. Aquat. Geochem. 4, 301–319.

Mukhin, L.M., Gerasimov, M.V., Safonova, E.N., 1989. Origin of precursors of organic molecules during evaporation of meteorites and mafic terrestrial rocks. Nature 340, 46–48.

Nakazawa, H., 2008. Origin and evolution of life: endless ordering of the Earth's light elements. In: Okada, H., Mawatari, S.F., Suzuki, N., Gautum, P. (Eds.), International Symposium on Origin and Evolution of Natural Diversity. Hokkaido University, Sapporo, pp. 13–19.

Nakazawa, H., Sekine, T., Kakegawa, T., Nakazawa, S., 2005. High yield shock synthesis of ammonia from iron, water and nitrogen available on the early Earth. Earth Planet. Sci. Lett. 235, 356–360.

Oparin, A.I., 1938. The Origin of Life. Macmillan.

Schaefer, L., Fegley, B., 2010. Chemistry of atmospheres formed during accretion of the Earth and other terrestrial planets. Icarus 208, 438–448.

Sekine, T., 1997. Shock wave chemical synthesis. Eur. J. Solid. State Inorg. Chem. 34, 823–833.

Shapiro, R., 1999. Prebiotic cytosine synthesis: acritical analysis and implications for the origin of life. Proc. Natl. Acad. Sci. U. S. A. 96, 4396–4401.

Smirnov, A., Hausner, D., Laffers, R., Strongin, D.R., Schoonen, M.A.A., 2008. Abiotic ammonium formation in the presence of Ni-Fe metals and alloys and its implications for the Hadean nitrogen cycle. Geochem. Trans. 9, 5.

Summers, D.P., 1999. Sources and sinks for ammonia and nitrite on the early Earth and the reaction of nitrite with ammonia. Orig. Life Evol. Biosph. 29, 33–46.

Summers, D.P., Basa, R.C.B., Khare, B., Rodoni, D., 2012. Abiotic nitrogen fixation on terrestrial planets: reduction of no to ammonia by FeS. Astrobiology 12, 107–114.

Summers, D.P., Chang, S., 1993. Prebiotic ammonia from reduction of nitrite by iron(II) on the early Earth. Nature 365, 630–632.

Valley, J.W., Peck, W.H., King, E.M., Wilde, S.A., 2002. A cool early Earth. Geology 30, 351–354.

Walker, J.C.G., 1985. Carbon-dioxide on the early earth. Orig. Life Evol. Biosph. 16, 117–127.

Wołos, A., Roszak, R., Żądło-Dobrowolska, A., Beker, W., Mikulak-Klucznik, B., et al., 2020. Synthetic connectivity, emergence, and self-regeneration in the network of prebiotic chemistry. Science 369, eaaw1955.

6

Life-related molecules on Venus

6.1 Life-related molecules on Venus atmosphere

6.1.1 Introduction

As we know, the average temperature and pressure of Venus today are 750K and 90 bar, respectively, which make the present Venus incompatible to the existence of life. However, billions of years ago, the surface was habitable for life because Venus was located at the habitable zone of the fainter younger sun (e.g., Sagan and Mullen, 1972; Claire et al., 2012; Westall et al., 2023).

6.1.2 Phosphine on Venus atmosphere

The Grinspoon group insisted that there is an "unknown ultraviolet absorber" (e.g., Grinspoon, 1999). The ultraviolet spectra of the clouds of Venus are obtained from the Earth and by spacecrafts of Galileo, Venus Express, Akatsuki, etc. Based on these data, the Venus clouds are made of particles of sub-millimeter to a few millimeters in radius, containing sulfuric acid of 75%~95%. The ultraviolet absorbing materials are SO_2 gas, sulfur aerosol, $FeCl_3$, and sulfur monoxide dimers [$(SO)_2$; OSSO]. $FeCl_3$ easily transforms into $Fe_2(SO_4)_3$ with sulfuric acid, therefore, the synthetic reactions of $FeCl_3$ should exist.

The Limaye group (e.g., Limaye et al., 2018; 2021) suggested that the temperature and pressure of the Venus clouds are $\sim 15°C$ and $\sim 10^5$ Pa, respectively, at 55 km height from the surface. In such conditions, the life which performs photosynthesis or lives by CO_2 assimilation can survive, as a result, oxidative or reductive reactions occur by which "unknown absorbers" are produced as byproducts, which work as unknown absorbers of ultraviolet lights.

Introductory Astrochemistry
DOI: https://doi.org/10.1016/B978-0-443-23938-0.00006-6

Since the phosphine (PH_3) signal in the clouds was found (Greaves et al., 2021), the lives in upper atmosphere (cloud) are strongly supported. As Venus is a rocky planet, and the atmospheric and surface conditions are very oxidative, there are no synthesis pathways to produce reductive phosphine. In order to produce phosphine continuously, some synthetic reactions (possibly some kinds of life) should exist in the clouds. Thus, the possibility of life on Venus has been hotly debated.

References

Claire, M.W., Sheets, J., Cohen, M., Ribas, I., Meadows, V.S., Catling, D.C., 2012. The evolution of solar flux from 0.1 nm to 160 μm: quantitative estimates for planetary studies. Astrophys. J. 757, 95. Available from: https://doi.org/10.1088/0004-637X/757/1/95.

Greaves, J.S., Richards, A.M.S., Bains, W., Rimmer, P.B., Sagawa, H., et al., 2021. Phosphine gas in the cloud decks of Venus. Nat. Astron. 5, 655−664.

Grinspoon, D.H., 1999. Venus Revealed: A New Look Below the Clouds of Our Mysterious Twin Planet. Perseus Publishing.

Limaye, S.S., Mogul, R., Smith, D.J., Ansari, A.H., Slowik, G.P., Vaishampayan, P., 2018. Venus' spectral signatures and the potential for life in the clouds. Astrobiology 18, 1181−1198.

Limaye, S.S., Rakesh, M., Baines, K.H., Bullock, M.A., Cockell, Cutts, J.A., et al., 2021. Venus, an astrobiology target. Astrobiology 21, 1163−1185.

Sagan, C., Mullen, G., 1972. Earth and Mars: evolution of atmospheres and surface temperatures. Science 177, 52−56.

Westall, F., Höning, D., Avice, G., Gentry, D., Gerya, T., et al., 2023. The habitability of Venus. Space Sci. Rev. 219, 17.

Materials on Mars

7.1 Exploration of Mars

7.1.1 Introduction

Since the resolution of the telescopes increased in 19th century, seasonal changes of the Mars surface and so-called "Martian canals" have been observed, and people imagined that there would be Life on Mars (see Fig. 7.1). As Mars is within the habitable zone of the solar system, and organic compounds were found in many Martian meteorites (see Section 7.2), there is a possibility that Life could exist on Mars. Thus, Mars has become a target for investigation of Life by spacecrafts and landing rovers. Life-related materials (LRMs) are observed in the Martian atmosphere and support the existence of Life on Mars. Today, the possibility of Life on Mars is hotly debated. The sample return program of the Mars will finally decide whether the Life exists or not on Mars.

7.1.2 Searching for life-related materials (LRMs) and Life on Mars

The space program of NASA focused on searching for Life on Mars primarily because of its similarities with the Earth. The first exploration of Mars was a flyby mission by the Mariner 4 probe in 1965. By this exploration, it was revealed that the surface of Mars was barren with the $\sim 1/100$ air pressure lower than that of the Earth. This suggested that there is no large amount of liquid or atmosphere on the Mars' surface (Dick, 2006; Martel et al., 2012). The Mariner 4 probe also suggested that Mars does not have a magnetic field, which would protect the planet against deadly solar and cosmic rays (see Section 10.2.6).

Introductory Astrochemistry
DOI: https://doi.org/10.1016/B978-0-443-23938-0.00007-8

FIGURE 7.1   **Mars.Mars taken by the Rosetta spacecraft in 2007.** A half of a white dot in the right down corner indicate the size of Phobos. *Source: ESA & MPS for OSIRIS Team MPS/UPD/LAM/IAA/RSSD/INTA/UPM/DASP/IDA. https://www.esa.int/ESA_Multimedia/ Images/2007/02/True-colour_image_of_Mars_seen_by_OSIRIS.*

After the Mariner 4 mission, two Viking spacecrafts landed on Mars in 1976. Although the experiments performed in these missions gave evidence for the presence of oxidizing materials in the Martian soil, absence of organic molecules in the analyzed samples led to the consensus that there is no Life on the surface of Mars (Dick, 2006; Martel et al., 2012).

At the end of 2022, three rovers, the Curiosity and Perseverance rovers which are operated by NASA as well as the Zhurong (祝融) rover (the part of the Tianwen [天問]-1 mission) by the China National Space Administration (CNSA) are still working. Seven orbiters, Mars Odyssey, Mars Express, Mars Reconnaissance Orbiter, MAVEN (NASA), the Hope Probe (UAE Space Agency), the ExoMars Trace Gas Orbiter (ESA), and the Tianwen-1 orbiter (CNSA) are surveying Mars.

The Curiosity rover was moving on the surface of Mars over 3 years from 2012. The rover's observations confirmed that once there was plenty of water on the surface making lakes 3.8–3.3 billion years ago. There were streams delivering sediments which were deposited as shown in Fig. 7.2. The rocks in the photograph are the finely laminated

FIGURE 7.2   A view of Mars by Curiosity Such views confirmed that water filled this place forming a lake. There were streams delivering sediments (mudstones) which deposited as layers. *Source: NASA/JPL-Caltech/MSSS. http://www.nasa.gov/sites/default/files/thumbnails/image/pia19839-galecrater-main.png.*

mudstones, which should be deposited as a lake (Clavin, 2015). If there was plenty of liquid water on Mars, it is not unreasonable to think that there was Life on Mars.

7.1.3 Mars Sample Return mission

The NASA-ESA Mars Sample Return is proposed. The Perseverance rover has collected about 40 pencil-sized samples since 2021. Then the Sample Retrieval Lander will put the samples up into space, and the Earth Return Orbiter will receive them and bring back to the Earth. If this mission is successful, we can know whether there are/were Life on Mars.

7.2 Life-related molecules (LRMs) in the atmosphere on Mars

7.2.1 Introduction

Mars has lost its magnetosphere four billion years ago, thus solar wind directly hit the Martian ionosphere. Then the density of atmosphere was lowered by stripping away atoms from the outer layer of ionosphere. Mars Global Surveyor (launched in 1996 by NASA), Mars Express (launched in 2003 by ESA), and MAVEN (launched in 2013 by NASA) detected ionized atmospheric particles. The atmospheric pressure of Mars is only 600 Pa compared to that of the Earth of 1.0×10^5 Pa, indicating that Mars is only 0.6% of that of the Earth. The composition was 96% of CO_2, 1.9% of Ar, 1.9% of N_2, and traces of O_2 and water.

7.2.2 Methane on Mars

The presence and state of LRMs, including the possible biosignatures, in Martian materials are unsolved questions, despite limited reports of the existence of organic matter on Mars. Webster et al. (2018) reported in situ measurements of methane at Gale crater over 5 years by the Curiosity rover. The background levels of methane had a mean value of 0.41 ± 0.16 parts per billion by volume (ppbv) (95% confidence interval) and showed a strong, repeatable seasonal variation (0.24−0.65 ppbv). This variation was greater than that predicted from either ultraviolet degradation of impact-delivered organics on the surface or from the annual surface pressure cycle. The large seasonal variation in the background and occurrences of higher temporary spikes (~7 ppbv) were consistent with small localized sources of methane released from Martian surface or subsurface reservoirs.

Eigenbrode et al. (2018) reported the in situ detection of organic matter preserved in lake mudstones at the base of the ~3.5-billion-year-old Murray formation at Pahrump Hills, Gale crater, by the Curiosity rover. Diverse pyrolysis products, including thiophenic, aromatic, and aliphatic compounds released at high temperatures (500°C−820°C), were directly detected by evolved gas analysis. Thiophenes (C_4H_4S) were also observed by gas chromatography−mass spectrometry. Their presence suggests that sulfurization aided organic matter preservation. At least 50 nanomoles of organic carbon existed, probably as macromolecules containing 5% carbon as organic sulfur molecules.

7.2.3 Liquid water beneath Mars ice cap

The presence of liquid water at the base of the Martian polar caps has long been suspected but not proved. Orosei et al. (2018) surveyed the Planum Australe region using the Mars Advanced Radar for Subsurface and Ionosphere Sounding (MARSIS) instrument that is a low-frequency radar on the Mars Express spacecraft. Radar profiles obtained between May 2012 and December 2015 contain evidence of liquid water trapped below the ice of the South Polar Layered Deposits. Anomalously bright subsurface reflections are evident within a 20-km-wide zone centered at 193°E, 81°S, which is surrounded by much less reflective areas. Quantitative analysis of the radar signals shows that this bright feature has high relative dielectric permittivity (>15), matching that of water-bearing materials. Orosei et al. (2018) interpret this feature as liquid water on Mars.

7.2.4 Hydrogen chloride was found in Mars's atmosphere

Finding new atmospheric gases is one of Mars' exploration purposes. Because we can know what activities of geophysical or biological origins are going on. Korablev et al. (2021) first detected a halogen gas, HCl, which could be generated by contemporary volcanic degassing or chlorine releasing from gas-solid reactions. They detected $\sim 3.2 - 3.8 \, \mu m$ by the Atmospheric Chemistry Suite, and confirmed with Nadir and Occultation for Mars Discovery instruments onboard the ExoMars Trace Gas Orbiter, which revealed a wide distribution of HCl in the 1- to 4-ppbv range, 20 times higher than previously reported upper limits. HCl increased during the 2018 global dust storm and declined soon after its end, indicating the exchange between the dust and the atmosphere. It is necessary to understand the origin and variability of HCl to build new models of Martian geo- and photochemistry.

7.2.5 High concentrations of O_2 in the Martian brines

Mars is assumed to have too low concentrations of O_2 in the Martian atmosphere to support aerobic respiration on Mars. Stamenković et al. (2018) presented a thermodynamic framework for the solubility of O_2 in brines under Martian near-surface conditions. Their estimation for modern Mars showed that dissolved O_2 values in liquid environments should range from $\sim 2.5 \times 10^{-6} \, mol/m^3$ to $2 \, mol/m^3$ across the planet. Particularly polar regions have high concentrations because lower temperatures at higher latitudes raise O_2 entry into brines. General circulation model simulations showed that variations of O_2 concentrations in near-surface environments are large both spatially and with time. The time variation is associated with secular changes in obliquity or axial tilt. Even at the limits of the uncertainties, their simulations suggest that sufficient O_2 was available for aerobic microbes to breathe in near-surface environments on Mars. Their findings can also explain the formation of highly oxidized phases such as manganese oxide minerals in Martian rocks found by Mars rovers. Furthermore, it is implied that aerobic Life could exist on modern Mars and on other planetary bodies with sources of O_2 independent of photosynthesis.

7.3 Organic compounds in the Martian meteorites

The Martian meteorites are divided into five types: shergottites (basalt), nakhlites (cumulate of clinopyroxenite), chassignites (cumulate of dunite), basaltic breccia (NWA 7034), and orthopyroxenite (ALH84001). The first three types are sometimes integrated and called as the SNC meteorites (see Section 1.4.12). The meteorites are finally concluded by the noble gas composition in the meteorite. They came to the Earth by the impact shock on Mars of another space body.

Steele et al. (2016) summarized the igneous carbon contents and the $\delta^{13}C$ (‰) values of the Martian meteorites. The compilation results are shown in Table 7.1. As organic carbon compounds are ubiquitous on the Earth, it is essential to separate the terrestrial contamination and the organic compounds from Mars both of biotic and abiotic origins. Although not all forms of reduced carbon are linked to Life (LRMs), they defined "organic carbon" as the carbon that has a C−H bond, therefore, methane is included in the organic carbon. A bulk sample is treated with acid to remove carbonates and measured by combustion or pyrolysis gas chromatography-isotope ratio mass spectrometry (GC-IRMS), and high-temperature component (> 600°C) is considered to be of a Martian origin carbon. From Table 7.1, the carbon of Martian origin was detected from all the types of Martian meteorites.

As the organic compounds were found in many Martian meteorites, there is a possibility that Life could exist on Mars. In addition, as related in Section 7.2, Life-related materials (LRMs) were observed in the Martian atmosphere and supported the existence of Life on Mars. McKay et al. (1996) reported that the fossils of the Martian Life were found in the Antarctic meteorite, ALH84001 (see Section 7.4). This report caused a sensation in the world, and Mars has become a target for investigation of Life. Furthermore, this study greatly contributed to the progress of astrobiology. Based on this meteorite, what was the evidence of Life and how the evidence should be checked are discussed. It seems that all the biological evidence in this meteorite can be explained by the chemical processes, and the fossils seemed to be formed by inorganic processes. After the hot discussions, the Martian meteorite, Tissint fell in 2011 and was analyzed as the least contaminated Martian meteorite (see Section 7.5). Even today, the possibility of Life on Mars is discussed.

7.4 Fossils of microorganisms were found in the Martian meteorite, ALH 84001?!

7.4.1 Introduction to finding microorganism fossils in the Martian meteorite, ALH 84001

In 1996, D.S. McKay and his colleagues presented a series of evidences that Life existed in the Martian meteorite Allan Hills 84001 (ALH84001) (McKay et al., 1996). This meteorite was an igneous orthopyroxenite discovered in Antarctica in 1984. The meteorite was originated from Mars because of both its mineralogy and the isotope composition of the air which was trapped inside the meteorite (Mittlefehldt, 1994; Gibson et al., 1997). From the fragments of ALH84001, McKay et al. found traces of polycyclic aromatic hydrocarbons (PAHs), magnetite

TABLE 7.1 Igneous carbon contained in martian meteorites.

Meteorite	Type	Age (Ga)	[C] (ppm)	$d^{13}C$ (‰)	References
ALH 84001	O	4.1−4.43	280	−11.0	Jull et al. (1998)
DaG 476	S	0.47	24	−15.6	Steele et al. (2012)
DaG 476	S	0.47	6	−22.1	Grady et al. (2004)
Los Angeles	S	0.17	6.5	−24.3	Grady et al. (2004)
QUE 94201	S	0.33	1.3	−23.1	Grady et al. (2004)
Shergotty	S	0.17	4.3	−19.4	Grady et al. (2004)
SAU 130	S	0.81	19	−16.9	Grady et al. (2004)
Nakhla	N	1.3	26	−15.0 ~ −33.0	Jull et al. (1998)
NWA 998	N	1.29	17	−24.7	Steele et al. (2012)
Zagami	S	0.18	4	−16.8	Steele et al. (2012)
Zagami	S	0.18	52.8	−22.7	Grady et al. (2004)
Dhofar 019	S	0.55	109	−21.7	Grady et al. (2004)
NWA 1183	S	0.19	19.7	−14.8	Grady et al. (2004)
SAU 005	S	—	8.2	−6.4	Grady et al. (2004)
ALH 77005	S	0.18	1.7	−21.4	Grady et al. (2004)
LEW 88516	S	0.18	1.8	−20.2	Grady et al. (2004)
Y 793605	S	0.21	2.6	−14.9	Grady et al. (2004)
Chassigny	C	1.34	2.1	−20.8	Grady et al. (2004)
Tissint	S	0.57	14	−17.8	Steele et al. (2012)
NWA 7034	B	1.5−4.4	22	−23.4	Agee et al. (2013), Humayun et al. (2013)
Total			18 ± 26	−19.1 ± 4.5	Jull et al. (1998)

Type: B, breccia; C, chassignite; N, nakhlite; O, orthopyroxenite; S, shergottite. *Data from Steele, A., McCubbin, F.M., Fries, M.D., 2016. The provenance, formation, and implications of reduced carbon phases in Martian meteorites. Meteorit. Planet. Sci. 51, 2203−2225.*

crystals, carbonate globules, and pseudo-microfossils of miniature bacteria (McKay et al., 1996).

This research made a lot of researchers burst into searching for extraterrestrial Life and spearheading the creation and expansion of NASA's astrobiology program. Furthermore, this report caused major scientific controversy about Life on other planets and exciting debates about what exactly constitutes Life and whether nanobacteria is a real Life or not (Arrhenius and Mojzsis, 1996; Grady et al., 1996; Gibson et al., 1997; Hamilton, 2000; Hogan, 2003; Young and Martel, 2010; Martel et al., 2012).

From the next Section 7.4.2 to Section 7.4.5, finding Life in the Martian meteorite and the evidences that supported the past Life of the ALH84001 meteorite are explained. In the following Sections 7.4.6 and 7.4.8, the evidences are alternatively explained by the chemical processes, resulting in denial of the Life in the meteorite. Especially, morphology is a poor indicator of Life, however, it played an important role again in the discussion of the oldest Life of the Apex Chert (Section 4.3.2).

7.4.2 Discovery of the Life in the Martian meteorite, ALH84001

McKay et al. (1996) studied an Antarctic meteorite, ALH84001 (Mittlefehldt, 1994), which is an igneous orthopyroxenite composed of coarse-grained orthopyroxene [$(Mg,Fe)SiO_3$] and minor maskelynite ($NaAlSi_3O_8$), olivine [$(Mg,Fe)SiO_4$], chromite ($FeCr_2O_4$), pyrite (FeS_2), and apatite [$Ca_3(PO_4)_2$]. It crystallized 4.5 Ga. It records at least two shock events separated by a period of annealing. The first shock was ~ 4.0 Ga.

The $\delta^{13}C$ values of the carbonate in ALH 84001 were $-17 \sim +42‰$. The value of $+42‰$ is higher than those of other SNC meteorites. This range of $\delta^{13}C$ exceeds the range made by most terrestrial inorganic processes. Alternatively, biogenic processes are known to produce negative values in $\delta^{13}C$ on the Earth.

In examining the Martian meteorite ALH84001 by the state-of-the-art methods, McKay et al. (1996) concluded that the following evidences were compatible with the existence of past Life on Mars:

- The igneous Mars rock (of unknown geologic context) was penetrated by a fluid along fractures and pore spaces, then such places became the sites of secondary mineral formation and possible biogenic activity.
- A formation age for the carbonate globules was younger than the age of the igneous rock.
- SEM (see Section 1.3.6.3) and transmission electron microscope (TEM) images of carbonate globules and features resembled terrestrial

microorganisms, terrestrial biogenic carbonate structures, or microfossils.

- Magnetite and iron sulfide particles could have resulted from oxidation and reduction reactions known to be important in terrestrial microbial systems.
- PAHs are associated with surfaces rich in carbonate globules.
- Although none of these observations is in itself conclusive for the existence of the past Life. However, when they are considered collectively, it becomes reasonable to conclude that they are evidences for primitive Life on early Mars.

7.4.3 Polycyclic aromatic hydrocarbons

PAHs (see Fig. 7.3) are molecules of aromatic rings made of carbon and hydrogen. On the Earth, PAHs are abundant as fossil molecules in ancient sedimentary rocks, coal, and petroleum. Although PAHs are ubiquitous, they are not produced by living organisms and do not play specific roles in living things.

McKay et al. (1996) analyzed freshly broken fracture surfaces on small chips of ALH84001 for PAHs using a microprobe two-step laser mass spectrometer (mL^{2}MS). The average PAH concentration on the surfaces was >1 ppm. Contamination checks and control experiments indicated that the observed organic materials were from ALH84001. No PAHs in ALH84001 with intact fusion crust are present, and the PAH signal increases with increasing depth, becoming constant at ~1200 μm within the interior, well away from the fusion crust.

The accumulation of PAHs on the Greenland ice sheet over the past 400 years had been studied in ice cores (Kawamura and Suzuki, 1994). The total concentration of PAHs in the cores varies from 10 ppt for pre-industrial times to 1 ppb for recent snow deposition. Concentrations of PAHs in the Antarctic ice were expected to lie between these two values. Analysis of Antarctic salt deposits on a heavily weathered meteorite (LEW 85320) by μ^{2}Ms did not show the presence of terrestrial PAHs within detection limits, which suggested that the terrestrial contamination of PAHs for ALH84001 is less than 1%.

The freshly broken but preexisting fracture surfaces rich in PAHs also typically displayed carbonate globules. The globules tend to be disk-shaped and flattened parallel to the fracture surface. Intact carbonate globules appeared orange in visible light and had a rounded appearance; many displayed alternating black and white rims. Under high magnification stereo light microscopy or SEM stereo imaging, some of the globules appear to be quite thin and pancake-like, suggesting that the carbonates formed in the restricted width of a thin fracture.

This geometry limited their growth perpendicular to, but not parallel to, the fracture.

The PAHs were highest in the carbonate-rich regions. From averaged mass spectra, two groups of PAHs were identified. A middle-mass group of 178–276 Da (Dalton) dominated, which were composed of simple 3- to 6-ring PAH skeletons. The alkylated homologs were <10% of the total signal intensity. Principal peaks at 178, 202, 228, 252, and 278 Da were assigned to phenanthrene ($C_{14}H_{10}$; Fig. 7.3A), pyrene ($C_{16}H_{10}$; Fig. 7.3B), chrysene ($C_{18}H_{12}$; Fig. 7.3C), perylene or benzopyrene ($C_{20}H_{12}$; Fig. 7.3D), and anthanthrene ($C_{22}H_{12}$; Fig. 7.3E).

A second weak high-mass group was from about 300 to 450 Da. The primary source of PAHs was anthropogenic emissions, which were characterized by the presence of abundant aromatic heterocycles, primarily dibenzothiophene ($C_{12}H_8S$; 184 Da; see Fig. 7.3F). The PAHs in ALH84001 were present at ppm levels but dibenzothiophene was not observed in any of the samples.

7.4.4 Carbonate globules and magnetite crystals

In analysis by an electron microprobe, carbonate globules were observed (Fig. 7.4). The larger globules ($>100\,\mu m$) had Ca-rich cores surrounded by alternating Fe and Mg-rich bands. Near the edge of the globule, several sharp thin bands were present. The first band was rich in Fe and S, the second was rich in Mg with no Fe, and the third was rich in Fe and S again. The Fe-rich rims were composed mainly of fine-grained magnetite ranging in size from ~ 10 to 100 nm and minor amounts of pyrrhotite (~ 5 vol%).

Magnetite crystals were cubic, teardrop, or irregular shape. Individual crystals had well-preserved structures with no lattice defects. The magnetite and Fe-sulfide were in a fine-grained carbonate matrix. Composition of the fine-grained carbonate matrix matched that of coarse-grained carbonates located adjacent to the rim.

7.4.5 Structures resembling the terrestrial Life

The occurrence of the fine-grained carbonate and magnetite phases could be explained by either inorganic or biogenic processes. Magnetite particles were similar (chemically, structurally, and morphologically) to terrestrial magnetite particles known as magnetofossils, remaining as bacterial magnetosomes. The sketch of extracellular precipitated superparamagnetic magnetite particles produced by terrestrial magnetotactic bacteria is shown in Fig. 7.5 which resembled some of magnetite crystals of ALH84001.

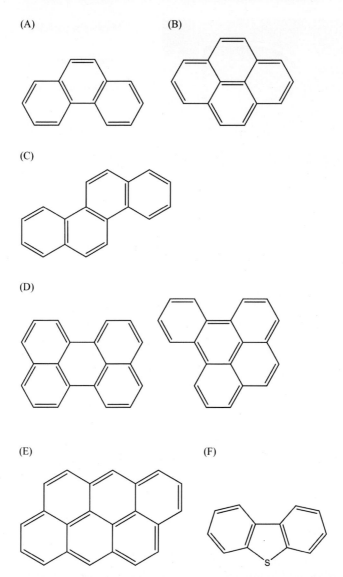

FIGURE 7.3 **Chemical structures of polycyclic aromatic hydrocarbons (PAHs) in ALH 84001.** (A) phenanthrene ($C_{14}H_{10}$); (B) pyrene ($C_{16}H_{10}$); (C) chrysene ($C_{18}H_{12}$); (D) perylene or benzopyrene ($C_{20}H_{12}$); (E) anthanthrene ($C_{22}H_{12}$); and (F) dibenzothiophene ($C_{12}H_8S$). *Source: Data from McKay, D.S., Gibson Jr, E.K., Thomas-Keprta, K.L., et al., 1996. Search for past life on Mars: possible relic biogenic activity in martian meteorite ALH84001. Science 273, 924–930.*

The elongated carbonates formed by the connection of tiny ovoid to cylindrical particles resembling the magnetite particles in Fig. 7.5A were observed in Fig. 7.5B. The object was about 400 nm in the longest

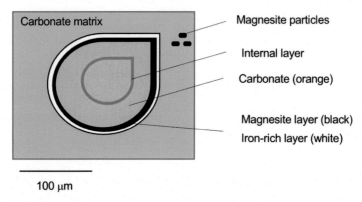

FIGURE 7.4 **A schematic diagram of a carbonate globule in a backscattered electron image of ALH84001**. The colors are artificial colors. *Source: Modified from Martel, J., Young, D., Peng, H.H., Wu, C.Y., Young, J.D., 2012. Biomimetic properties of minerals and the search for life in the Martian meteorite ALH84001. Ann. Rev. Earth Planet. Sci. 40, 167–193.*

dimension, and about 40 nm across. Such object was found on the surface of calcite concretions grown from Pleistocene ground water in southern Italy, which was interpreted as nanobacteria.

The origin of these textures on the surface of the ALH84001 carbonates was unclear. One possible explanation is that the textures were erosional remnants of the carbonate that happened to be in the shape of ovoids and elongate forms.

An alternative explanation was that these textures, as well as the nanosized magnetite and Fe-sulfides, were the products of microbiological activity. In general, authigenic secondary minerals in Antarctica were oxidized or hydrated. The lack of PAHs in the other analyzed Antarctic meteorites, the sterility of the sample, and the nearly unweathered nature of ALH84001 argued against an Antarctic biogenic origin.

The filamentous features were similar in size and shape to nanobacteria in limestone. The elongate forms resembled some forms of fossilized filamentous bacteria in the terrestrial fossil record. In general, the terrestrial bacteria microfossils were more than an order of magnitude larger than the forms seen in the ALH84001 carbonates.

The carbonate globules in ALH84001 shown in Fig. 7.5 were clearly a key element in the interpretation of this Martian meteorite. The origin of these globules was controversial. Harvey and McSween (1996) and Mittlefehldt (1994) argued, on the basis of microprobe chemistry and equilibrium phase relationships, that the globules were formed by high-temperature metamorphic or hydrothermal reactions. Alternatively, Romanek et al. (1992) argued on the basis of isotopic relationships that the carbonates were formed under low-temperature hydrothermal conditions. Finally, they interpreted that the carbonate globules had a

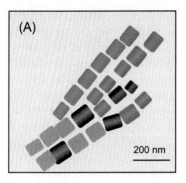

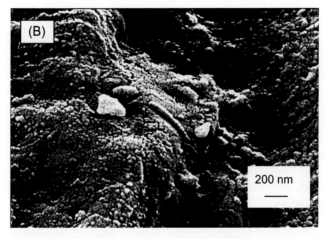

FIGURE 7.5   **Schematic diagrams showing morphologies of crystals.** (A) A sketch of magnetites observed in terrestrial magnetotactic bacteria. This sketch is drawn based on the photographs of Martel et al. (2012). (B) The tube-like structures observed in situ in the ALH84001 meteorite. *Source: (A) Modified from Martel, J., Young, D., Peng, H.H., Wu, C.Y., Young, J.D., 2012. Biomimetic properties of minerals and the search for life in the Martian meteorite ALH84001. Ann. Rev. Earth Planet. Sci. 40, 167–193. (B) NASA/JSC/Stanford University. PIA00288.jpeg. The scale bar is added by the author.*

biogenic origin and were likely formed at low temperatures. Pure Mg-carbonate (magnesite) was produced by biomineralization under alkaline conditions.

7.4.6 Magnetite crystals were not biomarkers

On the Earth, some species of bacteria contain magnetosomes, which consisted of membrane-delineated vesicles with crystals of magnetite

(Fe_3O_4) and/or greigite (Fe_3S_4) (Bazylinski and Frankel, 2004; Chang and Kirschvink, 1989). McKay et al. (1996) and the same group studies (Thomas-Keprta et al., 2000, 2001, 2002) reported that the magnetite crystals detected in ALH84001 were similar to those found in terrestrial magnetotactic bacteria.

Several characteristics of the magnetite particles such as crystal purity, structural perfection, lack of defects, etc., were proposed to be signs of biological processes (Thomas-Keprta et al., 2000). Some authors noted that only one-fourth of the magnetite in ALH84001 had sizes and shapes similar to those of the terrestrial magnetotactic bacteria (Buseck et al., 2001; Golden et al., 2004). This means that the sizes and shapes of most magnetite crystals observed in ALH84001 were completely different from those found in magnetotactic bacteria from the Earth crystals, and those found in ALH84001 were on average much smaller (Grady et al., 1996).

There existed many structural, morphological, and crystallographic variability in the magnetite crystals of magnetotactic bacteria, thus it was difficult to confirm a biological origin for magnetite particles simply by comparing them with the magnetite crystals observed in terrestrial bacteria (Buseck et al., 2001). Based on these results, it became obvious that the features of magnetite crystals such as shapes, sizes, purity, and crystalline structures did not indicate a biogenic origin (Buseck et al., 2001; Golden et al., 2004). Some researchers had also reported that magnetite chains could be produced in artificial condition (Golden et al., 2004; Nealson and Cox, 2002), insisting against the use of these chains as an evidence for biogenic origin.

These results gave doubt on the hypothesis that the magnetite crystals in ALH84001 were of biogenic origin. Several researchers were against the use of magnetite crystals as signs of past Life (Barber and Scott, 2002; Buseck et al., 2001; Frankel and Buseck, 2000; Golden et al., 2004). Based on this information, the hypothesis that magnetotactic microorganisms present in the ancient meteorite ALH84001 were suspected.

7.4.7 Nanobacteria are not Life

Many scientists were not convinced for the idea of nanobacteria (Abbott, 1999, 2000; Bradbury, 1998; Day, 1998; Deresinski, 2008; Hamilton, 2000; Hogan, 2003, 2004a, 2004b; Hopkins, 2008; Saey, 2008; Travis, 1998). Abbott and other authors (Abbott, 1999; Hogan, 2004b; Hopkins, 2008; Saey, 2008; Urbano and Urbano, 2007) considered the idea of nanobacteria as the difference of thinking grounds. Martel's group (Martel and Young, 2008; Martel et al., 2010; Wu et al., 2009; Young et al., 2009a; Young and Martel, 2010) and others (Barr et al., 2003; Benzerara et al., 2003; Cisar et al., 2000; Drancourt et al., 2003;

Kirkland et al., 1999; Raoult et al., 2008; Vali et al., 2001) tried to prove the nanobacteria to be experimentally mistaken idea.

Especially, the Martel's group showed that the nanobacteria specimens as studied in many publications were nonliving mineralo-organic nanoparticles. They had further demonstrated that the round shape of the bacteria-like forms was due to the presence of an amorphous mineral phase transiently stabilized by mineralization inhibitors that bind the growing mineral and prevented it from forming the faceted edges (Martel and Young, 2008; Martel et al., 2010; Wu et al., 2009; Young et al., 2009a; Young and Martel, 2010). Amorphous minerals represented random networks which was contrast against the orderly periodic crystal structure. This randomness formed the mineral to be rounded. The small initial size of the mineral nanoparticles was made by the presence of mineralization inhibitors which interfered with the particle growth.

Proteins can act as mineralization inhibitors (Martel and Young, 2008; Raoult et al., 2008; Wu et al., 2009), and other biological macromolecules such as phospholipids (Cisar et al., 2000), and inorganic ions such as magnesium (Martel and Young, 2008; Young et al., 2009b). The nanobacteria proponents, such as Kajander or McKay, claimed that the presence of both inorganic and organic inhibitors within a mineral particle resulted from a normal living process (Ciftcioglu and McKay, 2010; Ciftcioglu et al., 2006; Hjelle et al., 2000; Kajander and Ciftcioglu, 1998; Kajander et al., 1997; Kumar et al., 2006; Lieske, 2008; Shiekh et al., 2009; Sommer and Wickramasinghe, 2005).

In the case of ALH84001, various ions were present during the formation of the carbonate globules because of the existence of magnesite and iron-rich minerals on the surfaces of the globules. Several researchers have made bacteria-like structures similar to the ones found in ALH84001 by the precipitation of calcium carbonate in vitro (Kirkland et al., 1999; Martel et al., 2012; Vecht and Ireland, 2000). In addition, bacteria-like structures similar to the ALH84001 had been observed in other meteorites from the Moon, where Life was considered to be unlikely. This also supported that the bacteria-like structures could be produced from nonliving processes (Martel et al., 2012; Sears and Kral, 1998).

7.4.8 Ineffectiveness of morphological similarities

The nanobacteria hypothesis is formed on morphological grounds since no chemical evidences are found by any laboratory except the initial studies advanced by Kajander and the nanobacteria group (Martel et al., 2012). It is clear to us that nanoparticles are composed of carbonate and/or phosphate and that this chemical reactivity should be able to explain the presence of organic compounds which may be bound as contaminants or trapped bystanders to the putative nanobacteria. The lack of specific biomarkers other than morphology needs great care when searching for Life in unknown specimens (Martel et al., 2012).

In studies of morphological signatures of terrestrial fossils and extraterrestrial Life, Garcia-Ruiz and his colleagues have shown that morphology alone is a poor and ambiguous indicator of biogenicity (Garcia-Ruiz et al., 2002, 2003; Hyde et al., 2003). These authors demonstrated that the morphologies of common inorganic minerals such as barium carbonate and silica that precipitate in alkaline environments resemble primitive organisms and complex biological structures. Notably, they propose a mechanism to explain the formation of the smooth and curved surfaces in waves (Garcia-Ruiz et al., 2009). The waves of growing curl and form structures like cauliflowers. These cauliflower-like forms remind the colonies of nanobacteria observed in human body fluids and the ALH84001 meteorite.

In the case of the ALH84001 studies, some researchers argued that the bacteria-like structures could represent gold coating in sample preparation, which is used for electron microscopy to give the samples electronically conducive (Bradley et al., 1997). These authors showed that the formation of parallel lamellae on surface fractures as well as appropriate sample orientation can give the impression of bacteria-like formations in ALH84001 samples. However, using atomic force microscopy (AFM), another group showed even without using gold coating, that bacteria-like structures were apparent in ALH84001 samples, thus concluding that the bacteria-like morphologies are not just the result of sample preparation (Steele et al., 1998; Martel et al., 2012). Using various forms of optical and electron microscopy, abundant bacteria-like structures in mineral precipitates that were not coated were observed (Martel and Young, 2008; Martel et al., 2010; Peng et al., 2011; Wu et al., 2009; Young et al., 2009a,b). Coating artifacts are unlikely to explain all the bacteria-like structures found in the ALH84001 meteorite. Instead, they are more likely the result of simple chemical processes (Martel et al., 2012).

These observations indicate that bacteria-like structures observed in geological specimens cannot be considered as convincing evidence of past Life. The bacteria-like structures are more likely to appear from chemical processes that are ubiquitous and that occur whenever minerals precipitate from supersaturated solutions in the presence of mineralization inhibitors (Martel et al., 2012).

7.4.9 Summary: morphology is not a decisive factor

In the discussions of these sections, we have learned that the morphology is not a decisive factor for determination of living things from the ALH84001 case. Researchers, especially geologists sometimes forget this simple rule. Thus, confusions sometimes occur as shown in the previous sections. We must always pay attention to morphological discussion.

7.4.10 Organic synthesis associated with serpentinization and carbonation on early Mars

The Martian meteorite ALH84001 formed during the Noachian period on Mars (igneous crystallization age of ~ 4.09 Ga). ALH84001 was predominately composed of orthopyroxene (Opx) and contained carbonate globules related to aqueous processes on early Mars at ~ 3.9 Ga. As ALH84001 is one of the oldest rocks from Mars, the similar early planetary processes on Mars should have occurred on early Earth. As nitrogen-containing organic compounds were found in ALH 84001, these organics were formed by possibly abiotic processes, such as impact-related, igneous, and hydrothermal processes, or biological production or mere terrestrial contamination. To investigate the origin of the organic materials, Steele et al. (2022) applied nanoscale spectral, imaging, structural, and isotopic analysis techniques to ALH84001.

They found that the microdenticular texture of Opx indicates aqueous and hydrothermal alteration of Opx. Amorphous silica, talc-like phases, magnetite, and carbonates were observed at the outer edges of the altered Opx, indicating serpentinization and carbonation. Serpentinization is an abiotic organic synthesis mechanism, in which basaltic rocks react with and aqueous fluid, resulting in serpentine minerals, magnetite, and hydrogen (see equations below):

$$3MgFeSiO_4 + 3H_2O \rightarrow Mg_3Si_2O_5(OH)_4 + Fe_3O_4 + SiO_2 + H_2 \qquad (7.1)$$

7.4.10.1 Opx serpentine magnetite

The hydrogen can reduce aqueous CO_2 to methane, CO and other organics, which forms alkanes and other organic minerals. Aqueous alteration of Opx in ALH 84001 produced carbonate globules, amorphous silicates, and silica and talc-like phases.

$$2Mg_3Si_2O_5(OH)_4 + 3CO_2 \rightarrow Mg_3Si_4O_{10}(OH)_2 + 3MgCO_3 + 3H_2O \qquad (7.2)$$

7.4.10.2 Serpentine talc magnesite

Steele et al. (2022) found complex refractory organic materials associated with mineral assemblages that formed by mineral carbonation and serpentinization reactions. The organic molecules existed with nanophase magnetite; both formed in situ during water-rock interactions on Mars. Two potentially distinct mechanisms of abiotic organic synthesis operated on early Mars during the late Noachian period (3.9 ~ 4.1 Ga).

7.5 The least contaminated Martian meteorite, Tissint fell

7.5.1 The freshest Martian meteorite, Tissint

The Tissint meteorite fell in Morocco in 2011. The meteorite was broken into pieces, and more than 17 kg were collected after a few days from the fall, resulting in keeping very fresh condition with minimum terrestrial contamination. Tissint comprises an olivine-phyritic shergottite of highly composite nature, with similarities to Antarctic meteorite EETA 79001 and with a crystallization age of 665 ± 74 Ma, and a cosmic ray exposure age of 0.9 ± 0.2 Ma (Schultz et al., 2020). Tissint comprises all high-pressure phases (seven minerals and two mineral glasses), indicating that shock metamorphism was up to ~ 25 GPa and $2000°C$. Steele et al. (2016) and Lin et al. (2014) found heterogeneous distribution of carbonaceous components in Tissint which filled open fractures and enclosed shock melt veins using NanoSIMS, time-of-flight-SIMS, electron microscopy, and EDX.

7.5.2 The diversity of organic geochemistry on Mars glimpsed from the Tissint meteorite

Schmitt-Kopplin et al. (2023) found carbon phases with magnesium silicates in Tissint using electron microscopy and EDX, and showed the coexistence of reactive mineral and organic phases. Thermogravimetry—differential scanning calorimetry—mass spectrometric evolved gas analysis (TG-DSC-MSEGA) showed the very low mass loss from the sample up to $1000°C$ (Δm 1.52%). In the temperature interval up to $600°C$, CO_2 ($m/z = 44$) was formed from the intramolecular redox reaction of organics, and other masses resulted from the pyrolytic decomposition of some organics [$m/z = 15$ (CH_3^+), 39 ($C_3H_3^+$), and 41 ($C_3H_5^+$)]. In $600°C \sim 1000°C$, formation of CO_2 and SO_2 was detected, where CO_2 resulted from the decomposition of carbonates, while SO_2 from the degradation of sulfates. These results were consistent with those of the tunable laser spectrometer of the Sample Analysis at Mars on Curiosity rover (SAM), showing the production of CO_2, CH_4, CO, carbonyl sulfide, and CS_2 when pyrolyzing the samples from Gale crater on Mars and the Tissint sample (House et al., 2022), and with earlier results from the evolved gas analyzer of SAM that detected water, H_2, SO_2, H_2S, NO, CO_2, CO, O_2, and HCl (Sutter et al., 2017).

Schmitt-Kopplin et al. (2023) also analyzed methanol extracts from the Tissint meteorite by electrospray ionization (ESI), atmospheric photoionization (APPI), Fourier transform ion cyclotron resonance MS (FTICR-MS), and by proton-detected nuclear magnetic resonance (^{1}H NMR) spectroscopy. The methanol extracts are considered to preserve the original speciation of organic molecules. The Tissint methanol extracts demonstrated more than 5000 assigned elementary compositions (considering the volatile elements C, H, O, N, S, and Mg), with CHO (38%), CHON (30%), CHOS (20%) and CHONS (11%) compounds. More than a hundred separate compounds of [CHO(S)Mg] were also detected. The data showed a typical meteoritic profile reflecting hydration, alkylations, hydroxylation, and carboxylations. These observations were consistent with in situ observations of Martian organic phases by scanning transmission X-ray microscopy analyses (Steel et al., 2018). In addition, hydrogen isotope analyses of Tissint confirmed its Martian origin.

7.5.3 Conclusions by the Tissint meteorite

The large range of organics, such as C3−7 aliphatic branched carboxylic acids and aldehydes, olefins, and polyaromatics with/without heteroatoms were found in the Tissint meteorite. Olivine macrocrystals and the melt veins contained abundant organo-magnesium compounds, indicating organo-synthesis processes with the magnesium silicates, as previously proposed. The diverse chemistry and abundant complex molecules suggested organic heterogeneity within the minerals in the Martian mantle and crust. Furthermore, this diversity was obtained within a relatively short time of 4 Ga.

7.5.4 Electrochemistry of brines and metal-rich minerals on Mars reduced CO_2 to alcohols

The origin and synthesis mechanisms of abiotic organic carbons on Mars are important to the habitability potential of Mars. Both rover missions to Mars and analyses of Martian meteorites gave clues for the presence of volatile and refractory organic compounds in Martian rocks. Steele et al. (2016) reviewed and presented the four possible synthesis pathways: (1) impact generated; (2) impact induced; (3) primary igneous; or (4) secondary hydrothermal. The organic carbon in the Tissint meteorite was not associated with the presence of a carbonate phase, nor is it igneous in origin. The precise mechanism of formation of such an organic carbon pool remains unclear.

Steele et al. (2018) reported on the analysis of organic phases associated with titano-magnetite, magnetite, pyrite, and pyrrhotite in three Martian meteorites, Nakhla, Tissint, and NWA 1950. Nakhla and Tissint are falls, and NWA 1950 is a find. There were secondary Martian carbonate and clay phases (including saponite and serpentine) in cracks within Nakhla, while Tissint shows evidence of minimal secondary alteration processes. NWA 1950 had carbonates of terrestrial origin, although an indigenous macromolecular carbon (MMC) phase is documented in this meteorite. Steele et al. (2018) developed the hypothesis that interactions among spinel-group minerals, sulfides, and a brine enable the electrochemical reduction of aqueous CO_2 to organic molecules. A similar process should likely occur wherever igneous rocks with spinel-group minerals and/or sulfides encounter brines.

References

Abbott, A., 1999. Battle lines drawn between 'nanobacteria' researchers. Nature 401, 105.

Abbott, A., 2000. Researchers fail to find signs of life in 'living' particles. Nature 408, 394.

Agee, C.B., Wilson, N.V., McCubbin, F.M., Ziegler, K., Polyak, V.J., et al., 2013. Unique meteorite from Early Amazonian Mars: water-rich basaltic breccia Northwest Africa 7034. Science 339, 780–785.

Arrhenius, G., Mojzsis, S., 1996. Extraterrestrial life: life on Mars- then and now. Curr. Biol. 6, 1213–1216.

Barber, D.J., Scott, E.R.D., 2002. Origin of supposedly biogenic magnetite in the Martian meteorite Allan Hills 84001. Proc. Natl. Acad. Sci. U. S. A. 99, 655–656.

Barr, S.C., Linke, R.A., Janssen, D., et al., 2003. Detection of biofilm formation and nanobacteria under lon-term cell culture conditions in serum samples of cattle, goats, cats, and dogs. Am. J. Vet. Res. 64, 176–182.

Bazylinski, D.A., Frankel, R.B., 2004. Magnetosome formation in prokaryotes. Nat. Rev. Microbiol. 2, 217–230.

Benzerara, K., Menguy, N., Guyot, F., Dominici, C., Gillet, P., 2003. Nanobacteria-like calcite single crystals at the surface of the Tatouine meteorite. Proc. Natl. Acad. Sci. U. S. A. 100, 7438–7442.

Bradbury, J., 1998. Nanobacteria may lie at the heart of kidney stones. Lancet 352, 121.

Bradley, J.P., Harvey, R.P., McSween Jr, H.Y., 1997. No 'nanofossils' in Martian meteorite. Nature 390, 454–456.

Buseck, P.R., Dunin-Borkowski, R.E., Devouard, B., Frankel, R.B., McCartney, M.R., 2001. Magnetite morphology and life on Mars. Proc. Natl. Acad. Sci. U. S. A. 98, 13490–13495.

Chang, S.R., Kirschvink, J.L., 1989. Magnetofossils, the magnetization of sediments, and the evolution of magnetite biomineralization. Annu. Rev. Earth Planet Sci. 17, 169–195.

Ciftcioglu, N., McKay, D.S., Mathew, G., Kajander, E.O., 2006. Nanobacteria: fact of fiction? Characteristics detection, and medical importance of novel self-replicating, calcifying nanoparticles. J. Investig. Med. 54, 385–394.

Ciftcioglu, N., McKay, D.S., 2010. Pathological calcification and replicating calcifying-nanoparticles: general approach and correlation. Pediatr. Res. 67, 490–499.

Cisar, J.O., Xu, D.Q., Thompson, J., Swaim, W., Hu, L., Kopecko, D.J., 2000. An alternative interpretation of nanobacteria-induced biomineralization. Proc. Natl. Acad. Sci. U. S. A. 97, 11511–11515.

Clavin, W., 2015. NASA's Curiosity rover team confirms ancient lakes on Mars. http://www.nasa.gov/feature/jpl/nasas-curiosity-rover-team-confirms-ancient-lakes-on-mars.

Day, M., 1998. Mean microbes-hard little bugs could cause everything from tumours to dementia. New Sci. 159, 11.

Deresinski, S., 2008. Nan(non)bacteria. Clin. Infect. Dis. 47, v—vi.

Dick, S.J., 2006. NASA and the search for life in the universe. Endeavour 30, 71—75.

Drancourt, M., Jacomo, V., Lepidi, H., et al., 2003. Attempted isolation of *Nanobacterium* sp. microorganisms from upper urinary tract stones. J. Clin. Microbiol. 41, 368—372.

Eigenbrode, J.L., Summons, R.E., Steele, A., Freissinet, C., Millan, M., et al., 2018. Organic matter preserved in 3-billion-year-old mudstones at Gale crater, Mars. Science 360, 1096—1101.

Frankel, R.B., Buseck, P.R., 2000. Magnetite biomineralization and ancient life on Mars. Curr. Opin. Chem. Biol. 4, 171—176.

Garcia-Ruiz, J.M., Carnerup, A., Christy, A.G., Welham, N.J., Hyde, S.T., 2002. Morphology: an ambiguous indicator of biogenicity. Astrobiology 2, 353—369.

Garcia-Ruiz, J.M., Hyde, S.T., Carnerup, A.M., et al., 2003. Self-assembled silica-carbonate structures and detection of ancient microfossils. Science 302, 1194—1197.

Garcia-Ruiz, J.M., Melero-Carcia, E., Hyde, S.T., 2009. Morphogenesis of self-assembled nanocrystalline materials of barium carbonate and silica. Science 323, 362—365.

Gibson Jr, E.K., McKay, D.S., Thomas-Keprta, K., Romanek, C.S., 1997. The case for relic life on Mars. Sci. Am. 27, 58—65.

Golden, D.C., Ming, D.W., Morris, R.V., Brearley, A.J., 2004. Lauer HVJr: evidence for exclusively inorganic formation of magnetite in Martian meteorite ALH84001. Am. Mineral 89, 681—695.

Grady, M., Wright, I., Pillinger, C., 1996. Opening a martian can of worms? Nature 382, 575—576.

Grady, M.M., Verchovsky, A.B., Wright, I.P., 2004. Magmatic carbon in Martian meteorites: attempts to constrain the carbon cycle on Mars. Int. J. Astrobiol. 3, 117—124.

Hamilton, A., 2000. Nanobacteria: gold mine or minefield of intellectual enquiry? Microbiol. Today 27, 182—184.

Harvey, R., McSween Jr, H.P., 1996. A possible high-temperature origin for the carbonates in the martian meteorite ALH84001. Nature 382, 49—51.

Hjelle, J.T., Miller-Hjelle, M.A., Poxton, I.R., et al., 2000. Endotoxin and nanobacteria in polycystic kidney desease. Kidney Int. 57, 2360—2374.

Hogan, J., 2003. 'Microfossils' made in the laboratory. New Sci. 2422, 14—15.

Hogan, J., 2004a. Are nanobacteria alive or just strange crystals? New Sci. 2448, 6—7.

Hogan, J., 2004b. Nanobacteria revelations provoke new controversy. New Sci. http://www.newscientist.com/article/d5009-nanobacteria-revelations-provoke-new-controversy.html.

Hopkins, M., 2008. Nanobacteria theory takes a hit. Nat. News. http://www.nature.com/news/2008/080417/full/news.2008.762.html.

House, C.H., Wong, G.M., Webster, C.R., Flesch, G.J., Franz, H.B., et al., 2022. Depleted carbon isotope compositions observed at Gale crater. Mars. Proc. Natl. Acad. Sci. U. S. A. 119, e2115651119.

Humayun, M., Nemchin, A., Zanda, B., Hewins, R.H., Grange, M., Kennedy, A., et al., 2013. Origin and age of the earliest Martian crust from meteorite NWA 7533. Nature 503, 513—516.

Hyde, S.T., Carnerup, A.M., Larsson, A.K., Christry, A.G., Garcia-Ruiz, J.M., 2003. Self-assembry of carbonate-silica colloids: between living and non-living form. Phys. A 339, 24—33.

Jull, A.J.T., Courtney, C., Jeffrey, D.A., Beck, J.W., 1998. Isotopic evidence for a terrestrial source of organic compounds found in Martian meteorites Allan Hills 84001 and Elephant Moraine 79001. Science 279, 366—369.

Kajander, E.O., Ciftcioglu, N., 1998. Nanobacteria: an alternative mechanism for pathogenic intra-and extra-cellular calcification and stone formation. Proc. Natl. Acad. Sci. U. S. A. 95, 8274–8279.

Kajander, E.O., Kuronen, I., Akerman, K.K., Pelttari, A., Ciftcioglu, N., 1997. Nanobacteria from blood: the smallest culturable autonomously replicating agent on earth. Proc. SPIE 3111, 420–428.

Kawamura, K., Suzuki, I., 1994. Ice core record of polycyclic aromatic hydrocarbons over the past 400 years. Naturwissenschaften 81, 502–505.

Kirkland, B.L., Lynch, F.L., Rahnis, M.A., Folk, R.L., Molineux, I.J., McLean, R.J.C., 1999. Alternative origins for nannobacteria-like objects in calcite. Geology 27, 347–350.

Korablev, O., Olsen, K.S., Trokhimovskiy, A., Lefevre, F., Montmessin, F., et al., 2021. Transient HCl in the atmosphere of Mars. Sci. Adv. 7. Available from: https://doi.org/10.1126/sciadv.abe4386.

Kumar, V., Farell, G., Yu, S., Harrington, S., Fitzpatrick, L., et al., 2006. Cell biology of pathologic renal calcification contribution of crystal transcytosis, cell-mediated calcification, and nanoparticles. J. Investig. Med. 54, 412–424.

Lieske, J.C., 2008. Can biologic nanoparticles initiate nephrolithiasis? Nat. Clin. Pract. Nephrol. 4, 308–309.

Lin, Y., Goresy, A.E., Hu, S., Zhang, J., Gillet, P., et al., 2014. NanoSIMS analysis of organic carbon from the Tissint martian meteorite: evidence for the past existence of subsurface organic-bearing fluids on Mars. Meteorit. Planet. Sci. 49, 2201–2218.

Martel, J., Young, J.D., 2008. Purported nanobacteria in human blood as calcium carbonate nanoparticles. Proc. Natl. Acad. Sci. U. S. A. 105, 5549–5554.

Martel, J., Wu, C.Y., Young, J.D., 2010. Critical evaluation of gamma-irradiated serum used as feeder in the culture and demonstration of putative nanobacteria and calcifying nanoparticles. PLoS One 5, e10343.

Martel, J., Young, D., Peng, H.H., Wu, C.Y., Young, J.D., 2012. Biomimetic properties of minerals and the search for life in the Martian meteorite ALH84001. Ann. Rev. Earth Planet. Sci. 40, 167–193.

McKay, D.S., Gibson Jr, E.K., Thomas-Keprta, K.L., et al., 1996. Search for past life on Mars: possible relic biogenic activity in martian meteorite ALH84001. Science 273, 924–930.

Mittlefehldt, D.W., 1994. ALH84001, a cumulate orthopyroxenite member of the martian meteorite clan. Meteoritics 29, 214–221.

Nealson, K.H., Cox, B.L., 2002. Microbial metal-ion reduction and Mars: extraterrestrial expectations? Curr. Opin. Microbiol. 5, 296–300.

Orosei, R., Lauro, S.E., Pettinelli, E., Cicchetti, A., Coradini, M., et al., 2018. Radar evidence of subglacial liquid water on Mars. Science 361, 490–493.

Peng, H.-H., Martel, J., Lee, Y.H., Ojcius, D.M., Young, J.D., 2011. Serum-derived nanoparticles: de novo generation and growth in vitro, and internalization by mammalian cells in culture. Nanomedicine 6, 643–658.

Raoult, D., Drancourt, M., Azza, S., et al., 2008. Nanobacteria are mineralo fetuin complexes. PLoS Patbog. 4, e41.

Romanek, C.S., Grossman, E.L., Morse, J.W., 1992. Carbon isotopic fractionation in synthetic aragonite and calcite: effects of temperature and precipitation rate. Geochim. Cosmochim. Acta 56, 419–430.

Saey, T.H., 2008. Rest in peace nanobacteria, you were not alive after all. Sci. News 173, 6–7.

Schmitt-Kopplin, P., Matzka, M., Ruf, A., Menez, B., Aoudjehane, H.C., et al., 2023. Complex carbonaceous matter in Tissint Martian metian meteorites give insights into the diversity of organic geochemistry on Mars. Sci. Adv. 9. Available from: https://doi.org/10.1126/sciadv.add6439.

Schulz, T., Povinec, P.P., Ferrière, L., Jull, A.T., Kováčik, A., et al., 2020. The history of the Tissint meteorite, from its crystallization on Mars to its exposure in space: new geochemical, isotopic, and cosmogenic nuclide data. Meteorit. Planet. Sci. 55, 294–311.

Sears, D.W., Kral, T.A., 1998. Martian "microfossils" in lunar meteorites? Meteorit. Planet. Sci. 33, 791–794.

Shiekh, F.A., Miller, V.M., Lieske, J.C., 2009. Do calcifying nanoparticles promote nephrolithiasis? A review of the evidence. Clin. Nephrol. 71, 1–8.

Sommer, A.P., Wickramasinghe, N.C., 2005. Functions and possible provenance of primordial proteins – Part II: microorganism aggregation in clouds triggered by climate change. J. Proteome Res. 4, 180–184.

Stamenković, V., Ward, L.M., Mischna, M., Fischer, W.W., 2018. O_2 solubility in Martian near-surface environments and implications for aerobic life. Nat. Geosci. 11, 905–909.

Steele, A., Goddard, D., Beech, I.B., et al., 1998. Atomic force microscopy imaging of fragments from the Martian meteorite ALH 84001. J. Microsc. 189, 2–7.

Steele, A., McCubbin, F.M., Fries, M., Kater, L., Boctor, N., et al., 2012. A reduced organic carbon component in Martian basalts. Science 337, 212–215.

Steele, A., McCubbin, F.M., Fries, M.D., 2016. The provenance, formation, and implications of reduced carbon phases in Martian meteorites. Meteorit. Planet. Sci. 51, 2203–2225.

Steele, A., Benning, L.G., Wirth, R., Siljeström, S., Fries, M.D., et al., 2018. Organic synthesis on Mars by electrochemical reduction of CO_2. Sci. Adv. 4, eaat5118.

Steele, A., Benning, L.G., Wirth, R., Schreiber, A., Araki, T., et al., 2022. Organic synthesis associated with serpentinization and carbonation on early Mars. Science 375, 172–177.

Sutter, B., Mcadam, A.C., Mahaffy, P.R., Ming, D.W., Edgett, K.S., Rampe, E.B., et al., 2017. Evolved gas analyses of sedimentary rocks and eolian sediment in Gale crater, Mars: results of the Curiosity rover's sample analysis at Mars instrument from Yellowknife Bay to the Namib Dune. J. Geophys. Res. Planets 122, 2574–2609.

Thomas-Keprta, K.L., Bazylinski, D.A., Kirschvink, J.L., Clemett, S.J., McKay, D.S., 2000. Elongated prismatic magnetite crystals in ALH84001 carbonate globules: potential Martian magnetofossils. Geochim. Cosmochim. Acta 64, 4049–4081.

Thomas-Keprta, K.L., Clemett, S.J., Bazylinski, D.A., Kirschvink, J.L., McKay, D.S., 2001. Truncated hexaoctahedral magnetite crystals in ALH84001: presumptive biosignatures. Proc. Natl. Acad. Sci. U. S. A. 4, 2164–2169.

Thomas-Keprta, K.L., Clemett, S.J., Bazylinski, D.A., Kirschvink, J.L., McKay, D.S., 2002. Magnetofossils from ancient Mars: a robust biosignature in the Martian meteorite ALH84001. Appl. Environ. Microbiol. 68, 3663–3672.

Travis, J., 1998. The bacteria in the stone: extra-tiny microorganisms may lead to kidney stones and other diseases. Sci. News 154, 75–77.

Urbano, P., Urbano, F., 2007. Nanobacteria: facts or fancies? PLoS Patbog. 3, e55.

Vali, H., McKee, M.D., Ciftcioglu, N., et al., 2001. Nanoforms: a new type of protein-associated mineralization. Geochim. Cosmochim. Acta 65, 63–74.

Vecht, A., Ireland, T.G., 2000. The role of vaterite and aragonite in the formation of pseudo-biogenic carbonate structures: implication for Martian exobiology. Geochim. Cosmochim. Acta 64, 2719–2725.

Webster, C.R., Mahaffy, P.R., Atreya, S.K., Moores, J.E., Flesch, G.J., et al., 2018. Background levels of methane in Mars' atmosphere show strong seasonal variations. Science 360, 1093–1096.

Wu, C.Y., Martel, J., Young, D., Young, J.D., 2009. Fetuin-A/albumin-mineral complexes resembling serum calcium granules and putative nanobacteria: demonstration of a dual inhibition-seeding concept. PLoS One 4, e8058.

Young, J.D., Martel, J., Young, L., Wu, C.Y., Young, A., Young, D., 2009a. Putative nano-bacteria represent physiological remnants and culture by-products of normal calcium homeostasis. PLoS One 4, e4417.

Young, J.D., Martel, J., Young, D., Young, A., Hung, C.M., Young, L., et al., 2009b. Characterization of granulations of calcium and apatite in serum as pleomorphic mineralo-protein complexes and as precursors of putative nanobacteria. PLoS One 4, e5421.

Young, J.D., Martel, J., 2010. The rise and fall of nanobacteria. Sci. Am. 302, 52–59.

8

Comet and asteroid materials

8.1 Asteroid and comet explorations

8.1.1 Overview of asteroid exploration missions

In this chapter, asteroid and comet exploration are explained. Samples on the Earth, Mars, and Venus have undergone severe weathering by atmosphere and water. However, some asteroids and comets are considered to be the remnants of the solar nebula and keep the condition when they formed 4.56 Gy ago. Some evolved asteroids record evolutionary histories until today. Thus, asteroid exploration is the key to reveal the history of our solar system from the beginning to today.

Asteroid exploration started from its direct observation in the near distance. Since the NASA's Galileo spacecraft took the photographs of S-type asteroids in 1993 for the first time, NEAR-Shoemaker launched by NASA in 1996, Rosetta by ESA in 2004, and Dawn by NASA in 2007 went to near asteroids or comets and obtained many important information. In addition, the spacecraft going to the M-type asteroid was launched in 2023 by JPL (Section 8.1.2).

To ascertain the consistency between its IR spectrum and the composition of the asteroid, the direct method is to obtain the asteroid materials and take the IR spectra. In addition, meteorite-researchers wanted to get and analyze the asteroid samples by their hands. Thus, sample-return missions of asteroids were planned.

Before the asteroid sample-return, the first sample-return project taking the sample from the coma of the comet was performed by NASA in 1999. The second sample-return project was the Genesis project (Section 8.1.3), in which the solar wind was recovered (Section 8.2.1).

Technological advances made it possible to fetch the asteroid samples (sample-return of asteroid samples) (see Section 8.2.2). The first sample-return mission of the asteroid was Hayabusa (2003−10) by JAXA, which

Introductory Astrochemistry
DOI: https://doi.org/10.1016/B978-0-443-23938-0.00008-X

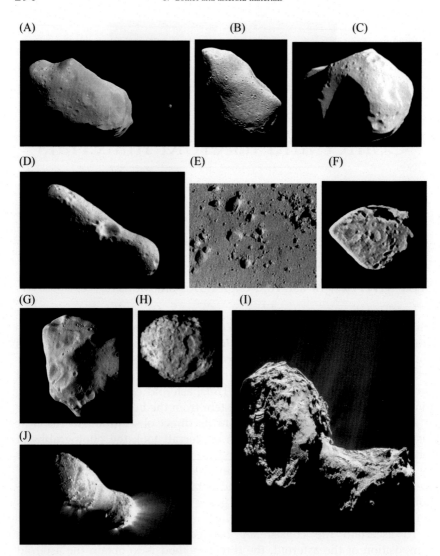

FIGURE 8.1   **Photographs of asteroids and the core of a comet.** (A) S-type asteroid, 243 Ida (60 km in the major axis) and its moon, Dactyl (1.6 km in diameter) were taken by Galileo in August 1991, from the distance of 10,300 km. (B) The S-type asteroid, 951 Gaspra (18 km in the major axis) taken by Galileo in October 1991 from a distance of 1600 km at a relative speed of 8 km/s. The colors were exaggerated. (C) C-type asteroid, 253 Mathilde (61 km in diameter) from 1200 km distance at 9.93 km/s. (D) S-type asteroid, Eros (33 km in the major axis). Eros is a peanut shape and belongs to the near-Earth asteroids (NEA) with an average distance from the sun is 1.46 au, and Eros approached the Earth only about 0.18 au in 2012. (E) The surface of Eros just before the landing of the spacecraft. It is covered with regolith like the surfaces of the moon and 25143 Itokawa. (F) E-type asteroid, 2867 Šteins (4.6 km in diameter), which resides in the main belt. Rosetta took this photograph from 1700 km at 9 km/s. A large crater of 1.7 km in diameter and

recovered several grains from an S-type asteroid, 25143 Itokawa, and the succeeding sample-return mission, Hayabusa2 successfully fetched ~5.4 g from a C-type asteroid, 162173 Ryugu. NASA also started the sample-return mission, OSIRIS-REx to collect 0.4–1 kg samples from a B-type asteroid, 101955 Bennu. The next sample-return mission going to the moon of Mars by JAXA has already started. In addition, as related in Section 6.2.3, the sample-return from Mars is also planned by NASA and ESA.

8.1.2 Asteroid and comet exploration missions

8.1.2.1 *The Galileo spacecraft*

In 1989, the Jupiter orbiter probe, Galileo was launched by JPL in NASA. In 1991, Galileo took photographs of an S-type asteroid, 234 Ida and 951 Gaspra (see Fig. 8.1A and B). Humankind first looked at the asteroids from the near distance. Many researchers were inspired by these photographs of near observation.

8.1.2.2 *The NEAR-Shoemaker spacecraft*

NEAR-Shoemaker was launched by NASA in 1996 to study the near-Earth asteroid 433 Eros from near orbital. It took photograph of the C-type asteroid, 253 Mathilde in 1997 (Fig. 8.1C). Mathilde belongs to the main-belt asteroid with a density of 1.3 g/cm^3, which suggests that Mathilde was made of rubble piles. Mathilde had a very large crater,

---

◀ seven craters in a line were found. The asteroid rotates backward. (G) The M-type large asteroid, 21 Lutetia, resides in the main belt (120 km in the major axis). Rosetta took this photograph from 3162 km at 15 km/s. Lutetia was the first M-type asteroid explored by a spacecraft. (H) The photograph of the core of Comet Wild-2. The diameter is 5 km. (I) The photograph of the coma of the comet 67 P/Churyumov_Gerasimenko. The photo was taken 31 km from the comet. The size is about ~6 km in the longest axis. (J) The photograph of the coma of Comet Hartley-2 (103 P) taken by the Deep Impact spacecraft. The size is ~2 km in length. *Source: (A) NASA/JPL/Processed by Kevin M. Gill. https://upload.wikimedia.org/wikipedia/commons/7/79/243_Ida_-_August_1993_%2816366655925%29.jpg. (B) USGS/NASA/JPL. https://upload.wikimedia.org/wikipedia/commons/thumb/8/81/951_Gaspra.jpg/681px-951_Gaspra. jpg. (C) NASA. http://nssdc.gsfc.nasa.gov/imgcat/html/object_page/nea_19970627s.html. (D) NASA/JPL/JHUAPL. http://photojournal.jpl.nasa.gov/catalog/PIA02475. (E) NASA. https:// upload.wikimedia.org/wikipedia/commons/a/a6/Erosregolith.jpg. (F) ESA ©2008 MPS for OSIRIS Team MPS/UPD/LAM/IAA/RSSD/INTA/UPM/DASP/IDA; processing by T. Stryk. https://www. esa.int/ESA_Multimedia/Images/2013/02/Steins_revisited. (G) ESA 2010 MPS for OSIRIS Team MPS/UPD/LAM/IAA/RSSD/INTA/UPM/DASP/IDA. https://www.esa.int/ESA_Multimedia/ Images/2010/07/Lutetia_at_closest_approach. (H) NASA. https://upload.wikimedia.org/wikipedia/ commons/4/40/Stardust_photo.jpg. (I) ESA/Rosetta/NAVCAM − CC BY-SA IGO 3.0. https:// www.esa.int/ESA_Multimedia/Images/2014/11/Comet_on_20_November_NavCam. (J) NASA. https://upload.wikimedia.org/wikipedia/commons/4/42/NASAHartley2Comet.jpg.*

the inside color of which was similar to that of the surface, it was inferred that the asteroid should be made of uniform materials from the surface to the inside. The rotation speed was too slow to take the photographs of all surfaces. Then NEAR-Shoemaker arrived the S-type asteroid, 433 Eros (Fig. 8.1D) in 2000. From the analysis by the X-ray and γ-ray spectrometer on board, Eros was made of ordinary chondrites containing silicates such as olivine and pyroxenes. Finally it collided to Eros with taking photographs just before landing on the surface of Eros (Fig. 8.1E).

8.1.2.3 The Rosetta spacecraft

Rosetta was a spacecraft launched by ESA in 2004, with a lander Philae. Rosetta was made to study the comet, 67 P/ Churyumov–Gerasimenko (67 P). During the journey to 67 P, it flew by the asteroids 2867 Šteins (Fig. 8.1F) in 2008 and 21 Lutetia in 2010 (Fig. 8.1G), respectively. Šteins has a rubble-pile structure and became the present shape by the YORP effects (see Section 1.4.13.6).

Rosetta took a photograph of a coma of 67 P (Fig. 8.1I). The jet activities of water and carbon dioxide gases were observed. The surface seemed to be made of hard organic materials and ice. The inside should be porous and amorphous ice. It was called as a "deep-fried ice cream".

Philae was a lander on the coma of 67 P, which was attached on Rosetta. It was detached from Rosetta in 2014, and landed on the surface of the coma. After the first touchdown, Philae rebounded on the surface of the comet several times. Unfortunately, Philae settled in the dark cleavage of the coma where sunshine is not enough to generate sufficient energy to communicate with Rosetta.

8.1.2.4 The Deep Impact spacecraft

Deep Impact was launched by NASA in 2005 to study the interior composition of the comet Tempel-1 (9 P) by hitting an impactor of 372 kg with an observer spacecraft. The impactor successfully hit the comet nucleus in 2005, forming an impact crater. The excavated debris indicated that the comet coma was less icy but more dusty, obscuring the view of the impact crater. Till then, the scientists thought that the comet coma was very porous and made of sand-like materials. Actually, the coma was made of a very fine powder, including clays, carbonates, sodium, and crystalline silicates. The observation was also done by the Stardust spacecraft (see Section 8.2.1.1), which studied Comet Wild 2 in 2004.

After the Deep Impact mission, the observer spacecraft was extended as EPOXI (Extrasolar Planet Observation and Deep Impact Extended Investigation). The image of Comet Hartley-2 (103 P) was sent in 2010 (Fig. 8.1J). It was found that the comet sprayed not water vapor but dry ice.

8.1.2.5 *The Dawn spacecraft*

Dawn spacecraft was launched by NASA in 2007, arrived the orbit around 4 Vesta (Fig. 8.2A) in 2011, and explored the asteroid. Then, Dawn entered the orbit around 1 Ceres (Fig. 8.2B) in 2015. In 2018, the mission of Dawn was ended by the lack of fuel, and the spacecraft stayed the orbit forever.

4 Vesta is considered to be an evolved asteroid, to have an iron−nickel core of diameter of ∼220 km with basaltic crust, and to be the source of HED (Howardite-Eucrite-Diogenite) achondrites (see Section 1.4.12 and Fig. 1.56). 1 Ceres is the largest protoplanet in the asteroid belt occupying one-third of the total asteroidal mass. It shows the IR spectrum of the C-type asteroid, but it is made of silicates and ice. Ceres is very carbon rich (∼20% at the surface) because it was made outside of the Jupiter orbit where carbon was richer.

8.1.2.6 *The Psyche spacecraft*

Psyche spacecraft (see Fig. 8.3A) was launched by JPL in NASA in September 2023 and will arrive at 16 Psyche in 2029. The 16 Psyche asteroid is an M-type asteroid with a diameter of 227 km, and is considered to be the core of the protoplanet with a diameter of ∼500 km, of

(A) (B)

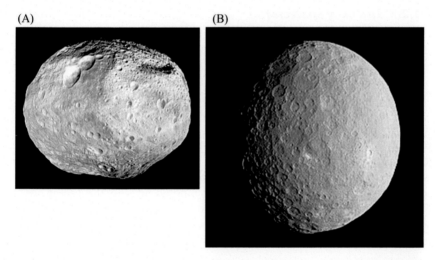

FIGURE 8.2   **4 Vesta and 1 Ceres.** (A) Composite grayscale image of V-type asteroid 4 Vesta. The mean diameter is 525 km. (B) C-type asteroid 1 Ceres in true color at a distance of 13,641 km. The mean diameter is 945 km, which is the largest asteroid in the asteroid belt. *Source: (A) NASA/JPL-Caltech/UCAL/MPS/DLR/IDA. https://upload.wikimedia.org/wikipedia/commons/thumb/1/14/Vesta_full_mosaic.jpg/280px-Vesta_full_mosaic.jpg. (B) NASA / JPL-Caltech / UCLA / MPS / DLR / IDA / Justin Cowart. https://upload.wikimedia.org/wikipedia/commons/thumb/7/76/Ceres_-_RC3_-_Haulani_Crater_%2822381131691%29_%28cropped%29.jpg/245px-Ceres_-_RC3_-_Haulani_Crater_%2822381131691%29_%28cropped%29.jpg.*

(A)   (B)

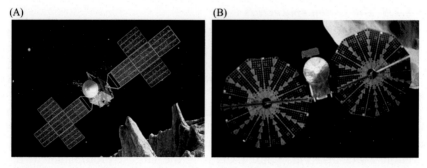

FIGURE 8.3   **Illustrations of Psyche and Lucy spacecrafts.** (A) Illustration of Psyche spacecraft on 16 Psyche asteroids. (B) Illustration of Lucy flying past the binary Trojan asteroids, 627 Patroclus-Menoetius. *Source: (A) NASA/JPL-Caltech/Arizona State Univ./Space Systems Loral/Peter Rubin. https://www.nasa.gov/sites/default/files/styles/full-width-feature/public/ thumbnails/image/pia21499−20170523.jpg. (B) NASA's Goddard Space Flight Center/Conceptual Image Lab/Adriana Gutierrez. https://upload.wikimedia.org/wikipedia/commons/thumb/0/0f/Lucy-PatroclusMenoetius-art.png/640px-Lucy-PatroclusMenoetius-art.png.*

which the mantle and crust were stripped off by violent collisions. It has high albedo, and 90% is made of iron−nickel and 6% is orthopyroxene. Psyche is the first mission to explore the M-type asteroids in detail.

8.1.2.7 The Lucy spacecraft

Lucy spacecraft was launched in October 2021 by NASA (Fig. 8.3B). It will flyby a main-belt asteroid as well as seven Jupiter trojans, which are at the Lagrange L4 and L5 orbits of the Sun and Jupiter (see Section 1.1.3). The mission is named Lucy hominin fossils, because the study of trojans is like that of the "fossils of planet formation".

Lucy carries three main instruments: (1) the color visible imager (0.4−0.85 μm) and infrared spectroscopic mapper (1−3.6 μm), (2) the high-resolution visible imager, and (3) the thermal infrared spectrometer (6−75 μm). Soon after the launch, NASA found that one of the solar arrays was stopped at 75%−95% deployment to the full. The cable to open the solar array may has lost its tension and tangled around the motor shaft. NASA stated that the stability can be kept during the mission, and the solar power is enough to continue the mission. The author hopes that the mission should complete the exploration with full success.

8.1.3 Some interesting topics on asteroids and comets

8.1.3.1 The D/H ratio on the comets and origin of water on the Earth

Rosetta spacecraft directly determined the deuterium to hydrogen ratio (D/H ratio) of the Jupiter family comet, 67 P, by a mass spectrometer for the D/H ratio determination and an enantiomer-sensitive gas chromatography-mass spectrometry (GC-MS) designed to decipher the chirality of LRMs on the comet (Thiemann and Meierhenrich, 2001). Surprisingly, the mass spectrometer on Rosetta reported a very high D/H ratio of 67 P of $\sim 5 \times 10^{-4}$, which is very different from those of the Earth of $\sim 1.6 \times 10^{-4}$, of Jupiter−Saturn−Uranus−Neptune of $2-6 \times 10^{-5}$, or other comets (possibly from Oort clouds) of $2-4 \times 10^{-4}$. As the D/H ratios of asteroids were $1.3-3 \times 10^{-4}$, it was suggested that the origin of ocean water of the Earth should have come from the asteroids, and not from such comets (Altwegg et al., 2017; Schroeder et al., 2019).

In contrast, the Herschel Space Observatory has shown that the comet 103 P/Hartley-2 had a similar (D/H ratio) to the Earth's water by the spectroscopic observation (Hartogh et al., 2011). This implies that the water on the Earth should come from such comets.

8.1.3.2 *The origin of volatiles on the Earth*

Poch et al. (2020) found that the nucleus of the comet 67 P showed an unidentified spectral reflectance feature around 3.2 μm. This absorption band was experimentally attributed to ammonium salts mixed with dust on the surface. Semivolatile ammonium salts are a substantial reservoir of N in the comet, potentially dominating over refractory organic matter and more volatile species. Similar absorption features were also observed in the spectra of some asteroids, implying a compositional relation among asteroids, comets, and the parent interstellar cloud.

8.1.3.3 *Identification of water emission from a main-belt comet*

Main-belt comets go around the Sun from the asteroid belt to the Jupiter orbit, repeatedly showing comet-like activity (dust comas or tails) when they approach the Sun (the perihelion passages), strongly suggesting ice sublimation (Hsieh and Jewitt, 2006; Jewitt et al., 2015). Although the main-belt comets are expected to have water ice, no gas has been detected around them even using the world's largest telescopes (Snodgrass et al., 2017).

Kelley et al. (2023) used the James Webb Space Telescope (JWST), and found that main-belt comet 238 P/Read has a coma of water vapor, but lacks a significant CO_2 gas coma. The findings demonstrated that the activity of comet Read is caused by water ice sublimation, and that main-belt comets are fundamentally different from the general cometary population.

Main-belt comets appear to be a source of volatile materials, and in understanding the early solar system's volatile inventory and its subsequent evolution.

8.1.3.4 Mystery of the bright spots on Ceres

Bright spots on the surface of Ceres (see the center of Fig. 8.2B) whose surface was covered by craters were found in the images taken by the Dawn spacecraft. The bright spots were observed both on smooth terrains and chaotically fractured terrains (Landaw, 2015). The bright spots were inferred as ice, salts, or vapor emission with high reflection. Such brightness indicated recent activity.

Various carbonates have been detected on Ceres, and their abundance and spatial distribution have been mapped using a visible and infrared mapping spectrometer on Dawn. Carbonates were abundant and ubiquitous across the surface. Mg–Ca carbonates were detected all over the surface, but localized areas showed Na carbonates, such as natrite (Na_2CO_3) and hydrated Na carbonates (e.g., $Na_2CO_3 \cdot H_2O$). Their geological settings and accessory NH_4-bearing phases suggested the upwelling, excavation, and exposure of salts formed from $Na–CO_3–NH_4–Cl$ brine solutions at multiple locations across Ceres. The presence of the hydrated carbonates indicated that their formation/ exposure on Ceres' surface was geologically recent and dehydration to the anhydrous form (Na_2CO_3) is ongoing, implying that Ceres is a still-evolving body (Carrozzo et al., 2018).

Ceres' craters often have smooth floors, which may perhaps contain impact melts and most are irregular in shape. In contrast, Vesta's craters are bowl-shaped. It is similar to the craters of a cold icy moon of Saturn with no melt but they are also irregular in shape (Krohn et al., 2016).

Dawn's exploration of Vesta and Ceres was testing long-held paradigms and giving new theories about the early solar system. Protoplanets were forming very early during the hot phase of the solar nebular and they delivered iron cores and rocky material to the forming terrestrial planets. Comets and wet asteroids delivered water to the inner solar system, as shown in the hydrated material on Vesta (Russel and Raymond, 2012).

8.1.3.5 The defense from the asteroidal collision with the Earth

The Double Asteroid Redirection Test (DART) is a NASA's space mission, which is not science but a test of the Earth's defense against collision with near-Earth objects (NEOs). In this mission, the double asteroids were chosen, and a spacecraft will intentionally crash into the minor asteroid (Dimorphos). The radius of Dimorphos orbit around the major asteroid 65803, Didymos will decrease by the impact (see Fig. 8.4). LICIACube (Light Italian CubeSat for Imaging of Asteroid) made by the Italian Space Agency (ASI) acquires images of the impact and ejecta from the asteroid. This project was supported by ESA, JAXA, and ASI, etc. The DART spacecraft has a 610 kg iron impactor. ESA is

(A)　　　　　　　　　　　　　　　　　　　　　　　　(B)

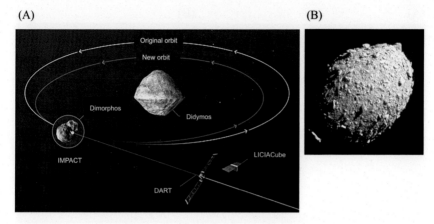

FIGURE 8.4 **The DART project.** (A) Illustration of the effect of DART impact on the orbit of the minor asteroid, Dimorphos around the major asteroid, Didymos. LICIACube is piggy-backed with DART and separated before the impact. It will take images of the impact and ejecta. (B) The image of Dimorphos before impact. The last complete image of Dimorphos (the length of 160 m) from 12 km just before 2 s before impact. *Source: NASA/Johns Hopkins APL. (A) https://upload.wikimedia.org/wikipedia/commons/5/52/Infographic_showing_the_effect_of_DART% 27s_impact_on_the_orbit_of_Didymos_B.jpg. (B) https://www.nasa.gov/sites/default/files/thumb- nails/image/all_dimorphos_dart_0401930040_12262_01_iof_imagedisplay-final.png.*

developing a spacecraft Hera, which will be launched in 2024 and arrive to Didymos in 2027. Such an asteroid impact project has already been tried by Deep Impact, although the purpose is completely different. In 2005, a 370 kg probe was collided on the core of comet Tempel-1, produced 19 GJ energy (4.8 tons of TNT), and changed the perihelion by 10 m.

The spacecraft impact that has the potential to deflect an asteroid on a collision course to the Earth was successful. The produced energy was 10.9 GJ (Daly et al., 2023). The DART succeeded to change the orbit of Dimorphos from 11.921 h to that of 11.372 h (-33.0 ± 1.0 min, 3 s; Thomas et al., 2023).

8.1.3.6 The Martian Moons eXploration mission

Martian Moons eXploration (MMX) mission is the third Japanese sample-return mission by JAXA. In this mission, MMX spacecraft will be launched in 2024, explore the Martian moons, Phobos, and Deimos (Fig. 8.5), land on Phobos, collect samples on Phobos, and return to the Earth in 2029. The purpose of this mission is to clarify the origin of Phobos and Deimos whether they are captured asteroids or made by giant impacts like our Moon. The former is supported by the IR spec-trum of Phobos which resembles to that of D-type asteroids formed by

(A)   (B)

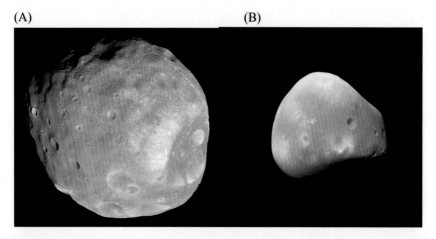

FIGURE 8.5   **The Martian moons: Phobos and Deimos.** (A) Phobos and (B) Deimos. Mean radii of Phobos and Deimos are 22.2 and 6.2 km, respectively. The sizes of each photograph are proportional to the actual sizes. The photographs were obtained by the Mars Reconnaissance Orbiter in 2008 and 2009, respectively. *Source: NASA/JPL-Caltech/University of Arizona. (A) http://photojournal.jpl.nasa.gov/catalog/PIA10368. (B) http://marsprogram.jpl.nasa. gov/mro/gallery/press/20090309a.html.*

colder conditions than C-type asteroids. However, the orbits of Phobos and Deimos are circular on the equatorial plane, which denies the capturing model.

Lots of problems remain to achieve success in this mission. The first one is the electric power problem because Mars is farther from the Sun than those of previous sample-return missions. The second one is the requirement of more fuel than the previous missions because the gravity of Mars is greater, and distances to and from Mars are farther. Furthermore, the surface condition of Phobos is not known. If Phobos is a D-type asteroid origin, its composition is ice and silicates. If the giant impact origin, Phobos is made of a mixture of melts of the proto-Phobos and Mars. In both cases, the sampling is very difficult.

8.2 Sample-return missions from the comets and asteroids

8.2.1 Early sample-return missions from space

NASA planned two sample-return missions from space. One mission was the Stardust mission, in which samples from the coma of comet Wild 2 were collected and returned to the Earth. The other is the Genesis mission, where the solar wind was collected and recovered

(A)   (B)

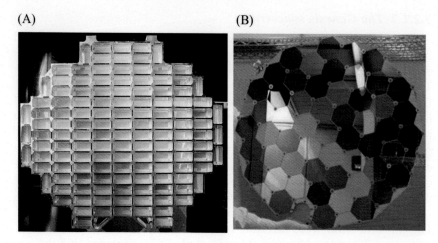

FIGURE 8.6 **Photographs of particle collectors from the comet and the solar wind.** (A) A photograph of the particle collector on Stardust made from aerogel to capture dust grains. The size is (B) A photograph of the solar wind collector array on Genesis. The collector is composed of hexagon panels made of semiconductor-grade pure wafers such as silicon, silicon carbide, corundum, gold on sapphire, diamond thin film, copper, etc. *Source: (A) NASA. https://upload.wikimedia.org/wikipedia/commons/1/1e/ Stardust_Dust_Collector_with_aerogel.jpg. (B) NASA/JPL. https://upload.wikimedia.org/wikipedia/ commons/3/37/Genesis_Collector_Array.jpg.*

to the Earth. These two sample-return missions from space are explained here.

8.2.1.1 *The Stardust spacecraft*

The Stardust mission was planned to collect samples from Comet Wild 2 (81 P) at 237 km in 2004 and to take images of the comet's coma. The particle collector was made of aerogel which is a low-density, inert, microporous, silica-based material to capture dust grains. The collector tray contained 90 blocks of aerogel, forming >1000 cm^2 of surface area (Fig. 8.6A). The porous sponge-like structure of 99.8% volume was empty.

Stardust successfully took images of the coma of the comet (Fig. 8.1H), went into the tail of the comet, and opened the collector at a planned velocity of 6.1 km/s to the comet. After sample collection, the sample collector was placed into the re-entry capsule, which fell in the desert in Utah and the collector was safely recovered.

Olivine grains and glycine were found, which meant that the silicates made at high temperatures (olivine) must be transferred to the cold places where the comets formed. In addition, the comet contained amino acids which can be a source of protein. Westphal et al. (2014) reported that seven possible interstellar dust particles were found from the Stardust sample collector. The samples have been analyzed until today.

8.2.1.2 The Genesis spacecraft

Genesis was launched by NASA in 2001, and collected the solar wind from 2001 to 2004 (totally for 27 months) at Lagrange point L_1 (see Section 1.1.3) by the four solar wind collector arrays (see Fig. 8.6B). Then, the arrays were placed into the re-entry capsule and returned to the Earth. The purpose was to determine the isotopic composition of light elements, especially of O and N of the sun. This was the first sample-return mission beyond the Moon.

In the plan, the return capsule opened the parachute and was recovered by the helicopter. However, the parachute did not open and the capsule directly hit the ground at 311 km/h. Everyone was disappointed when they saw the cracked capsule.

Fortunately, most wafers were intact because they fell on dry desert. The dirt and sand were carefully removed from the wafers, and lots of important data were obtained. Meshik et al. (2007) reported that the isotopic ratios of Ne and Ar of the Sun were consistent with the results obtained from the lunar rocks. McKeegan et al. (2011) reported the surprising results that the $\delta^{18}O$ and $\delta^{17}O$ of the sun are -58.5 and -59.1‰, respectively, which indicate that the abundance of ^{16}O of the sun is ~6% more than that of the Earth. This requires some early processes in the protoplanetary disk that caused the ^{16}O enrichment of ~6% of the Earth. Marty et al. (2011) reported that the nitrogen isotopic ratio of $^{15}N/^{14}N$ was 40% lighter than that of the Earth. This is consistent with the result of the oxygen.

8.2.2 Sample-return missions from asteroids

8.2.2.1 The Hayabusa spacecraft

Hayabusa was launched in 2003 by JAXA as a sample-return mission of an S-type asteroid, 25143 Itokawa, and returned to the Earth in 2010. The size of 25143 Itokawa was $535 \times 294 \times 209$ m (Fig. 8.7A), and the density was 1.9 ± 0.13 g/cm^3. The existence of large boulders suggests that an early collisional breakup of a pre-existing parent asteroid followed by a re-agglomeration into a rubble-pile object (Fujiwara et al., 2006). Hundreds of mineral particles of the asteroid were collected. The mineral chemistry of the recovered sample showed that the composition of Itokawa was similar to LL ordinary chondrite, which made a direct link between S-type asteroids and ordinary chondrites (Nakamura et al., 2011). On 50 μm-sized micrograins, μm to sub-μm sized adhered melts and sub-μm-sized craters were observed, indicating a hostile environment at an asteroid's surface (Nakamura et al., 2012).

FIGURE 8.7 **Photographs of asteroids targeted by sample-return missions.** The photograph sizes are proportional to the actual sizes. (A) 25143 Itokawa. The longest axis is 535 m. (B) 162173 Ryugu. The diameter (the horizontal axis) is ∼700 m. (C) 101955 Bennu. The diameter (the horizontal axis) is ∼492 m. *Source: (A) JAXA. https://s3-ap-north-east-1.amazonaws.com/jaxa-jda/http_root/photo/P-043-12077/d46e2b0e2980a8637b05f76265f1a804. jpg. (B) JAXA and University of Tokyo, etc. https://jda.jaxa.jp/result.php?lang = j&id = e18e0c63 b5111f065158aef7e2dbce0d. (C) NASA. https://www.nasa.gov/sites/default/files/styles/full_width_-feature/public/thumbnails/image/twelve-image_polycam_mosaic_12-2-18.png.*

8.2.2.2 The Hayabusa2 spacecraft

Hayabusa2 targeted the C-type asteroid, 162173 Ryugu. The scientific purposes of Hayabusa2 included the origin of water and organic materials that composed life as well as how the planet formed by the repetition of collision, breakup, and accretion of planetesimals. Furthermore, the technological purposes are as follows: (1) sampling from several positions, (2) making an artificial crater to remove surface materials, and (3) sampling the sub-surface materials from the artificial crater. The artificial crater was an important technology to remove the surface materials which is space-weathered by the solar wind.

The Hayabusa2 spacecraft (Fig. 8.8A) was launched in 2014, arrived at 162173 Ryugu in 2018, and touched down on the surface and collected samples (the first samples named A-series) in February 2019. Then, the spacecraft successfully made the crater by hitting an impactor, and collected the interior samples (the second sample named C-series). The samples in A- and C-series are shown in Fig. 8.9A and B,

(A)   (B)

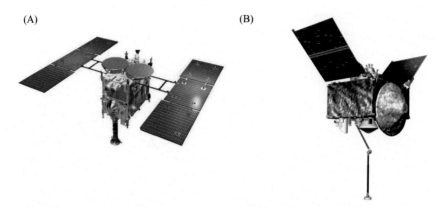

FIGURE 8.8   Illustrations of the sample-return spacecrafts. (A) The JAXA spacecraft,
Hayabusa2 for sample-return mission of the C-type asteroid, 162173 Ryugu was launched in
2014, touched down in 2019, and returned to the Earth in 2020 with collecting samples of
>5.4 g. (B) NASA spacecraft, OSIRIS-REx. OSIRIS-REx was launched in 2016 to collect samples
from the carbonaceous near-Earth asteroid (B-type), 101955 Bennu. In 2020 it touched down at
Bennu, and collected 0.4−1 kg samples, and coming back to the Earth in September 2023.
*Source: (A) ©Akihiro Ikeshita. https://www.hayabusa2.jaxa.jp/galleries/cg/pages/single3.html.
(B) NASA. https://commons.wikimedia.org/wiki/File:OSIRIS-REx_spacecraft_model.png.*

(A)   (B)

FIGURE 8.9   Photographs of the returned samples from the C-type asteroid, 162173
Ryugu. (A) The samples of "A series" (3.10 g), which were collected from the surface.
(B) The sample of "C series" (1.51 g), which were collected from the artificial crater. These
samples are expected to be mainly composed of the sub-surface materials of the asteroid.
The yellow arrow indicates artificial material possibly aluminum scraped off from the
sampler horn. *Source: (A) JAXA. https://www.hayabusa2.jaxa.jp/en/topics/20201225-samples/
img/fig6.png. (B) JAXA. https://www.hayabusa2.jaxa.jp/en/topics/20201225-samples/img/fig7.png.*

respectively. Somehow the C-series samples look bigger than the A-series samples,although the former includes the newly crushed samples by the explosion of the impactor.

From the remote-sensing, Ryugu had an oblate "spinning top" shape, with a prominent circular equatorial ridge (Fig. 8.7B). The bulk density was 1.19 ± 0.02 g/cm^3, indicating a high porosity of $>50\%$. Large surface boulders suggested a rubble-pile structure. The surface slope of Ryugu's shape indicated that it spun at twice the rate than that of the present. This suggested that Ryugu was reshaped during a period of rapid rotation (Watanabe et al., 2019).

The remote-sensing data showed that the parent body was partially dehydrated due to internal heating. The proto-asteroidal materials formed at ≤ 150K (the H_2O condensation temperature) were heated either by radiogenic isotopes like ^{26}Al or by the Sun (Sugita et al., 2019).

The reflectance spectra of Ryugu's surface acquired with the near-infrared spectrometer on board showed a weak, narrow absorption centered at $2.72\,\mu$m, indicating that hydroxyl (OH)-bearing minerals were ubiquitous. The intensity of the OH feature and low albedo were like thermally and/or shock-metamorphosed carbonaceous chondrites. Few variations in the OH-band position suggested that Ryugu was a compositionally homogeneous rubble-pile object generated from impact fragments of an undifferentiated aqueously altered parent body (Kitazato et al., 2019).

The analytical results for collected samples by the Hayabusa2 mission, especially for organic materials and LRMs, are summarized in Section 8.3 in detail.

Hayabusa2 left Ryugu in November 2019, and the sample capsule was safely arrived on the Australian Desert in December 2020. The collected samples were totally 5.4 g. The Hayabusa2 spacecraft itself was further extended to explore the L-type asteroid, 2001 CC_{21} in 2026 and 1998 KY_{26} in 2031.

8.2.2.3 OSIRIS-REx sample-return mission

OSIRIS-REx (Fig. 8.8B) was launched in 2016 to collect the samples from a carbonaceous near-Earth B-type asteroid, 101955 Bennu. OSIRIS-REx is named after the initial letters of Origins, Spectral Interpretation, Resource Identification, Security, Regolith Explorer. In 2017, it searched for NEOs at the Sun—Earth L_4 Lagrange point (see Section 1.1.3). From 2019, it began investigating Bennu at ~ 1.75 km, and was trying to find a sampling site. Compared to Ryugu, Bennu (Fig. 8.7C) was composed of larger boulders, resulting in a more difficult mission to collect samples (Lauretta et al., 2019). In 2020, OSIRIS-REx touched down Bennu, and collected 0.4—1 kg samples, which safely returned to the Earth on

September 24, 2023. The sample was moved to a clean room in the Astromaterials Curation Facility at NASA's Johnson space Center in Houston. After the sample-return mission, the spacecraft will go to 99942 Apophis as OSIRIS-APEX (Apophis Explorer) in 2029.

Chesley et al. (2014) described that Bennu was a half-kilometer near-Earth asteroid with a well-constrained orbit. The data set of optical astrometry from 1999 to 2013 and high-quality radar delay measurements in 1999, 2005, and 2011 revealed the action of the Yarkovsky effect (see Section 1.4.13.5), with a mean semi-major axis drift rate of 284 ± 1.5 m/year. The Yarkovsky effect and the thermal emissions yield a bulk density of 1.26 ± 0.07 g/cm^3, which indicates a macroporosity in the range of $40\% \pm 10\%$, suggesting a rubble-pile internal structure. The associated mass estimate was $7.8 \pm 0.9 \times 10^{10}$ kg.

The close approach to the Earth in 2135 is likely within the lunar distance and leads to strong scattering and numerous potential impacts in subsequent years in 2175–2196. The highest individual impact probability is 9.5×10^{-5} in 2196, and the cumulative impact probability is 3.7×10^{-4}, which corresponds to a cumulative Palermo Scale of -1.70 (see Section 1.4.13.8).

Bottke et al. (2015) used a Monte-Carlo method to search the escape routes of Bennu-like orbits from the main belt. They found the most likely parent families for Bennu were Eulalia and New Polana. The Eulalia and New Polana families had a $70\%–30\%$ probability of producing Bennu.

Barnouin et al. (2020) made digital terrain models (DTMs) of the asteroid which was a key for understanding the origin and evolution of the asteroid, providing geological and geophysical context for the sample, maximizing the amount of sample returned, navigating the spacecraft, and ensuring the safety of the spacecraft during sampling. To produce the DTMs, a camera- and a lidar (laser altimetry)-based approach were used. DTMs were produced at global and local scales. The global shape models were 1–0.1 m accuracy. Products such as geopotential elevation, slope, and tilt could be derived from the DTMs.

To return a pristine sample of carbonaceous asteroid Bennu, Dworkin et al. (2018) maintained the OSIRIS-REx spacecraft sampling hardware at a level of <180 ng/cm^2 of amino acids and hydrazine on the sampler head through precision cleaning, control of materials, and vigilance. The cleanliness of the spacecraft was achieved through communication among scientists, engineers, managers, and technicians.

Bates et al. (2020) collected visible and near-infrared and mid-infrared reflectance spectra from well-characterized CM1/2, CM1, and CI1 chondrites and identified trends related to their mineralogy and degree of secondary processing. The spectral slope between

0.65 and 1.05 µm decreased with increasing total phyllosilicate abundance and increasing magnetite abundance, both of which were related to more extensive aqueous alteration.

NASA and JAXA signed an agreement to exchange the samples of 0.3 g of Bennu from NASA to JAXA, and 10% (actually 0.54 g) of Ryugu from JAXA to NASA. Thus, after 2023 when the samples from Bennu are safely recovered, the samples of Bennu can be analyzed in Japan. The comparison of the samples of Bennu with those of Ryugu in the USA and Japanese researchers will give us important knowledge about the origins of carbonaceous asteroids, organic materials, water, and life on the Earth.

8.3 Organic materials and life-related materials in asteroids and comets

8.3.1 Organic materials and life-related materials in comets

If a comet is made of pure ice, no organic materials are found or formed on it. However, the comas of comets are made of lots of organic materials (see the next paragraph), and reactions on a comet or collision between a comet and another asteroid or comet can produce organic materials and LRMs. Laboratory experiments shown in Section 8.6.4 support that such reactions should occur.

LRMs of the amino acid precursors, such as ammonia, methanol and carbonyl compounds, have been observed in actual comets; for example, Halley, Hyakutake, Tempel-1, Giacobini-Zinner, Hartley-2 and Hale-Bopp (Crovisier and Bockelée-Morvan, 1999; Bockelée-Morvan et al., 2000; Ehrenfreund and Charnley, 2000; Ehrenfreund et al., 2002; Mumma et al., 2003; Festou et al., 2005; DiSanti et al., 2013). Glycine, one of the simplest amino acids, was detected on comet 81 P/Wild from samples returned by NASA's Stardust mission (Elsila et al., 2009; see Section 8.2.1.1).

8.3.2 Organic materials and life-related materials in carbonaceous chondrites

The carbonaceous chondrites are considered to be a chip of C-type asteroids. Various LRMs of amino acids and nucleobases are found in the carbonaceous chondrites (Kvenvolden et al., 1970; Cronin and Moore, 1971; Stoks and Schwartz, 1979, 1981; Martins et al., 2008; Callahan et al., 2011). In addition to amino acid, precursors (ammonia, methanol, and carbonyl compounds which correspond to LRMs) were observed in actual carbonaceous chondrites (Crovisier and Bockelée-

Morvan, 1999; Bockelée-Morvan et al., 2000; Ehrenfreund and Charnley, 2000; Ehrenfreund et al., 2002; Mumma et al., 2003; Festou et al., 2005; DiSanti et al., 2013).

Furukawa et al. (2019) found sugars in three carbonaceous chondrites (NWA 801 [CR2], NWA 7020 [CR2], and Murchison [CM2]), and showed evidence of extraterrestrial ribose and other bioessential sugars in the carbonaceous meteorites. The ^{13}C-enriched stable carbon isotope compositions (high $\delta^{13}C_{VPDB}$) of the detected sugars indicate that the sugars can be of extraterrestrial origin.

8.4 Mineralogy and inorganic chemistry of samples recovered from the asteroid Ryugu

8.4.1 Systematic analysis of the Ryugu samples

The Hayabusa2 spacecraft collected samples from the asteroid Ryugu, and returned to the Earth in 2022. The samples are the least contaminated by the terrestrial surface, especially for organic materials. In this section, the initial analytical results mainly on inorganic chemistry are summarized. The systematic analysis of organic materials in the asteroid samples from Ryugu is summarized in the next section, Section 8.5.

Each research team combined various analysis methods and made their own specific analytical flows. Every team separated the samples into several groups:

- The sample is used to make thin sections for mineralogical description by spot analytical methods such as EPMA, SEM-EDX, SIMS, HR-SIMS, etc.
- The sample is used for the determination of the carbon (C) and sulfur (S) abundances by the combustion method.
- The sample is used for the oxygen (O) concentration and oxygen isotopic analyses. Some teams used the laser-fluorination technique, in which the oxygen isotopic analysis of the spot area including each mineral analysis can be performed.
- The sample is used for the major and non-volatile trace element analysis. The sample is acid-digested and made into solution, and major and trace element concentrations are analyzed by ICP-QMS and HR-ICP-MS, etc. After column chromatography, isotopic analyses of non-volatile elements are performed using MC-ICP-MS or TIMS.
- For organic material analysis, the sample is stepwise leached with water and organic solvents and separated into soluble organic materials (SOM), and insoluble organic materials. Each SOM is analyzed for the detection and quantification of organic materials and LRMs using GC-MS, MS/MS, etc.

In this section, analyses of the first four are called as the inorganic analysis, and the last analysis is called as the organic analysis.

8.4.2 Inorganic analysis of the Ryugu samples

8.4.2.1 Abundances and speciation analyses of C and S by pyrolysis and combustion

Yokoyama et al. (2023) described analytical methods for C and S. C and S abundances were determined at a HORIBA EMIA-Step instrument for pyrolysis and combustion analysis. For C analysis, each sample was loaded on a quartz boat, and introduced into a quartz furnace at 800°C with continuous flow of N_2. Then, the temperature was increased linearly from 800°C to 1000°C. The gas flow was then changed to O_2. The generated gases were transported to a copper oxide converter at 600°C with N_2 or O_2 flow where carbon species were converted to CO_2. The gas was dehydrated, and the time variation of CO_2 intensity was measured with a non-dispersive infrared detector (NDIR), of which signals were converted to C concentrations (wt.%) using a $NaHCO_3$ standard. The carbons in different chemical forms in the sample were released at different times, producing a time spectrum. The timing of CO_2 from carbonates, organic matter, and graphite was calibrated using reference carbon-containing materials. For S analysis, each sample was loaded on an alumina boat and introduced into a ceramic furnace under a continuous flow of O_2 at 1450°C. The SO_2 gas was measured using the NDIR detector, and calibrated to S concentration (wt.%) using a K_2SO_4 standard.

8.4.2.2 Details of mineralogy and inorganic analytical results

Yokoyama et al. (2023) measured the mineralogy and the bulk chemical and isotopic compositions of Ryugu samples of both A- and C-series (see Fig. 8.5) using totally ~ 95 mg. The samples contained 15.38 ± 0.50 wt.% of H_2O and CO_2. Organic carbon was $74\% \pm 3\%$. The samples consisted predominantly of minerals formed in aqueous fluid on a parent planetesimal. The primary minerals were altered by fluids at a temperature of $37°C \pm 10°C$ from the oxygen isotope ratios at a few Myrs by the $^{53}Mn-^{53}Cr$ system after the formation of the first solids in the solar system. After aqueous alteration, the Ryugu samples were likely never heated above $\sim 100°C$. Major, trace and volatile element abundances were similar to the CI (Ivuna-type) carbonaceous chondrites. The $\delta^{18}O-\delta^{17}O$ plots of the Ryugu samples were almost on the terrestrial fractionation (TF) line. The $\delta^{18}O-\Delta^{17}O$ plots were also on the TF line with only a few exceptions. The chemical composition resembled more closely to the Sun's photosphere than any other natural samples.

Nakamura et al. (2023) analyzed 17 Ryugu samples of 1−8 mm. They found CO_2-bearing water inclusions within a pyrrhotite crystal, indicating that Ryugu's parent asteroid formed in the outer solar system. The abundances of chondrules and CAIs which are formed at high temperatures were low, and those of phyllosilicates and carbonates which are formed by aqueous alteration reactions at low temperature, high pH, and small water/rock ratios of <1 (by mass) were high. Less altered fragments made of olivine, pyroxene, amorphous silicates, calcite, and phosphide were found. Numerical simulations based on the mineralogical and physical properties of the samples indicated that the Ryugu's parent body was formed $\sim$2 Myrs after the beginning of solar system formation.

Nakamura et al. (2022) used representative 7 and 9 particles of A- and C-series, respectively, of totally 55 mg for comprehensive analysis. On average, Ryugu particles consisted of 50% phyllosilicate matrix, 41% porosity, and 9% minor phases including organic matter. The abundances of 70 elements from the particles were in close agreement with those of CI chondrites. Bulk Ryugu particles showed higher $\delta^{18}O$, $\Delta^{17}O$, and $\varepsilon^{54}Cr$ values than CI chondrites. Thus, Ryugu samples were the most primitive and least-thermally processed protosolar nebula reservoirs. This finding was consistent with that the origins of Ryugu organic matter were the protosolar nebular and the interstellar medium from the H−C−N isotopic compositions. The analytical data obtained here, suggests that complex soluble organic matter formed during aqueous alteration on the Ryugu progenitor planetesimal (several tens of km), <2.6 My after CAI formation. The Ryugu progenitor planetesimal was fragmented and evolved into the current asteroid Ryugu through sublimation.

Naraoka et al. (2023) used aggregated grains of <1 mm diameter with a total weight of 38.4 mg (named A0106). The A0106 sample contained 3.76 ± 0.14 wt.% of total carbon (C), 1.14 ± 0.09 wt.% of total hydrogen (H), 0.16 ± 0.01 wt.% of total nitrogen (N), and 3.3 ± 0.7 wt.% of total sulfur (S). The concentration of pyrolyzed oxygen (O), liberated at 1400°C under a helium gas flow, was 12.9 ± 0.42 wt.%. The total C, H, O, N, and S contents were $\sim$21.3 wt.%, which comprised hydrous minerals, carbonates, sulfides, and organics, including macromolecular insoluble organic matter and SOM. The stable isotopic compositions were as follows: $\delta^{13}C_{PDB} = -0.58 \pm 2.0‰$, $\delta D_{VSMOW} = +252 \pm 13‰$, $\delta^{15}N_{Earth\ atmosphere} = +43.0 \pm 9.0‰$, $\delta^{34}S_{VCDT} = -3.0 \pm 2.3‰$, and $\delta^{18}O_{VSMOW} = +12.6 \pm 2.0‰$. The effective heating temperature was estimated to be <150°C.

Ito et al. (2022) analyzed eight Ryugu particles (~ 60 mg in total) and found that they were chemically unfractionated, but aqueously altered, CI (Ivuna-type) chondrites, which are widely used as a proxy for the bulk solar system composition, but are biased as the strongest types survived atmospheric entry and are then modified by interaction with the terrestrial environment.

In summary, the Ryugu samples showed an intricate spatial relationship between aliphatic-rich organics and phyllosilicates and indicated maximum temperatures of $\sim 30°C$ during aqueous alteration. Heavy hydrogen and nitrogen abundances are consistent with an outer solar system origin. Ryugu particles are the most uncontaminated and unfractionated extraterrestrial materials studied so far, and provide the best available match to the bulk solar system composition so far estimated from the Ivuna CI chondrite.

8.5 Organic materials and life-related materials in the recovered samples from the asteroid Ryugu

In this section, details of analytical results for organic materials and LRMs are shown. These are the first data for the least polluted organic materials from terrestrial contamination. Two groups mainly performed detailed and systematic organic analyses: one group was the Okayama University group (Nakamura et al., 2022; Potiszil et al., 2023), and the other was the Kyushu and Hokkaido University group (Naraoka et al., 2023; Oba et al., 2023). This section explains how the organic analysis was performed in each group, and what analytical results were obtained by the two groups as well as other research groups (Yabuta et al., 2023) for the analyses of organic materials for the asteroid Ryugu samples.

8.5.1 Systematic analyses of organic materials for the asteroid Ryugu by Nakamura et al. (2022) and Potiszil et al. (2023)

Nakamura et al. (2022) applied two types of organic material analyses for the asteroid Ryugu samples. For the distribution analysis of organic materials, Desorption Electrospray Ionization-Orbitrap-Mass Spectrometry (DESI-OT-MS) was applied. And for the bulk analysis, Ultra-High-Performance Liquid Chromatography-Orbitrap-Mass Spectrometry (UHPLC-OT-MS) was applied.

8.5.1.1 Organic material study using Desorption Electrospray Ionization-Orbitrap-Mass Spectrometry

Using DESI-OT-MS, ambient mass spectrometry imaging can be performed. The C-series sample, C0008 contained many homolog series, including six ($C_nH_{2n-7}N$, $C_nH_{2n-9}N$, $C_nH_{2n-11}N$, $C_nH_{2n-13}N$, $C_nH_{2n-15}N$, and $C_nH_{2n-17}N$) that was not reported by DESI-OT-MS in meteorite samples. The A-series sample, A0048 contained fewer members of the series found in C0008 and one series was not recorded. The surface of A0048 yielded a less dense distribution of pixels that gave a response, compared to that of C0008.

Although Orgueil contained some of the $C_nH_{2n-7}N$ homologs, the response is very weak and no other homologs were detected in Orgueil. However, Orgueil contained some non-homolog compounds that have not been reported elsewhere. Their results showed that some compounds can be found in Orgueil that are not present in C0008 and A0048. The spatial distribution of homologs in C0008 and A0048 was clearly demonstrated due to the higher resolution of 10 μm. The distribution varies between the homolog series detected, indicating that SOM was heterogeneously distributed within the matrix and that its composition is also heterogeneous throughout the sample. In the case of C0008, the majority of SOM is concentrated in the top half of the sample, especially in the case of the homologs. Furthermore, SOM homologs and non-homologs demonstrated high-intensity areas within the sample, which are positively correlated in some cases but negatively correlated. This observation suggested that individual components of the SOM may be sourced from distinct reservoirs. In the case of Orgueil, many of the SOM compounds produce weak signals that make a comparison to C0008 and A0048 very difficult. However, for the SOM compounds that record strong responses, the overall relationships concerning the distribution of SOM are similar to those observed in C0008.

All homologs detected by DESI-OT-MS for C0008 were already found in Murchison by UHPLC-OT-MS, such as alkylpyridines ($C_nH2_{n-5}N$), alkylquinolines ($C_nH_{2n-11}N$), alkylcarbazoles ($C_nH_{2n-15}N$) and unsaturated alkylpyridines ($C_nH_{2n-5}N$) using tandem mass spectrometry (MS/MS).

8.5.1.2 Organic material study using Ultra-High-Performance Liquid Chromatography-Orbitrap-Mass Spectrometry

First, the asteroid samples were heated with water for 20 h at 110°C. The water was dried by freeze-drying. The freeze-dried residues were then heated at 110°C with 2 M HCl/isopropyl alcohol to convert the amino acids to isopropyl esters. The mixture was reduced under a flow of N_2, freeze-dried to complete dryness, and dissolved with ethyl acetate.

UHPLC-OT-MS (Thermo Scientific) analysis was performed in reversed phase using a UHPLC unit. A binary pump system installed in the UHPLC unit enabled a dynamically changing gradient between the

C#		Code
1	Urea	Ure*
2	Glycine	Gly
3	Sarcosine+Alanine	Sar*+Ala
3	β-Alanine	β-Ala*
3	Serine	Ser
4	Threonine	Thr
4	α-Aminobutyric acid	AABA*
4	α-Aminoisobutyric acid	AIB*
4	β-Aminoisobutyric acid	BAIBA*
	+β-Aminobutyric	+BABA*
4	DL-Aspartic Acid	Asp
5	L-valine	Val
5	Norvaline	Nva*
5	Glutamic acid	Glu
5	Proline	Pro
5	Hydroxyproline	Hyp*
6	Leucine	Leu
6	Isoleucine	Ile
6	Alloleucine	AABA*
6	α-Aminoadipic acid	α-Adp*
6	Cycloleucine	Cle*
6	Pipecolic acid	Pip*
7	α-Aminopimelic acid	α-Apm*
7	Homocycloleucine	Hcl*
9	Phenylalanine	Pha

FIGURE 8.10 Organic molecules found in the water-soluble fraction of the Ryugu sample. The left side box is the organic molecules detected in the water-soluble fraction of the Ryugu asteroid samples. C# is the number of carbon atoms. Tyrosine was not found in Ryugu. The right side shows the chemical structures of urea and amino acids with the asterisk (*) in the left side box, which are not used in the protein synthesis. *Source: Data from Nakamura, E., Kobayashi, K., Tanaka, R., Kunihiro, T., Kitagawa, H., et al., 2022. On the origin and evolution of the asteroid Ryugu: a comprehensive geochemical perspective. Proc. Jpn. Acad. Ser. B 98, 227−282.*

two phases, A (10 mM ammonium formate in water) and B (100% acetonitrile).

The data for UHPLC-OT-MS was compared to two standards: the standard with 27 amino acids and the commercial standard with an additional 2 amino acids (proline and hydroxyproline) and urea. Assignments were initially made by the retention times. When the response was enough, tandem mass spectrometry (MS/MS) was performed, and the fragmentation patterns were compared between the samples and the standards.

Nakamura et al. (2022) found 23 amino acids and urea from the water-soluble fraction of a C0008 (see Fig. 8.10) using UHPLC-OT-MS and compared with those of Orgueil (CI1). Orgueil also contained at least 24 amino acids and urea, however, tyrosine was not detected in C0008. Alanine and sarcosine and O-aminobutyric and O-aminoisobutyric acid (AIB) were found in both Ryugu and Orgueil samples.

Many peaks in both Ryugu and Orgueil were observed but could not be identified because of a lack of standards for all the amino acid isomers for a given mass. The unidentified peaks and their ratios to each other were different between Ryugu and Orgueil. The identified amino acids had also clear differences in the ratios between amino acid isomers. For example, valine and norvaline were almost 1:1 in Ryugu, but in Orgueil the peak of valine is much larger than that of norvaline (~ 8:1).

Similarly, in Ryugu the ratio between α-aminobutyric acid (ABBA) and α-amino-isobutyric acid was 0.34, whereas in Orgueil the ratio was 0.87 and significantly higher. Another difference between Ryugu and Orgueil was the normalized intensity of their amino acids. Ryugu records a lower normalized intensity among all amino acids compared to Orgueil. However, the overall distribution of amino acids was similar between Ryugu and Orgueil.

8.5.1.3 Comparison between amino acids in the A- and the C-series samples, and their implications

Potiszil et al. (2023) analyzed A0022 using UHPLC-OT-MS, and compared the analytical results with those of C0008. The Raman spectrum caused by the organic materials (called as "G-band") of A0022 showed a lower and weaker energy shift with larger FWHM (Full width at half maximum) compared to those of C0008, indicating the effects of irradiation at the asteroid surface. It is reasonable that the analysis by DESI-OT-MS of C0008 showed greater intensity and the concentrated distribution of SOM than those of another A-series sample, A0048 (see Section 8.5.1.1), because the SOM in the A-series samples should have destructed by irradiation.

Although the abundances of amino acids in the two particles were different, no uniform difference in the amino acids was found. It is against the expectation that the amino acids in the A-series sample should have broken than in the C-series samples, because the A-series sample should be irradiated by greater solar wind and cosmic rays.

Concentrations of glycine and valine in C0008 were higher than those in A0022, whereas β-alanine, serine, and glutamic acid in A0022 were higher than those in C0008. Various amino acids were quantified in one particle, but not in the other particle. N,N-dimethylglycine, which was the most abundant amino acid in A0022, was below the detection limit in C0008. The weight difference (A0022 and C0008 were 0.988 and 1.900 mg, respectively) could not explain this phenomenon.

Another interesting result was the different β-alanine/glycine ratio; C0008 $= 0.57$ and A0022 $= 1.03$. Orgueil (CI1) had $1.71-3.16$ and

Murchison (CM2) had 0.26−0.43. The β-alanine/glycine ratio was previously thought to be sensitive to the degree of aqueous alteration: higher alteration shows a higher ratio. Thus, although the implication is given by only two grains, Ryugu experienced higher aqueous alteration than Murchison (CM2), but less intense than Orgueil (CI1). In addition, A0022 suffered from higher aqueous alteration than C0008. Furthermore, from the large difference of β-alanine/glycine ratio, the aqueous alteration is very heterogeneous in the sub-mm scale.

Glycine, β-alanine, and serine are the major components in residues from the UV processing on comet-like ices (Meinert et al., 2012; Oba et al., 2016; Modica et al., 2018). However, there are no experiments to synthesize all amino acids in carbonaceous chondrites. It is likely that such amino acids as well as *N,N*-dimethylglycine could be produced in the ISM (interstellar medium) or outer PSN (protosolar nebular) before accretion as Ryugu. In this scenario, A0022 is required to accrete large quantities of ice to collect *N,N*-dimethylglycine. The abundances of carbonates (6.9 vol% for A0022, and 1.65 vol% for C0008) would have formed from CO_2 (Fujiya et al., 2019). The higher porosity (51 vol% for A0022 and 44 vol% for C0008) may be filled with ice, and the high concentration of β-Alanine and serine in A0022 supports this hypothesis.

However, glycine which is a common product of comet-like ice irradiation experiments (Oba et al., 2016; Modica et al., 2018) is not higher in A0022 than C0008. As both formaldehyde and formic acid were observed in the comet 67 P/Churyumov−Gerasimenko (Altwegg et al., 2017), and such ice could be expected to be the ice accreted by Ryugu.

In summary, if the amino acids of Ryugu were synthesized on the proto-planetesimal of Ryugu, this would have occurred ∼2.6 Ma after the formation of CAI, when the accreted ice had melted into liquid water (Nakamura et al., 2022). The differences in the amino acids indicate different water/rock ratios accreted on Ryugu at a very small scale.

8.5.2 The systematic study of organic materials by Naraoka et al. (2023) and Oba et al. (2023)

8.5.2.1 Overview

Naraoka et al. (2023) analyzed the soluble organic molecules (SOM) of the aggregate Ryugu samples, A0106 (38.4 mg). Fig. 8.11 shows the analytical scheme they used. Samples were divided into three groups.

The first group sample was consumed to determine total carbon, hydrogen, oxygen, nitrogen, and sulfur (CHONS) contents and their isotopic ratios. Analytical results were already described in Section 8.4.2.2.

The second group sample was stepwise leached by organic solvents. The samples were sequentially extracted with hexane, dichloromethane

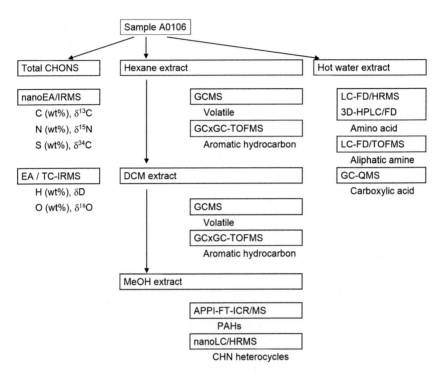

FIGURE 8.11   **Analytical scheme of organic materials by Naraoka et al. (2023).** The sample was separated into three groups; (1) total CHONS analysis, (2) the leaching experiments with organic solvents (hexane, dichloromethane, and methanol), and (3) the leaching experiments with hot water. Each abbreviation of analytical methods was explained in the text. *Source: Modified from Naraoka, H., Takano, Y., Dworkin, J.P., Oba, Y., Hamase, K., et al., 2023. Soluble organic molecules in samples of the carbonaceous asteroid (162173) Ryugu: samples returned from the asteroid Ryugu are similar to Ivuna-type carbonaceous meteorites. Science 379, eabn7850.*

(DCM), and methanol (MeOH). The hexane and DCM extracts were analyzed by 2D gas chromatography with time-of-flight mass spectrometry (GC × GC-TOFMS) for polycyclic aromatic hydrocarbons (PAHs). The methanol extract was analyzed by electrospray ionization (ESI) and atmospheric pressure photoionization (APPI) coupled with Fourier transform-ion cyclotron resonance mass spectrometry (FT-ICR/MS). The methanol extract was also analyzed by nano-liquid chromatography/high-resolution mass spectrometry (nanoLC/HRMS) to detect N-containing heterocyclic molecules.

The third group sample was leached with hot water. The hot water extracts were measured by liquid chromatography with fluorescence and HRMS (LC-FD/HRMS) and three-dimensional high-performance liquid chromatography with a high-sensitivity fluorescence detector (3D-HPLC/FD) for amino acid detection. The hot water extracts were

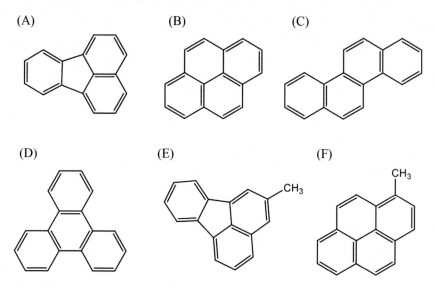

FIGURE 8.12 **Polycyclic aromatic hydrocarbons (PAHs) in the Ryugu sample.** (A) Fluoranthene. (B) Pyrene. (C) Chrysene. (D) Triphenylene. (E) Methylated fluoranthene. The position of the methyl group is variable. (F) Methylated pyrene. The position of the methyl group is variable. *Source: From Naraoka, H., Takano, Y., Dworkin, J.P., Oba, Y., Hamase, K., et al., 2023. Soluble organic molecules in samples of the carbonaceous asteroid (162173) Ryugu: samples returned from the asteroid Ryugu are similar to Ivuna-type carbonaceous meteorites. Science 379, eabn7850.*

also analyzed by liquid chromatography with fluorescence detection and TOFMS (LC-FD/TOFMS) for aliphatic amine detection and gas chromatography quadrupole mass spectrometry for monocarboxylic acids detection, respectively.

8.5.2.2 The hexane and dichloromethane extracts

The hexane and DCM extracts were analyzed by GC × GC-TOFMS for PAHs. The highest abundance of PAHs were fluoranthene (Fig. 8.12A) and pyrene (Fig. 8.12B), followed by chrysene (Fig. 8.12C)/triphenylene (Fig. 8.12D) and methylated fluoranthene and pyrene (Fig. 8.12E and F). Two rings of PAH of naphthalene and three rings of phenanthrene and anthracene were detected at low abundances (<ppm levels). The presence of alkylated PAHs (including alkylbenzenes) in the organic solvent extracts was confirmed by Fourier-transform infrared (FTIR) spectroscopy. Large alkylated PAHs were also identified by APPI FT-ICR/MS in the methanol extracts. The GC × GC-TOFMS analysis of hexane and DCM extracts did not find any common volatiles in carbonaceous chondrites such as methanol (CH_3OH), ethanol (C_2H_5OH), methyl formate ($HCOOCH_3$), acetone (CH_3COCH_3), diethyl ether ($C_2H_5OC_2H_5$), or acetonitrile (CH_3CN).

8.5.2.3 The methanol extracts

ESI-APPI-FT-ICR/MS was used for the methanol extract analysis. Relative m/z errors were lower than 100 ppb in a range of 92 < m/z < 1000. The average mass resolving power was about 400,000 at nominal mass 400. Data were internally calibrated to produce m/z lists. Almost 20,000 compounds consisting of CHONS were detected in the methanol extract by ESI-APPI-FT-ICR/MS. The most intense signals in the negative ions were polythionates ($NaS_3O_6^-$, $NaS_4O_6^-$, $NaS_5O_6^-$, $NaS_6O_6^-$, $NaS_7O_6^-$, $NaS_8O_6^-$, and $NaS_9O_6^-$). Abundant series of signals with repetitive mass differences were observed, indicating an evidence for a systematic reaction network including methylation, hydration, hydroxylation, and sulfurization. Mg-containing organic compounds such as CHOMg or CHOSMg were not detected.

The methanol extracts were also analyzed by nanoLC/HRMS. Alkylated N-containing heterocyclic molecules were identified, and confirmed by ESI FT-ICR/MS. These alkylated N-heterocycles included pyridine (C_5H_5N), piperidine ($C_5H_{11}N$), pyrimidine ($C_4H_4N_2$), imidazole ($C_3H_4N_2$), or pyrrole (C_4H_5N) rings with various amounts of alkylation. Alkylpyridines and alkylimidazoles (aromatic N-heterocycles) have previously been found in CM chondrites, whereas alkylpiperidines (aliphatic N-heterocycles) are more abundant in CR chondrites. The difference in relative abundances might reflect differing redox conditions on the meteorite parent bodies.

8.5.2.4 The hot water extract

The hot water extracts were analyzed by LC-FD/TOFMS, which indicated to contain methylamine (CH_3NH_2), ethylamine ($C_2H_5NH_2$), isopropylamine [$(CH_3)_2CHNH_2$] and n-propylamine ($C_3H_7NH_2$), but no other volatile compounds, such as methanol (CH_3OH), ethanol (C_2H_5OH), methyl formate ($HCOOCH_3$), acetone (CH_3COCH_3), diethyl ether ($C_2H_5OC_2H_5$), or acetonitrile (CH_3CN) which were detected in carbonaceous chondrites, were all below the detection limits.

The acid-hydrolyzed hot water extract (hot water extract was further treated with 6 M HCl at 105°C for 20 h in a flame-sealed glass ampoule) contained 15 amino acids such as glycine ($C_2H_5NO_2$), D,L-alanine ($C_3H_7NO_2$), D,L-valine ($C_5H_{11}NO_2$), and nonproteinogenic amino acids of β-alanine ($C_3H_7NO_2$), D,L-α-amino-n-butyric acid ($C_4H_9NO_2$), D,L-β-amino-n-butyric acid ($C_4H_9NO_2$), and several isomers of valine (D,L-norvaline, D,L-isovaline, δ-amino-n-valeric acid), which are rare in terrestrial biology (see Fig. 8.10). D/L was ~1, also indicating nonbiological origin. The amino acid content in A0106 was lower than Orgueil (CI), resulting from the different parent body and alteration conditions.

In summary, SOM concentrations in the A0106 sample were less than those of Murchison, and closer to those of Ivuna and Orgueil.

8.5.2.5 A nucleic acid, uracil was found in the Ryugu samples, A0106 and C0107

Oba et al. (2022) have developed a small-scale detection and identification method of nucleobases at parts per billion (ppb) to parts per trillion (ppt) levels using high-performance liquid chromatography coupled with ESI high-resolution mass spectrometry (HPLC/ESI-HRMS). From the aqueous extracts from the Murchison (CM2) meteorite, they successfully determined all five canonical nucleobases (adenine, guanine, cytosine, thymine, and uracil) to be 4−72 ppb.

Oba et al. (2023) applied this method to the hydrolyzed water extracts from the Ryugu samples, A0106 and C0107, and that of Orgueil and found uracil in all the hydrolyzed water extracts (see Section 8.5.2.4 for "hydrolyzing"). In addition, they found nicotinic acid (niacin, a B3 vitamer) and its derivatives, and imidazoles as nitrogen heterocyclic molecules (see Table 8.1 and Fig. 8.13). The difference in the uracil concentrations in A0106 and C0107 may be related to the degree of alteration by energetic particles such as ultraviolet photons and cosmic rays.

This study strongly suggests that these LRMs of prebiotic interest are commonly formed in carbonaceous asteroids and were delivered to the early Earth.

8.5.3 Macromolecular organic matter in the Ryugu samples

Yabuta et al. (2023) found the macromolecular organic matter in Ryugu samples by micro-Raman spectroscopy. They found that the Ryugu samples showed similar spectra to those of CI1 and CM2

TABLE 8.1   Concentrations (ppb) of uracil and other N-heterocyclic molecules identified in the hydrolyzed Ryugu samples (A0106 and C0107), and the hydrolyzed Orgueil meteorite.

	A0106	C0107	Orgueil
Uracil	11 ± 6	32 ± 9	140
Imidazole-2-carboxylic acid	6	9 ± 1	12
Imidazole-4-carboxylic acid	17 ± 3	19 ± 3	218
Nicotinic acid	49 ± 1	99 ± 4	715
Picolinic acid	n.d.	n.d.	n.d.
Isonicotinic acid	49 ± 2	62 ± 23	203

See Fig. 8.13 for the chemical structures of each molecule. *Data from Oba, Y., Koga, T., Takano, Y., Ogawa, N.O., Ohkouchi, N., Sasaki, K., et al., 2023. Uracil in the carbonaceous asteroid (162173) Ryugu. Nat. Commun. 14, 1292.*

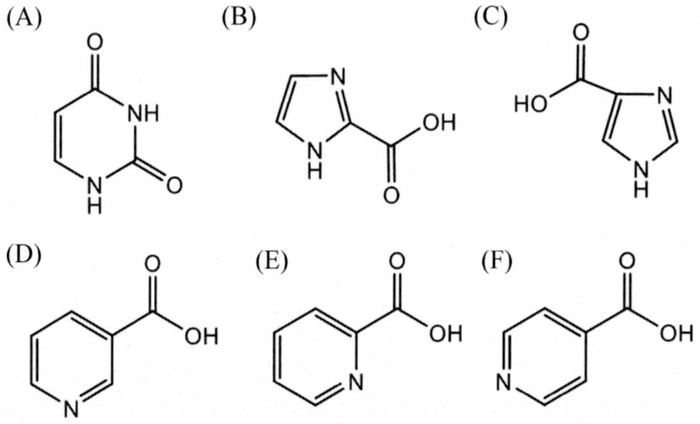

FIGURE 8.13   Chemical structure of molecules in Table 8.1. (A) Uracil. (B) Imidazole-2-carboxylic acid. (C) Imidazole-4-carboxylic acid. (D) Nicotinic acid. (E) Picolinic acid. (F) Isonicotinic acid. *Source: Data from Oba, Y., Koga, T., Takano, Y., Ogawa, N.O., Ohkouchi, N., Sasaki, K., et al., 2023. Uracil in the carbonaceous asteroid (162173) Ryugu. Nat. Commun. 14, 1292.*

carbonaceous chondrites. Micro-FTIR spectroscopy showed the existence of aromatic and aliphatic carbon, ketone, and carboxyl functional groups. Such organic matters in Ryugu probably suggest that LRMs were formed in the early stage of the solar system, and then modified by heterogeneous alteration in the parent body (and fell on the early Earth).

8.6 Origin of life-related materials in asteroids and comets by laboratory experiments

8.6.1 Introduction to impact-shock experiments for the synthesis of life-related materials

In this section, the origin of LRMs such as proteins, nucleic acids, sugars, and heteroatom (N, P, and S) containing organic materials in asteroids and comets are discussed based on the laboratory experiments. There are a few proposals for the origin of LRMs in space, and three cases were experimentally simulated:

- The icy comet was irradiated by strong UV light in space (Meinert et al., 2016).
- The icy comet hit the icy comet or silicate planetesimals including the Earth, which was experimentally simulated by a gas gun (Martins et al., 2013).

TABLE 8.2   Summary of experiments for synthesizing LRMs.

Researcher	Simulated reaction	Starting materials	Experimental condition	Detection
Meinert et al. (2016)	Icy grains at the molecular cloud stage	Ice Methanol NH_3	Ultraviolet light 10^{-5}Pa, 78K	GC × GC-TOFMS
Martins et al. (2013)	Icy comet/Icy comet collision	NH_3 solution Dry ice (CO_2) Methanol	Gas gun > 50 GPa	GC-MS
Ferus et al. (2015)	Icy comet/Icy comet or Earth collision (LHB)	Formamide ($HCONH_2$) Olivine chondrite	High energy laser >4500K	GC-MS

- The collisions of ice−ice or ice−silicate planet like the Earth were simulated by high-energy laser (Ferus et al., 2015), because the chemical reactions are driven by the collisional energy, which is also simulated by the high-energy laser.

To prove these hypotheses for the origin of LRMs, three experiments were selected. In Table 8.2, the simulated reaction, starting materials, irradiation medium, experimental condition, and their detection methods in each experiment are summarized.

8.6.2 Syntheses of life-related materials by ultra-violet irradiation on interstellar ice

Sugars are essential LRMs for all terrestrial biota. Ribose is essential as a building block of RNA. Both ribose and RNA could have stored information and catalyzed reactions in primitive life on the Earth. Meteorites contain many organic compounds including key LRMs, i.e., amino acids, nucleobases, and phosphates. The amino acids are identified in a cometary sample. However, the presence of sugars in extraterrestrial materials remains unclear.

Meinert et al. (2016) performed an experiment that ultraviolet light was irradiated on the ice composed of water (H_2O), methanol (CH_3OH), and ammonia (NH_3) in the interstellar condition (vacuum and low temperature). The experiment targeted whether aldoses, ketoses (see Section 1.5.19), and finally ribose, which is a backbone of RNA, are synthesized or not.

In the experiment of Meinert et al. (2016), the starting material gases were deposited on a MgF_2 substrate glass, irradiated by Lyman-α photons, and further irradiated with right-hand circularly polarized

(A)   Aldoses

		ppm (w/w)		ppm (w/w)
C-2	Ethylene glycol	550	C-5	
	Glycolaldehyde	2390	Ribitol	560
	Glycolic acid	6330	Ribose	260
C-3	Glycerol	2860	Ribonic acid	82
	Glyceraldehyde	302	Arabitol	1150
	Glyceric acid	2440	Arabinose	200
C-4	Erythritol	5070	Arabinoic acid	165
	Erythrose	*	Xylitol	630
	Erythronic acid	960	Xylose	240
	Threitol	7200	Xylonic acid	67
	Threose	**	Arabitol	1150
	Threonic acid	840	Lyxose	145
			Lyxonic acid	140

(*) and (**) indicate below quantification and detection limits, respectively.

(B)   Ketoses

		ppm (w/w)
C-3	Dihydroxyacetone	540
C-4	Erythrulose	37
C-5	Ribulose	2010
C-5	Xylulose	470

FIGURE 8.14 **Products in the UV irradiation experiments of the icy comet.**
(A) Aldoses. (B) Ketoses. The concentration is ppm (weight/weight). Lager molecules than C-5 are not shown. *Source: Data from Meinert, C., Myrgorodska, I., de Marcellus, P., et al., 2016. Ribose and related sugars from ultraviolet irradiation of interstellar ice analogs. Science 352, 208–212.*

synchrotron radiation (CPSR) from the Source Optimisée de Lumière d'Énergie Intermédiaire du Lure in French to observe stereochemical effects, and analyzed by GC × GC-TOFMS.

The experiment products of the ice samples contained aldoses and ketoses are shown in Fig. 8.14. Monosaccharides, ribose, arabinose, xylose, and lyxose, which belong to aldopentoses, and ribulose and xylulose, which belong to ketopentoses, were observed. However, no stereochemical effects of CPSR were observed.

Sugar molecules can be explained by the photochemically initiated formose-type reaction in which formaldehyde and glycolaldehyde are changed into hydroxyl aldehydes and hydroxyl ketones (Fig. 8.15). These photo-synthesized products are summed up to >3.5% by mass, therefore, these sugars, sugar alcohols, and sugar acids are synthesized

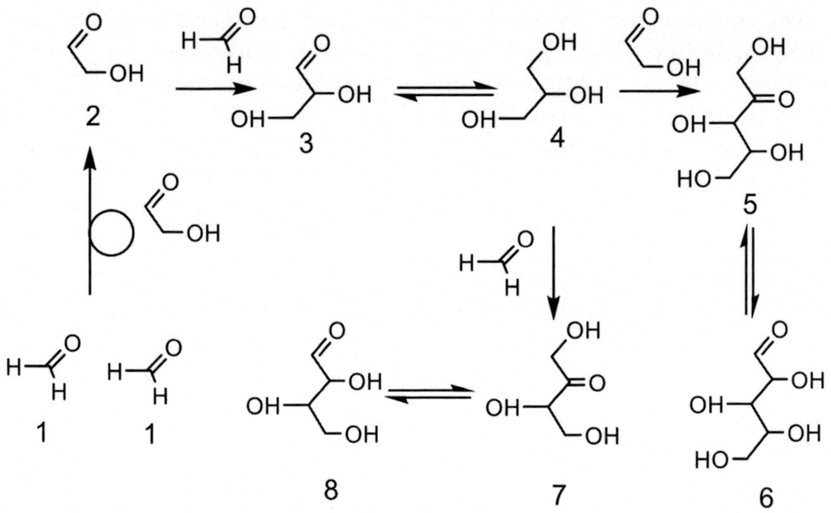

FIGURE 8.15 **The formose reactions.** Formaldehyde 1 reacts under autocatalytic reaction kinetics to form glycolaldehyde 2, which undergoes an aldol reaction by forming glyceraldehyde 3. Dihydroxyacetone 4 is formed by the aldoseketose isomerization of 3 and reacts with 2, making pentulose 5, which isomerizes to an aldopentose 6 such as ribose. In another pathway, dihydroacetone 7 and formaldehyde 1 react forming ketotetrose 7 and aldotetrose 8. The flow of high energy synthesis from formamide to four canonical nucleobases.

in major constituent levels. Ribose was also synthesized to a major-constituent level. It should also be noted that no N-containing materials were synthesized, although nitrogen source was added as ammonia. To make N-containing LRMs, another starting material as formamide like Section 8.6.3 would be required.

8.6.3 Synthesis of life-related materials when the icy comet hits another icy comet in space

Martins et al. (2013) performed an experiment where the icy comet hit another icy comet in space using a gas gun, in which LRMs can be synthesized. Methanol (CH_3OH) is one of the simplest compounds of carbon available in space. The ice containing ammonia solution, CO_2, and methanol was impacted at 7.15 and 7.00 km/s by the steel projectile (ices #1 and #2, respectively). The two pieces of ice ($\sim$100 g each) were used in one collision experiment (one is control), but <1 mg was at the peak pressure of >50 GPa. The ice piece was recovered, dried, and analyzed by gas chromatography-mass spectrometry (GC-Ms). The detection limits for amino acids were $\sim$10 pg.

The impact shock by the ice—ice collision produced several amino acids, including linear and methyl α-amino acids such as glycine, D- and L-Ala, α- AABA, D- and L-Nva, isovaline, and α-AIB (see Fig. 8.10). A racemic mixture of alanine was detected, with a D/L ratio of 0.99 ± 0.05 (ice #1) and 0.99 ± 0.02 (ice #2). A racemic mixture of norvaline with a D/L ratio of 0.97 ± 0.04 (ice #1) and 0.97 ± 0.02 (ice #2) was also detected. This clearly indicates no contamination of terrestrial materials.

A suggested synthetic pathway to produce the detected linear and methyl α-amino acids included a two-step process. The first reaction is the synthesis of the α-amino acid precursors (carbonyl compounds R-CHO, NH_3, and hydrogen cyanide H-CN).

$$CH_3OH \text{ (methanol)} \rightarrow HCHO \text{ (formaldehyde)} \rightarrow HCOOH \text{ (formic acid)}$$

$$(8.1)$$

$$HCOOH \text{ (formic acid)} + NH_3 \rightarrow H-CN \text{ (hydrogen cyanide)} + 2H_2O$$

$$(8.2)$$

The second reaction is the Strecker-cyanohydrin synthesis where:

$$R-CHO + H-CN + NH_4OH \rightarrow R-C(NH_2)-CN \rightarrow R-C(NH_2)-COOH \text{ (amino acid)}$$

$$(8.3)$$

Further support for these reactions is indicated by an ice mixture containing only ammonia and carbon dioxide without methanol. Even with an impact velocity of 7.12 km/s, no detectable quantities of amino acids were produced, indicating the importance of the initial methanol.

It is important to note that the impact shock experimentally occurs on the timescales of nanoseconds to milliseconds (depending on the size of the impactor), which is a much shorter timescale than that required for the two-step Strecker-cyanohydrin process. A high shock pressure results in the formation of ions and radicals, which will then be involved in the post-shock reaction to form amino acids.

These results are a significant step forward in our understanding of the origin of LRMs. The ice—ice collisions should occur in space. It is also known that comets contain significant quantities of the compounds used in this study, and that these compounds are found on the impacted surfaces of the icy bodies in the outer solar system. Therefore, it is highly probable that there are conditions on the surfaces of Saturnian bodies where ammonium compounds, simple alcohols, CO_2, and water ice coexist in an intimately mixed solid form. An icy body with high velocity (5—20 km/s) would give enough energy to promote shock synthesis of more complex LRMs from the ice comets.

As high concentrations of amino acids (Kvenvolden et al., 1970) and NH_3 (Pizzarello et al., 2011) have been detected in CR2 chondrites, these carbonaceous meteorites and comets may share a common condition for the formation of LRMs. When scaling between laboratory scales (millimeters) and planetary scales (kilometers) is taken into account, planetary-scale impacts would lead to significant, long-term, melting and evaporative loss of volatiles may affect the final abundance of LRMs, and the results should be treated with care. However, as there is no atmosphere to stop impacts from millimeter-sized projectiles, the experimental study may be directly applicable and provide the formation mechanism of LRMs.

8.6.4 The icy planet and icy/silicate planet (like the Earth) collisions

8.6.4.1 Introduction

An icy planet or silicate planet like the Archean Earth collisions are assumed to be the origin of the source of LRMs. To test this assumption and to simulate the collision, Ferus et al. (2015) heated the sample with a high-power laser and analyzed run products by GC-MS.

8.6.4.2 From the birth of the sun to late heavy bombardment, and life-related materials

Ferus et al. (2015) summarized the Earth's history from the birth of the Sun to the late heavy bombardment (LHB). The origin of LHB was linked to the dynamic instability of the outer solar system in the Nice model (Section 2.3.3; Tsiganis et al., 2005; Nesvorny and Morbidelli, 2012). The theoretical constraints indicated that the impactor flux on the Earth became ~ 10 times higher at LHB than in the period before LHB, and that this flux slowly decayed afterward (Koeberl, 2006; Morbidelli et al., 2012; Geiss and Rossi, 2013). At the peak, LHB most likely attained an impact frequency of 10^9 tons of materials per year. The typical impact speeds were estimated to have increased from ~ 9 to $\sim 21 \, km/s$, once the LHB began. The ratio of the gravitational cross-sections of the Earth and the Moon is $\sim 17:1$. Thus, for every lunar basin, such as Orientale or Imbrium, larger basins of ~ 17 times should have formed on the Earth (Bottke et al., 2012).

The atmosphere was partly eroded and transformed (Ferus et al., 2009; De Niem et al., 2012), and the hydrosphere was enriched by water (Morbidelli et al., 2000; Cavosie et al., 2005). These impact-related processes also transformed LRMs and their precursors on the Earth's surface, which would have played an important role as the origin of life (Chyba et al., 1990; Chyba and Sagan, 1992).

Although the impact energies were most likely not large enough to produce ocean evaporation or globally sterilizing events (Valley, 2005), they could have served as local energy sources for the synthesis of LRMs (Ferus et al., 2012, 2014; Jeilani et al., 2013). The formation of the nucleobases, as well as purine and glycine, were reported during the dielectric breakdown induced by the high-power laser Asterix in the presence of catalytic materials (meteorites, TiO_2, clay) (Ferus et al., 2012, 2014). Therefore, the high-impact activity may not have been harmful for the formation of LRMs and the first living structures. In contrast, it may have been the source of energy required to initiate chemical reactions, such as the synthesis of LRMs (Zahnle and Sleep, 2006).

One of the progresses in prebiotic chemistry is that formamide could be the parent compound of the components of the first informational polymers (Saladino et al., 2001). Saladino et al. (2012a, 2012b, 2013) extensively studied the formamide-based chemistry that can lead to the synthesis of nucleobases and nucleotides and their metabolic products.

By choosing the appropriate catalyst, purine, adenine, guanine, and cytosine (catalyzed by limestone, kaolin, silica, alumina, or zeolite), thymine (irradiated by sunlight and catalyzed by TiO_2), and hypoxanthine and uracil (in the presence of montmorillonites) were obtained (Saladino et al., 2012a,b,c, 2013; Senanayake and Idriss, 2006).

Formamide-based synthesis in the high-density energy event (impact plasma) can solve the enigma of the simultaneous formation of four nucleobases. The main object of this section is to demonstrate a mechanism of the formation of the nucleobases through the reaction of formamide and its dissociation products in a high-energy impact event relevant to LHB.

The plasma formed by the impact of an extraterrestrial body was simulated using the high-power chemical iodine Prague Asterix Laser System. During the dielectric breakdown in gas (Laser-Induced Dielectric Breakdown or LIDB) generated by a laser pulse of energy of 150 J (time interval of ~ 350 ps, wavelength of 1.315 μm, and output density of 10^{14} to 10^{16} W/cm^2), the outcomes for a high-energy density event occurred. The shock makes temperature to 4500K (Babankova et al., 2006), and generates secondary hard radiation (UV and X-ray).

The unstable radicals produced in the formamide dissociation have been identified and quantified using time-resolved discharge emission spectroscopy (Civiš et al., 2006, 2012).

8.6.4.3 Experiments of laser-induced dielectric breakdown

Saladino et al. (2006) proposed the prebiotic scenario that formamide could be concentrated in lagoons and was exposed to hard UV radiation at high temperatures and in the presence of various catalysts. After the Nice model, during LHB, such a lagoon must have been exposed to plasma during an extraterrestrial body impact event, which could initiate a cascade of chemical reactions, eventually leading to the formation of nucleobases.

In the LIDB plasma experiments by Ferus et al. (2015), formamide is first decomposed into radicals. Radicals of CN• and NH• are the most abundant species. ("•" indicates a radical, which is a free bond. For example, NH• shows a state that one electron of a lone electron pair of N is removed. Therefore, this nitrogen is very reactive.). The absorption gas-phase spectra showed the presence of HCN, CO, NH_3, CO_2, CH_3OH, and N_2O.

Ferus et al. (2014) reported that total yields of the radicals of CN•, NH•, and stable CO were 55, 4, and 41% in the impact plasma. Due to the very rigid structure of CN• from the strong triple bond, this molecule can adopt a series of excited electron configurations. These properties make this transient species an ideal reactant in the high-energy plasma environment (Civiš et al., 2008; Ferus et al., 2011; Horka et al., 2004). Therefore, CN• radical was also discovered in interstellar space in envelopes of giant stars (Shimizu, 1973). Ferus et al. (2012) showed that the stepwise addition of CN• radicals to formamide with atomic H can give rise to the formation of 2,3-diaminomaleonitrile (DAMN) in a highly exergonic reaction. DAMN (see Fig. 8.16) is generally considered to be the common precursor of all nucleobases.

8.6.4.4 Four canonical nucleobase syntheses from DAMN by Laser-Induced Dielectric Breakdown

Except cytosine, all canonical nucleobases were detected in the samples of formamide exposed to LIDB. Irradiation of formamide and DAMN using a high-power laser-produced adenine, guanine, and uracil, whereas DAMN suspensions produced all of the bases (Fig. 8.16).

When formamide was irradiated in the presence of clay, all four bases were detected. The role of clay is to protect the adsorbed cytosine against deamination to uracil in a further reaction step. The photoisomerization of DAMN produces 2,3-diaminofumaronitrile (DAFN), which binds to another CN• radical to readily cyclize into a trisubstituted pyrimidinyl radical. This moiety serves as the precursor for cytosine and uracil. The other pathway includes an additional reaction step in which DAFN cyclizes to a 4-amino-precursor of guanine.

8.6.4.5 Conclusion of LIDB experiments

Ferus et al. (2015) demonstrated that during the era of LHB, nucleobases may have been synthesized in an impact plasma via reactions of the dissociation products of formamide, such as CN• and NH• radicals, with the formamide parent molecule without a catalyst. Their proposal extends the original idea of Saladino et al. (2006), who suggested formamide to be

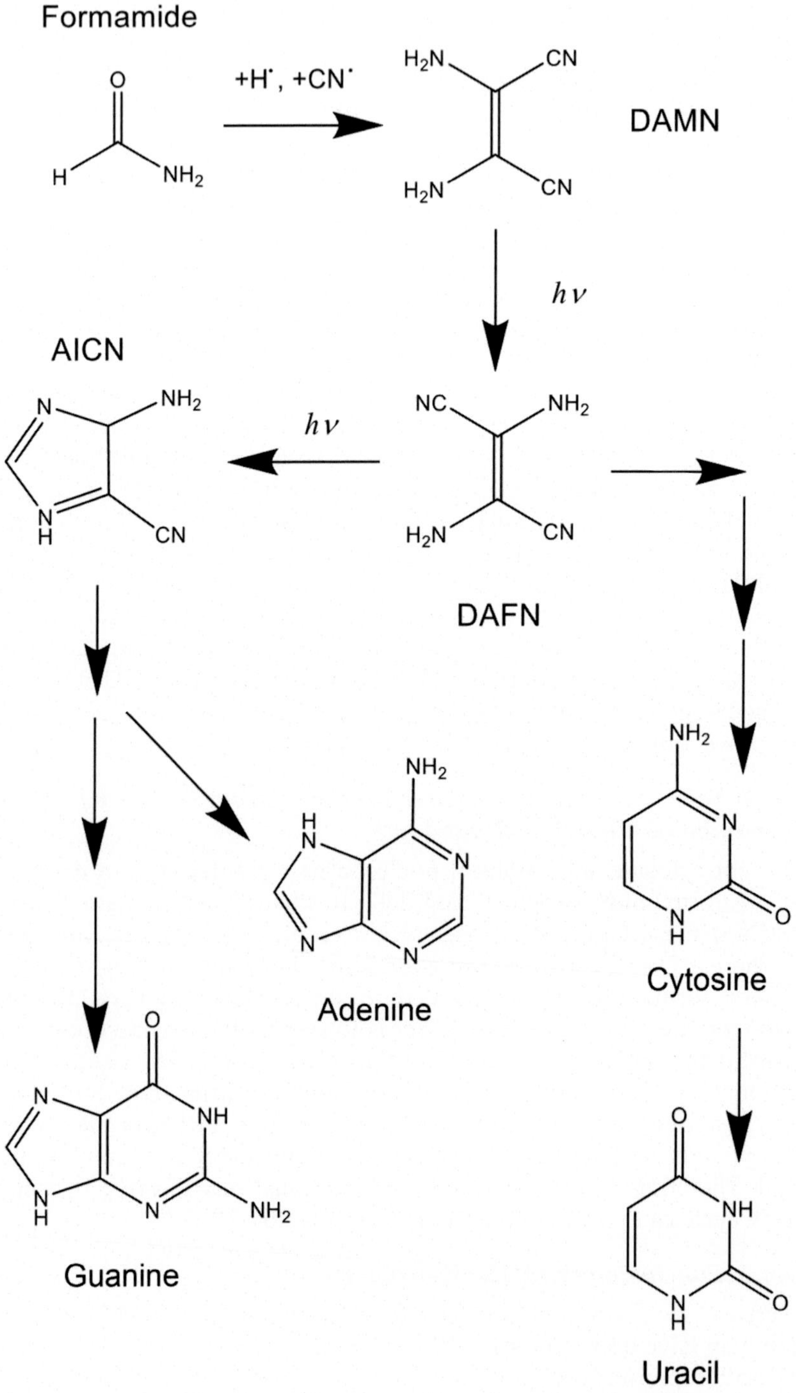

FIGURE 8.16 Flow of high energy synthesis from formamide to four canonical nucleo-bases. Formamide with H and CN forms DAMN. DAMN transforms into DAFN, and then AICN by light. AICN with two steps forms adenine, and with three steps guanine. DAFN with three steps forms cytosine, which becomes with one step uracil. *Source: Data from Ferus, M., Nesvorný, D., Sponer, J., et al., 2015. High-energy chemistry of formamide: a unified mechanism of nucleobase formation. Proc. Natl. Acad. Sci. U. S. A. 112, 657–662.*

the precursor of nucleobases in a prebiotic environment. Ferus et al. (2015) suggested that during the LHB period, the environment influenced by extremely frequent impact events was potentially favorable for nucleobase synthesis. The first biosignatures of life are dated to roughly coincide with the LHB or near the end of it. In conclusion, it is suggested that the emergence of life is not the result of an accident but a direct consequence of the condition of the primordial Earth.

8.6.5 Selection of enantiomers of life-related materials

The next step is how to explain the selective use of enantiomers of LRMs by living things, which is so-called "homochirality." Glucose is one of the LRMs and the main sugar on the Earth with a chemical formula of $C_6H_{12}O_6$. All sugars like glucose which are utilized by living things are a D-isomer.

A typical example of LRMs is an amino acid, which is a major constituent of proteins. The most biologically important molecules are made of proteins. The amino acid has also two types of enantiomers, but all living things use L-amino acid.

Three reasons are proposed: (1) life should have started from selected enantiomers because the abundances of the selected enantiomers were high by the reason during their syntheses in space or on the Earth; (2) the early life should have used selected enantiomers because of presently unknown advantages; and (3) the enantiomer-favored living thing won the early stages of the battle of survival for life by chance, and the present selectivity for enantiomers is not related to neither the abundance nor the advantageous reasons.

Hence, the next step is to successfully explain the selectivity for enantiomers of LRMs, which might give us clues for constraints on the initial life-related molecules. The LIDB (laser-induced dielectric breakdown; Ferus et al., 2015; Meinert et al., 2016) should have affected the L/D ratio, however, as related, no effects on LIDB were observed.

Hawbaker and Blackmond (2019) used experimental data from the model Soai autocatalytic reaction system to evaluate the energy required for symmetry breaking and chiral amplification in molecular self-replication (Shibata et al., 1996; see Fig. 8.17). One postulate for the source of the original imbalance is the tiny difference in energy between enantiomers due to parity violation in the weak force. The plausibility of parity violation energy difference coupled with asymmetric autocatalysis as a rationalization for absolute asymmetric synthesis and the origin of the homochirality of biological molecules are discussed. Their results allowed us to identify the magnitude of the energy imbalance that gives rise to directed symmetry breaking and asymmetric amplification in this autocatalytic system.

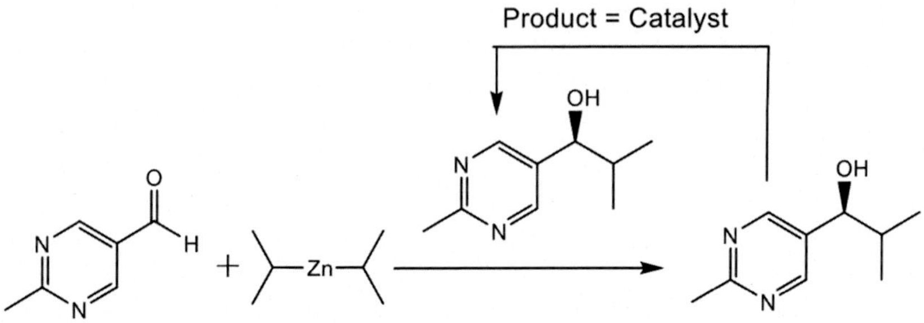

FIGURE 8.17 The Soai reaction. Asymmetric amplification occurs via autocatalysis. *Source: From Shibata, T., Morioka, H., Hayase, T., Choji, K., Soai, K., 1996. Highly enantioselective catalytic asymmetric automultiplication of chiral pyrimidyl alcohol. J. Am. Chem. Soc. 118, 471–472.*

References

Altwegg, K., Balsiger, H., Berthelier, J.J., Bieler, A., Calmonte, U., Fuselier, S.A., et al., 2017. Organics in comet 67P—a first comparative analysis of mass spectra from ROSINA–DFMS, COSAC and Ptolemy. Mon. Not. R. Astron. Soc. 469, S130–S141.

Babankova, D., Civis, S., Juha, L., et al., 2006. Optical and X-ray emission spectroscopy of high-power laser-induced dielectric breakdown in molecular gases and their mixtures. J. Phys. Chem. A 110, 12113–12120.

Barnouin, O.S., Daly, M.G., Palmer, E.E., Johnson, C.L., Gaskell, R.W., et al., 2020. Digital terrain mapping by the OSIRIS-REx mission. Planet. Space Sci. 180, 104764.

Bates, H.C., King, A.J., Hanna, K.L.D., Bowles, N.E., Russell, S.S., et al., 2020. Linking mineralogy and spectroscopy of highly aqueously altered CM and CI carbonaceous chondrites in preparation for primitive asteroid sample return. Meteorit. Planet. Sci. 55, 77–101.

Bockelée-Morvan, D., Lis, D.C., Wink, J.E., et al., 2000. New molecules found in cometC/1995 O1(Hale-Bopp). Investigating the link between cometary and interstellar material. Astron. Astrophys. 353, 1101–1114.

Bottke, W.F., Vokrouhlicky, D., Minton, D., et al., 2012. An Archaean heavy bombardment from a destabilized extension of the asteroid belt. Nature 485, 78–81.

Bottke, W.F., Vokrouhlicky, D., Walsh, K.J., Delbo, M., Michel, P., et al., 2015. In search of the source of asteroid (101955) Bennu: applications of the stochastic YORP model. Icarus 247, 191–217.

Callahan, M.P., Smith, K.E., Cleaves, H.J., et al., 2011. Carbonaceous meteorites contain a wide range of ex-traterrestrial nucleobases. Proc. Natl. Acad. Sci. U. S. A. 108, 13995–13998.

Carrozzo, F.G., De Sanctis, M.C., Raponi, A., Ammannito, E., Castil, J., et al., 2018. Nature, formation, and distribution of carbonates on Ceres. Sci. Adv. 4, 31701645.

Cavosie, A.J., Valley, J.W., Wilde, S.A., 2005. Magmatic delta O-18 in 4400-3900Ma detrital zircons: a record of the alteration and recycling of crust in the Early Archean. Earth Planet. Sci. Lett. 235, 663–681.

Chesley, S.R., Farnocchia, D., Nolan, M.C., Vokrouhlicky, D., Chodas, P.W., et al., 2014. Orbit and bulk density of the OSIRIS-REx target Asteroid (101955) Bennu. Icarus 235, 5–22.

Chyba, C., Sagan, C., 1992. Endogenous production, exogenous delivery and impact-shock synthesis of organic molecules: an inventory for the origins of life. Nature 355, 125–132.

Chyba, C.F., Thomas, P.J., Brookshaw, L., Sagan, C., 1990. Cometary delivery of organic molecules to the early. Earth Planet. Sci. Lett. 249, 366–373.

Civiš, S., Kubat, P., Nishida, S., Kawaguchi, K., 2006. Time-resolved Fourier transform infrared emission spectroscopy of H_3+ molecular ion. Chem. Phys. Lett. 418, 448–453.

Civiš, S., Kubelík, P., Ferus, M., 2012. Time-resolved Fourier transform emission spectroscopy of He/CH_4 in a positive column discharge. J. Phys. Chem. A 116, 3137–3147.

Civiš, S., Sedivcova-Uhlikova, T., Kubelik, P., Kawaguchi, K., 2008. Time-resolved Fourier transform emission spectroscopy of $A^2\Pi-X^2\Sigma^+$ infrared transition of the CN radical. J. Mol. Spectrosc. 250, 20–26.

Cronin, J.R., Moore, C.B., 1971. Amino acid analyses of Murchison, Murray, and Al-lende carbonaceous chondrites. Science 172, 1327–1329.

Crovisier, J., Bockelée-Morvan, D., 1999. Remote observations of the composition of cometary volatiles. Space Sci. Rev. 90, 19–32.

Daly, R.T., Ernst, C.M., Barnouin, O.S., Chabot, N., Rivkin, A.S., Cheng, A.F., et al., 2023. Successful kinetic impact into an asteroid for planetary defence. Nature 616, 443–447.

De Niem, D., Kuehrt, E., Morbidelli, A., Motschmann, U., 2012. Atmospheric erosion and replenishment induced by impacts upon the Earth and Mars during a heavy bombardment. Icarus 221, 495–507.

DiSanti, M.A., Bonev, B.P., Villanueva, G.L., Mumma, M.J., 2013. Highly depleted ethane and mildly depleted methanol in Comet 21P/Giacobini-Zinner: application of a new empirical 2-band model for CH3OH near 50 K. Astrophys. J. 763, 1–15.

Dworkin, J.P., Adelman, L.A., Ajluni, T., Andronikov, A.V., Aponte, J.C., Bartels, A.E., et al., 2018. OSIRIS-REx contamination control strategy and implementation. Space Sci. Rev. 214, 19.

Ehrenfreund, P., Charnley, S.B., 2000. Organic molecules in the interstellar medium, comets, and meteorites: a voyage from dark clouds to the early Earth. Annu. Rev. Astron. Astrophys. 38, 427–483.

Ehrenfreund, P., Irvine, W., Becker, L., et al., 2002. Astrophysical and astrochemical insights into the origin of life. Rep. Prog. Phys. 65, 1427–1487.

Elsila, J.E., Glavin, D.P., Dworkin, J.P., 2009. Cometary glycine detected in samples returned by Stardust. Meteorit. Planet. Sci. 44, 1323–1330.

Ferus, M., Civis, S., Mladek, A., Sponer, J., Juha, L., Sponer, J.E., 2012. On the road from formamide ices to nucleobases: IR-spectroscopic observation of a direct reaction between cyano radicals and formamide in a high-energy impact event. J. Am. Chem. Soc. 134, 20788–20796.

Ferus, M., Kubelík, P., Civiš, S., 2011. Laser spark formamide decomposition studied by FTIR spectroscopy. J. Phys. Chem. A 115, 12132–12141.

Ferus, M., Matulkova, I., Juha, L., Civis, S., 2009. Investigation of laser-plasma chemistry in CO-N_2-H_2O mixtures using O-18 labeled water. Chem. Phys. Lett. 472, 14–18.

Ferus, M., Michalcikova, R., Shestivska, V., Sponer, J., Sponer, J.E., Civis, S., 2014. High-energy chemistry of formamide: a simpler way for nucleobase formation. J. Phys. Chem. A 118, 719–736.

Ferus, M., Nesvorný, D., Sponer, J., et al., 2015. High-energy chemistry of formamide: a unified mechanism of nucleobase formation. Proc. Natl. Acad. Sci. U. S. A. 112, 657–662.

Festou, M., Uwe-Keller, H., Weaver, H.A., 2005. Comets-II. University Arizona Press, Tuscon.

Fujiwara, A., Kawaguchi, J., Yeomans, D.K., et al., 2006. The rubble-pile asteroid Itokawa as observed by Hayabusa. Science 312, 1330–1334.

Fujiya, W., Hoppe, P., Ushikubo, T., Fukuda, K., Lindgren, P., Lee, M.R., et al., 2019. Migration of D-type asteroids from the outer Solar System inferred from carbonate in meteorites. Nat. Astron. 3, 910–915.

Furukawa, Y., Chikaraishi, Y., Ohkouchi, N., Ogawa, N.O., Glavin, D.P., et al., 2019. Extraterrestrial ribose and other sugars in primitive meteorites. PNAS 116, 24440–24445.

Geiss, J., Rossi, A.P., 2013. On the chronology of lunar origin and evolution Implications for Earth, Mars and the Solar System as a whole. Astron. Astrophys. Rev. 21, 1–54.

Hartogh, P.D., Lis, C., Bockelee-Morvan, D., et al., 2011. Ocean-like water in the Jupiter-family comet 103P/Hartley 2. Nature 478, 218–220.

Hawbaker, N.A., Blackmond, D.G., 2019. Energy threshold for chiral symmetry breaking in molecular self-replication. Nat. Chem. 11, 957–962.

Horka, V., Civis, S., Spirko, V., Kawaguchi, K., 2004. The infrared spectrum of CN in its ground electronic state. Collect. Czech Chem. Commun. 69, 73–89.

Hsieh, H.H., Jewitt, D.A., 2006. Population of comets in the main asteroid belt. Science 312, 561–563.

Ito, M., Tomioka, N., Uesugi, M., Yamaguchi, A., Shirai, N., et al., 2022. A pristine record of outer solar system materials from asteroid Ryugu's returned sample. Nat. Astron. . Available from: https://doi.org/10.1038/s41550-022-01745-5.

Jeilani, Y.A., Nguyen, H.T., Newallo, D., Dimandja, J.-M.D., Nguyen, M.T., 2013. Free radical routes for prebiotic formation of DNA nucleobases from formamide. Phys. Chem. Chem. Phys. 15, 21084–21093.

Jewitt, D., Hsieh, H., Agarwal, J., 2015. The active asteroids. In: Michel, P., et al., (Eds.), Asteroids IV. University of Arizona Press, pp. 221–241.

Kelley, M.S.P., Hsieh, H.H., Bodewits, D., Saki, M., Villanueva, G.L., Milam, S.N., et al., 2023. Spectroscopic identification of water emission from a main-belt comet. Nature 619, 720–723.

Kitazato, K., Milliken, R.E., Iwata, T., Abe, M., Ohtake, M., Matsuura, S., et al., 2019. The surface composition of asteroid 162173 Ryugu from Hayabusa2 near-infrared spectroscopy. Science 364, 272–275.

Koeberl, C., 2006. Impact processes on the early Earth. Elements 2, 211–216.

Krohn, K., Jaumann, R., Tosi, F., Nasu, A., et al., 2016. Geologic mapping of the Ac-H-6 quadrangle of Ceres from Nasa's Dawn mission: compositional changes. Geophys. Res. Abstr. 18, EGU2016–7848.

Kvenvolden, K., Lawless, J., Pering, K., et al., 1970. Evidence for extraterrestrial amino-acids and hydrocarbons in the Murchison meteorite. Nature 228, 923–926.

Landaw, L., 2015. New clues to Ceres' bright spots and originsAvailable from: http://www.jpl.nasa.gov/news/news.php?feature = 4785 (2015).

Lauretta, D.S., Bartels, A.E., Barucci, M.A., Bierhaus, E.B., Binzel, R.P., Bottke, W.F., et al., 2019. The OSIRIS-REx target asteroid (101955) Bennu: constraints on its physical, geological, and dynamical nature from astronomical observations. Meteoritics Planet Sci 50, 834–849.

Martins, Z., Botta, O., Fogel, M.L., et al., 2008. Extraterrestrial nucleobases in the Murchison meteorite. Earth Planet. Sci. Lett. 270, 130–136.

Martins, Z., Price, M.C., Goldman, N., Sephton, M.A., Burchell, M.J., 2013. Shock synthesis of amino acids from impacting cometary and icy planet surface analogues. Nat. Geosci. 6, 1045–1049.

Marty, B., Chaussidon, M., Wiens, R.C., Jurewicz, A.J.G., Burnett, D.S., 2011. A ^{15}N-poor isotopic composition for the solar system as shown by genesis solar wind samples. Science 332, 1533–1536.

McKeegan, K.D., Kallio, A.P.A., Heber, V.S., Jarzebinski, G., Mao, P.H., et al., 2011. The oxygen isotopic composition of the sun inferred from captured solar wind. Science 332, 1528–1532.

Meinert, C., Filippi, J.J., de Marcellus, P., 2012. Le Sergeant d'Hendecourt L, Meierhenrich UJ: N-(2-Aminoethyl)glycine and amino acids from interstellar ice analogues. Chempluschem 77, 186–191.

Meinert, C., Myrgorodska, I., de Marcellus, P., et al., 2016. Ribose and related sugars from ultraviolet irradiation of interstellar ice analogs. Science 352, 208–212.

Meshik, A., Mabry, J., Hohenberg, C., Marrocchi, Y., Pravdivtseva, O., et al., 2007. Constraints on neon and argon isotopic fractionation in solar wind. Science 318, 433–435.

Modica, P., Martins, Z., Meinert, C., Zanda, B., D'Hendecourt, L.L.S., 2018. The amino acid distribution in laboratory analogs of extraterrestrial organic matter: a comparison to CM chondrites. Astrophys. J. 865, 41.

Morbidelli, A., Chambers, J., Lunine, J.I., et al., 2000. Source regions and timescales for the delivery of water to the Earth. Meteorit. Planet. Sci. 35, 1309–1320.

Morbidelli, A., Marchi, S., Bottke, W.F., Kring, D.A., 2012. A sawtooth-like timeline for the first billion years of lunar bombardment. Earth Planet. Sci. Lett. 355, 144–151.

Mumma, M.J., Disanti, M.A., Russo, N.D., et al., 2003. Remote infrared observations of parent volatiles in comets: a window on the early solar system. Adv. Space Res. 31, 2563–2575.

Nakamura, E., Kobayashi, K., Tanaka, R., Kunihiro, T., Kitagawa, H., et al., 2022. On the origin and evolution of the asteroid Ryugu: a comprehensive geochemical perspective. Proc. Jpn. Acad. Ser. B 98, 227–282.

Nakamura, E., Makishima, A., Moriguti, T., et al., 2012. Space environment of an asteroid preserved on micrograins returned by the Hayabusa spacecraft. Proc. Natl. Acad. Sci. U. S. A. 109, E624–E629.

Nakamura, T., Matsumoto, M., Amano, K., Enokido, Y., Zolensky, M.E., Mikouchi, T., et al., 2023. Formation and evolution of carbonaceous asteroid Ryugu: direct evidence from returned samples. Science 379, eabn8671.

Nakamura, T., Noguchi, T., Tanaka, M., et al., 2011. Itokawa dust particles: a direct link between S-type asteroids and ordinary chondrites. Science 333, 1113–1116.

Naraoka, H., Takano, Y., Dworkin, J.P., Oba, Y., Hamase, K., et al., 2023. Soluble organic molecules in samples of the carbonaceous asteroid (162173) Ryugu: samples returned from the asteroid Ryugu are similar to Ivuna-type carbonaceous meteorites. Science 379, eabn7850.

Nesvorny, D., Morbidelli, A., 2012. Statistical study of the early solar system's instability with four, five, and six giant planets. Astron. J. 144, 20–68.

Oba, Y., Koga, T., Takano, Y., Ogawa, N.O., Ohkouchi, N., Sasaki, K., et al., 2023. Uracil in the carbonaceous asteroid (162173) Ryugu. Nat. Commun. 14, 1292.

Oba, Y., Takano, Y., Frukawa, Y., Koga, T., Glavin, D.P., Dworkin, J.P., et al., 2022. Identifying the wide diversity of extraterrestrial purine and pyrimidine nucleobases in carbonaceous meteorites. Nat. Commun. 13, 2008.

Oba, Y., Takano, Y., Watanabe, N., Kouchi, A., 2016. Deuterium fractionation during amino acid formation by photolysis of interstellar ice analogs containing deuterated methanol. Astrophys. J. 827, L18.

Pizzarello, S., Williams, L.B., Lehman, J., Holland, G.P., Yarger, J.L., 2011. Abundant ammonia in primitive asteroids and the case for a possible exobiology. Proc. Natl. Acad. Sci. U. S. A. 108, 4303–4306.

Poch, O., Istiqomah, I., Quirico, E., Beck, P., Schmitt, B., et al., 2020. Ammonium salts are a reservoir of nitrogen on a cometary nucleus and possibly on some asteroids. Science 367, eaaw7462.

Potiszil, C., Ota, T., Yamanaka, M., Sakaguchi, C., Kobayashi, K., Tanaka, R., et al., 2023. Insights into the formation and evolution of extraterrestrial amino acids from the asteroid Ryugu. Nat. Commun. 14, 1482.

Russel, C., Raymond, C., 2012. The Dawn Mission to Minor Planets 4 Vesta and 1 Ceres. Springer Science & Business Media, New York.

Saladino, R., Botta, G., Delfino, M., Di Mauro, E., 2013. Meteorites as catalysts for prebiotic chemistry. Chem. Eur. J. 19, 16916—16922.

Saladino, R., Botta, G., Pino, S., Costanzo, G., Di Mauro, E., 2012a. From the one-carbon amide formamide to RNA all the steps are prebiotically possible. Biochimie 94, 1451—1456.

Saladino, R., Botta, G., Pino, S., Costanzo, G., Di Mauro, E., 2012b. Genetics first or metabolism first? The formamide clue. Chem. Soc. Rev. 41, 5526—5565.

Saladino, R., Crestini, C., Ciciriello, F., Costanzo, G., Di Mauro, E., 2006. About a formamide based origin of informational polymers: syntheses of nucleobases and favourable thermodynamic niches for early polymers. Orig. Life Evol. Biosph. 36, 523—531.

Saladino, R., Crestini, C., Costanzo, G., Negri, R., Di, Mauro, E., 2001. A possible prebiotic synthesis of purine, adenine, cytosine, and 4(3H)-pyrimidinone from formamide: implications for the origin of life. Bioorg. Med. Chem. 9, 1249—1253.

Saladino, R., Crestini, C., Pino, S., Costanzo, G., Di Mauro, E., 2012c. Formamide and the origin of life. Phys. Life Rev. 9, 84—104.

Schroeder, I.R.H.G., Altwegg, K., Balsiger, H., Berthelier, J.J., Combi, M.R., Keyser, J.D., et al., 2019. A comparison between the two lobes of comet 67P/Churyumov—Gerasimenko based on D/H ratios in H_2O measured with the Rosetta/ROSINA DFMS. Monthly Not. R. Astron. Soc. 489, 4734—4740.

Senanayake, S.D., Idriss, H., 2006. Photocatalysis and the origin of life: synthesis of nucleoside bases from formamide on $TiO_2(001)$ single surfaces. PNAS 103, 1194—1198.

Shibata, T., Morioka, H., Hayase, T., Choji, K., Soai, K., 1996. Highly enantioselective catalytic asymmetric automultiplication of chiral pyrimidyl alcohol. J. Am. Chem. Soc. 118, 471—472.

Shimizu, M., 1973. Interstellar dust and related topics. In: Greenberg, J.M., van de Hulst, H.C. (Eds.), International Astronomical Union Symposium. International Astronomical Union, Dordrecht, pp. 405—412.

Snodgrass, C., Agarwal, J., Combi, M., Fitzsimmons, A., Guilbert-Lepoutre, A., et al., 2017. The main belt comets and ice in the solar system. Astron. Astrophys. Rev. 25, 5.

Stoks, P.G., Schwartz, A.W., 1981. Nitrogen-heterocyclic compounds in meteorites: significance and mechanisms of formation. Geochim. Cosmochim. Acta 45, 563—569.

Stoks, P.G., Schwartz, A.W., 1979. Uracil in carbonaceous meteorites. Nature 282, 709—710.

Sugita, S., Honda, R., Morota, T., Kameda, S., Sawada, H., et al., 2019. The geomorphology, color, and thermal properties of Ryugu: implications for parent-body processes. Science 364. Available from: https://doi.org/10.1126/science.aaw0422.

Thiemann, W.H.P., Meierhenrich, U., 2001. ESA mission ROSETTA will probe for chirality of cometary amino acids. Orig. Life Evol. Biol. 31, 199—200.

Thomas, C., Naidu, S.P., Scheirich, P., Moskovitz, N.A., Pravec, P., Chesley, S.R., et al., 2023. Nature 616, 448—451.

Tsiganis, K., Gomes, R., Morbidelli, A., Levison, H.F., 2005. Origin of the orbital architecture of the giant planets of the Solar System. Nature 435, 459—461.

Valley, J.W., 2005. A cool early Earth? Sci. Am. 293, 58—65.

Watanabe, S., Hirabayashi, M., Hirata, N., Hirata, N.A., Noguchi, R., et al., 2019. Hayabusa2 arrives at the carbonaceous asteroid 162173 Ryugu—a spinning top—shaped rubble pile. Science 364, 268—272.

Westphal, A.J., Stroud, R.M., Bechtel, H.A., Brenker, F.E., et al., 2014. Evidence for interstellar origin of seven dust particles collected by the Stardust spacecraft. Science 345, 786—791.

Yabuta, H., Cody, G.D., Engrand, C., Kebukawa, Y., De Gregorio, B., et al., 2023. Macromolecular organic matter in samples of the asteroid (162173) Ryugu. Science 379, eabn9057.

Yokoyama, T., Nagashima, K., Nakai, I., Young, E.D., Abe, Y., et al., 2023. Samples returned from the asteroid Ryugu are similar to Ivuna-type carbonaceous meteorites. Science 379, eabn7850.

Zahnle, K.J., Sleep, N.H., 2006. Impacts and the Early evolution of life. In: Thomas, P.J., Hicks, R.D., Chyba, C.F., McKay, C.P. (Eds.), Comets and the Origin and Evolution of Life, second ed. Springer, Berlin, pp. 207−252.

Liquid water on the moons of Jupiter and Saturn

9.1 Exploration of Jupiter and Saturn

9.1.1 Introduction

Since Jupiter is farther away than Mars from the Earth, it becomes more difficult to explore Jupiter. The largest problem is the energy production, because the photovoltaic system generates far less energy than the case of Mars. One way to overcome the energy problem is to use the Radioisotope Thermoelectric Generators (RTGs) using plutonium-238 (^{238}Pu). However, RTG is difficult to build anymore because it has become difficult to attain enough amounts of ^{238}Pu, and the SiGe semiconductor used in RTG is not produced anymore. Another way is to use very large photovoltaic panels of $\sim 100 \, \text{m}^2$.

Although the spacecraft arrives at Jupiter or Saturn, as both planets are so heavy that the decelerating or turning the direction of the spacecraft requires lots of propellant, resulting in allocating less weight to scientific instruments. The communication for maneuvering and transferring data requires special techniques and experience. This is why every spacecraft has a large parabola antenna.

Table 9.1 summarizes the exploration of Jupiter and Saturn performed so far. The illustrations of the spacecraft are shown in Fig. 9.1. The Jupiter and Saturn explorations were mainly done by flyby of other exploration for other purposes. The Jupiter exploration is only Galileo and Juno, and the Saturn exploration is only Cassini—Huygens. The Saturn moon exploration is only done by Cassini—Huygens who landed on the Saturn moon, Titan. The Jupiter moon exploration, Juice (Jupiter icy moons explorer) has just launched in 2023 and will arrive at Jupiter in 2031 (see Section 9.1.2 in detail).

Introductory Astrochemistry
DOI: https://doi.org/10.1016/B978-0-443-23938-0.00009-1

TABLE 9.1 List of Jupiter and Saturn explorer missions.

Explorer	Launch and shutdown years	Organization	Type	Power	Main scientific target
Pioneer 10	1972–2003	NASA, Ames Research Center	Flyby	RTGs[a]	First Jupiter flyby to outer Solar System, Galilean moons observation
Pioneer 11	1973–95	NASA, Ames Research Center	Flyby	RTGs[a]	Jupiter & Saturn flyby to outer Solar System, Callisto, Ganymede Europa & Io, Enceladus & Titan
Voyager 1	1977–2025?	NASA, Jet Propulsion Laboratory	Flyby	RTGs[a]	Jupiter & outer Solar System flybys, Rings, magnetic fields
Voyager 2	1977–2025?	NASA, Jet Propulsion Laboratory	Flyby	RTGs[a]	Jupiter & outer Solar System flybys, Volcanic activity of Io, plate tectonics of Ganymede
Galileo	1989–2003	NASA, Jet Propulsion Laboratory	Orbiter	RTGs[a]	Jupiter orbiter, Jupiter entry probe
Ulysses	1990–2008	ESA, NASA	Flyby	RTGs[a]	Solar polar observer, Jupiter swing-by
Cassini–Huygens	1997–2017	NASA, ESA, ASI[b], Jet Propulsion Laboratory	Flyby	RTGs[a]	Jupiter flyby & Saturn orbiter, Titan lander
New Horizons	2006–	NASA, Applied Physics Laboratory	Flyby	RTGs[a]	Jupiter & Pluto flybys, New Galilean moon photos
Juno	2011–	NASA, Jet Propulsion Laboratory	Orbiter	Solar	Jupiter polar orbiter, Magnetic field, polar magnetosphere
Juice	2023–	ESA, NASA	Orbiter	Solar	Jupiter Icy Moons Explorer, Ganymede, Callisto, Europa

[a]RTG, Radioisotope Thermoelectric Generators.
[b]Italian Space Agency.

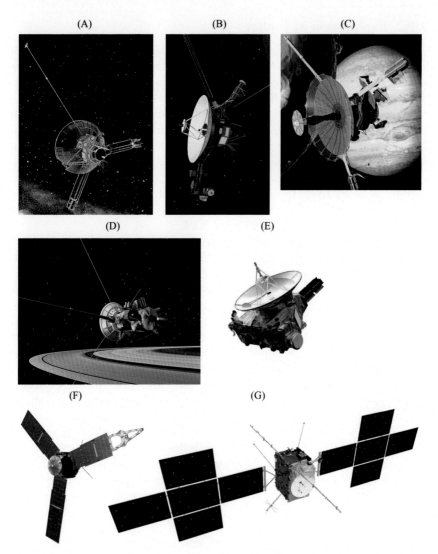

FIGURE 9.1 Explorers of Jupiter and Saturn. (A) An artist's impression of Pioneer 10. Pioneer 11 has the same appearance. (B) An artist's impression of Voyager 1. Voyager 2 has the same appearance. (C) An artist's illustration of Galileo spacecraft. (D) An illustration of Cassini on the Saturn Rings. (E) An image of New Horizons. (F) An illustration of Juno. (G) An artist's impression of Juice. *Source: (A) NASA/Don Davis. http://www.nasa.gov/ centers/ames/news/2013/pioneer11-40-years.html. (B) NASA/JPL-Caltech. https://www.nasa.gov/ sites/default/files/styles/full_width_feature/public/pia17462main_8k.jpg. (C) NASA. https://upload. wikimedia.org/wikipedia/commons/thumb/9/9c/Artwork_Galileo-Io-Jupiter.JPG/1280px-Artwork_ Galileo-Io-Jupiter.JPG. (D) NASA/JPL. https://upload.wikimedia.org/wikipedia/commons/thumb/b/b2/ Cassini_Saturn_Orbit_Insertion.jpg/800px-Cassini_Saturn_Orbit_Insertion.jpg?20220808161016. (E) NASA, Applied Physics Laboratory. https://upload.wikimedia.org/wikipedia/commons/thumb/e/ee/ New_Horizons_spacecraft_model_1.png/643px-New_Horizons_spacecraft_model_1.png?20150715130657. (F) NASA. https://upload.wikimedia.org/wikipedia/commons/f/f6/Juno_spacecraft_model_1.png. (G) ESA/ ATG medialab. https://www.esa.int/ESA_Multimedia/Images/2017/07/Juice_spacecraft.*

9.1.2 Exploration of Jupiter

The first spacecraft which visited Jupiter was Pioneer 10 in 1973, followed by Pioneer 11 (Fig. 9.1A) in 1974. The first close-up photographs were taken, and the spacecraft found the magnetosphere of Jupiter, indicating the liquid flow in Jupiter.

The next spacecrafts were Voyager 1 and Voyager 2 (Fig. 9.1B), which visited the planet in 1979, and studied the moon and the ring system. The volcanic activity of Io and the presence of water ice on the surface of Europa were found.

Ulysses, which was the solar polar orbiter, used Jupiter's gravity to change its orbit from the horizontal plane of the Solar System to the inclination orbit for observing the poles of the Sun. Ulysses studied the magnetosphere of Jupiter in 1992 and 2004.

Galileo (Fig. 9.1C) that arrived Jupiter in 1995 was the first spacecraft to enter the orbit around Jupiter and studied until 2003. In 1995, Galileo sent the atmospheric probe into the Jovian atmosphere. Galileo collected a lot of information on the Jovian system. Galileo closely approached four large Galilean moons ("Galilean moons" indicate four Jovian satellites [Callisto, Europa, Ganymede, and Io] which Galileo Galilei found) and found the thin atmospheres on three of them and found the possibility of the existence of liquid water under their surfaces. Galileo also found the magnetic field of Ganymede.

Cassini–Huygens (Fig. 9.1D), who was also the Saturn explorer, approached Jupiter in 2000 and took a very detailed images of the atmosphere of Jupiter and Saturn.

Juno (Fig. 9.1F) is the orbiter of Jupiter and arrived at Jupiter in 2016. Juno is powered by solar panels. Juno is measuring Jupiter's composition, gravitational and magnetic fields, and polar magnetosphere. Juno is searching for clues how Jupiter formed, whether the rocky core exists or not, the amount of water in the deep atmosphere, and its deep winds up to 620 km h^{-1}.

Juno's microwave radiometer provided a groundbreaking look beneath the water–ice crust of Ganymede and Europa down to 24 km from the surface. JunoCam, which took the visible-light image on Juno, indicated the complex Ganymede's surface, older dark terrain, younger bright terrain, and bright craters, and linear features, which is related to the tectonic activity.

The Jupiter icy moons explorer (Juice; Fig. 9.1G) was launched in 2023, will arrive at Jupiter in 2031, and enter the orbit of Ganymede in 2034. Juice will make a map of the surface, characterize the icy crust, study the magnetic field, the possible interior seas, and the chemistry of organic materials, especially LRMs of Ganymede as well as Europa and Callisto, all of which have internal liquid water oceans, and perhaps Life exists. Then it observes the volcanic activity of Io. Finally, Juice will impact Ganymede at the end of 2035.

NASA is planning the Europa Clipper mission, which will be launched in October 2024. The main goal of Europa Clipper is to determine whether there are places below the ice surface of Europa that could support life or not.

9.1.3 Exploration of Saturn

Saturn missions were mainly flybys by the outer Solar System exploration. In 1979, Pioneer 11 made the first close look at Saturn. In 1980 and 1981, Voyagers 1 and 2 made flybys of Saturn. In 2004, Cassini was the first orbiter of Saturn, which studied the Saturn system over 13 years, and finally plunged into Saturn in 2017. Cassini carried the Huygens probe of ESA which landed on the Saturn moon, Titan in 2005. The Hubble space telescope has been observing from the Earth's orbit.

9.2 Four contrasting moons of Jupiter

9.2.1 Introduction

The four largest moons of Jupiter: Io, Europa, Ganymede, and Callisto are shown in Fig. 9.2. Astonishingly, the appearances of each Jovian satellite are so different.

Cutaway views of the possible internal structures of the Galilean satellites are shown in Fig. 9.3. Except Callisto, all the satellites have metallic (iron, nickel) cores (shown in gray) drawn to the correct relative size. Except Callisto, all the cores are surrounded by rock (shown in brown) shells.

Io's rock or silicate shell extends to the surface, while the rock layers of Ganymede and Europa (drawn to correct relative scale) are in turn surrounded by shells of water in ice or liquid form (shown in blue and white and drawn to the correct relative scale).

Callisto is shown as a relatively uniform mixture of comparable amounts of ice and rock. Recent data, however, suggests a more complex core as shown here.

The surface layers of Ganymede and Callisto are shown as white to indicate that they may differ from the underlying ice/rock layers in the percentage of rock they contain. The white surface layer on Europa could be similar, although it could also suggest an ice layer overlying a liquid water ocean.

Europa might have a liquid water ocean under a surface ice layer several to 10 km thick; however, this evidence is also consistent with the

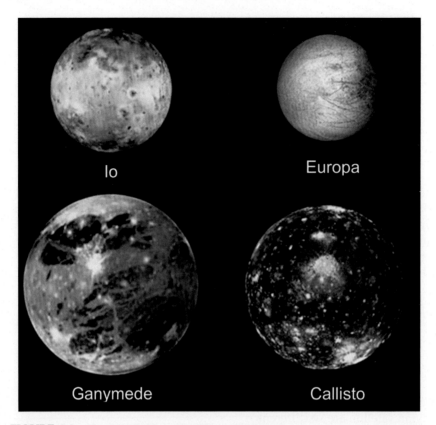

FIGURE 9.2 The four Galilean Satellites of Jupiter. The appearances of these neigh-boring satellites are amazingly different even though they are relatively close to Jupiter (0.35 Gm for Io; 1.8 Gm for Callisto). These images were acquired at very low "phase" angles (the Sun-moon-spacecraft angle) so that the Sun is illuminating the Jovian moons from completely behind the spacecraft. The colors have been enhanced to bring out subtle color variations of surface features. North is at the top of all the images which were taken by the Galileo spacecraft. The original image of NASA was slightly rearranged. *Source: NASA/JPL/DLR. http://photojournal.jpl.nasa.gov/jpegMod/PIA01400_modest.jpg.*

existence of a liquid water ocean in the past. Therefore, it is not certain whether there is a liquid water ocean on Europa at present.

The four Jovian moons are emphasized not only because they have interesting characters, but also they have the possibility of Life in under-lying "liquid" water, or a hot environment as Io. The possibility of Life in these moons is higher than that in Mars!

Why liquid water can exist in the cold moons? It is said that the heat is generated by the tidal force of the giant mother planet. When the icy planets exist with internal water as shown in Europa or Ganymede (see Sections 9.2.3 and 9.2.4), the core can move inside the moon. Thus the energy can be transferred as the tidal force, friction, or potential energy

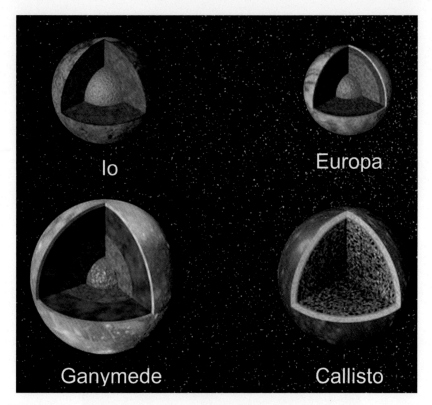

FIGURE 9.3 **Cutaway views of the possible internal structures of the Galilean satellites.** Gray, metallic core; brown, silicates; blue, liquid water. *Source: NASA/JPL. http:// photojournal.jpl.nasa.gov/jpegMod/PIA01082_modest.jpg.*

from the giant mother planet, Jupiter. However, once all water becomes ice or solid such as Callisto, it becomes difficult to transfer energy by such way. The inner liquid in the moon is a prerequisite to energy transfer, but without energy transfer, the inner liquid does not exist (it is like a chicken-and-egg debate).

9.2.2 Io

Io is the nearest moon to Jupiter in the four Jovian moons. Io, which is slightly larger than the Earth's moon, is the most colorful in the Galilean satellites. Its surface is covered by deposits of actively erupting volcanoes, and hundreds of lava flows. The volcanic vents are visible as small dark spots in Fig. 9.4. Several of these volcanoes are very hot, at least one reached 2000°C in 1997. Prometheus, a volcano located slightly right of center in Io's image, was active in 1979 and was still active in

FIGURE 9.4   **Active volcanic plume of Io.** This color image, acquired by Galileo, shows two volcanic plumes on Io. The plume is 140 km high. The second plume, seen near the boundary between day and night, is called Prometheus. The shadow of the airborne plume can be seen extending to the right of the eruption vent. The vent is near the center of the bright and dark rings. Plumes on Io have a blue color, so the plume shadow is reddish. It is possible that this plume has been continuously active for more than 18 years. North is at the top of the picture. The resolution is about 2 km per picture element. *Source: NASA/JPL/ University of Arizona. http://photojournal.jpl.nasa.gov/jpegMod/PIA01081_modest.jpg.*

1996. The active volcanic eruption of Io is shown in Fig. 9.4. The bright, yellowish, and white materials located at equatorial latitudes in Fig. 9.4 are believed to be composed of sulfur and sulfur dioxide. The polar caps are darker and covered by a redder material (Lopes, 2015).

9.2.3 Europa

Europa is the second nearest moon to Jupiter. Europa has a very different surface from its rocky neighbor, Io. A Galileo image of

Fig. 9.5A and B hints the possibility of liquid water beneath the icy crust of this moon. The bright white and bluish part of Europa's surface are composed of water ice. In contrast, the brownish regions on the right side of the image may be covered by salts (such as hydrated magnesium-sulfate) and an unknown red component. The yellowish terrain on the left side of the image is caused by some other, unknown contaminant. This global view was obtained in 1997; the finest details that can be discerned are 25 km across (Castillo-Roges, 2015).

Indications of possible plume activity were reported in 2013 by researchers using NASA's Hubble Space Telescope. Hubble Space Telescope observed water vapor above the frigid south polar region of Europa, providing the first strong evidence of water plumes erupting off the moon's surface (Fig. 9.5C). However, the Cassini spacecraft during its 2001 flyby of Jupiter did not find the plume activity. The plume activity was infrequent, or the plumes were smaller than that at Enceladus.

Europa has become one of the most exciting destinations in the solar system for future exploration because it shows strong indications of having an ocean beneath its icy crust and, thus, has a possibility of Life.

Therefore NASA plans the Europa Clipper mission, which will be launched in October 2024, and arrive in 2030 at Jupiter. The main goal of the mission is to determine whether there are places below the ice surface of Europa that could support Life or not. Europa Clipper is the largest spacecraft NASA has made for a planetary mission. It has two large solar arrays of 14.2 m long × 4.1 m height to collect enough power, and its dry mass is 3241 kg. The electronics and other payload is covered with vault walls which protect the electronics from the strong radiation of Jupiter's magnetic field. It equips an ice-penetrating radar to search for the subsurface to get clues about its ocean and deep interior, a thermal instrument to pinpoint locations of warmer ice, and instruments to measure the compositions.

9.2.4 Ganymede

Ganymede is the third nearest moon among the Galilean moons. Ganymede, larger than the planet Mercury, is the largest Jovian satellite. Its distinctive surface is characterized by the patches of dark and light terrain (see Fig. 9.6). Bright frost is visible at the north and south poles. Ganymede's surface is characterized by a high degree of crustal deformation. Much of the surface is covered by water ice, with a higher amount of rocky material in the darker areas. The global view of

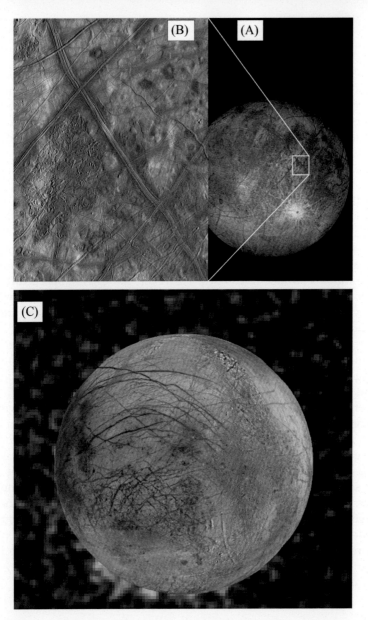

FIGURE 9.5   Jupiter's icy moon Europa. (A) Jupiter's moon Europa has a crust made of blocks, which are thought to have broken apart. (B) An enlarged photograph of (A). Europa may have had a subsurface ocean at some time. The presence of a magnetic field leads scientists to believe an ocean is present at Europa today. Reddish-brown areas represent non-ice material resulting from geologic activity. White areas are material ejected during the formation of the impact crater. Icy plains are shown in blue tones to distinguish coarse-grained ice (dark blue) from fine-grained ice (light blue). Long, dark lines are ridges and fractures in the crust. These images were obtained by the Galileo spacecraft. (C) The composite image of Europa and a background photograph taken by the Hubble Space Telescope. The white plumes from Europa surface may be the water plume coming from the sea under the surface ice. *Source: (A) and (B) NASA/JPL/University of Arizona. http://www.nasa.gov/sites/default/files/images/ 337344main_image_1339_full.jpg (C) NASA/ESA/W. Sparks (STScI)/USGS Astrogeology Science Center.*

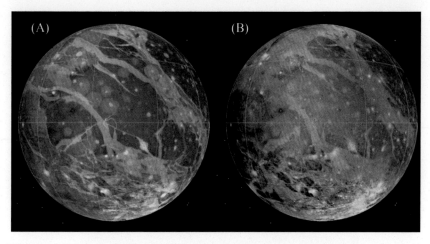

FIGURE 9.6 Ganymede. To present the best information on Jupiter's moon Ganymede, global images obtained by the Voyagers 1 and 2 and Galileo spacecrafts were assembled. (A) The geologic map of Ganymede. (B) The base map for the geologic map (A). *Source: USGS Astrogeology Science Center/Wheaton/NASA/JPL-Caltech. http://photojournal. jpl.nasa.gov/jpegMod/PIA17901_modest.jpg.*

Fig. 9.6B was taken in September 1997 when Galileo was 1.68 Gm from Ganymede; the finest details that can be discerned are about 67 km across (Cook, 2014).

Fig. 9.6B is an image map to make a geological map of Fig. 9.6A. The geological map was made with data from Voyagers 1 and 2 and Galileo spacecrafts. The geology of Ganymede is very complex.

In Fig. 9.7, an image of aurora around Ganymede taken by Hubble Space Telescope is shown. If a saltwater ocean was present, Jupiter's magnetic field would create a secondary magnetic field in the ocean. This friction of the magnetic fields would suppress the reaction of two auroras. This ocean resists Jupiter's magnetic field so strongly that it reduces the aurora to 2 degrees, instead of 6 degrees, if the ocean was not present. Thus scientists estimated that the ocean is 10 km thick which means 10 times deeper ocean than that of the Earth, and is buried under a 150 km crust of ice.

Identifying liquid water is crucial in the search for habitable worlds beyond the Earth and in the search of life. A deep ocean under the icy crust of Ganymede indicates the possibility of life beyond the Earth. The subterranean ocean is thought to have more water than the water on Earth's surface.

NASA's scientists made a sandwich model, shown in Fig. 9.8. This artist's concept of Ganymede illustrates its interior oceans. Scientists suspect Ganymede has a massive ocean under an icy crust. Ganymede's oceans may have 25 times the volume of those on the Earth. This model,

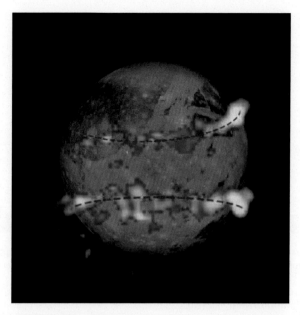

FIGURE 9.7   **Auroral belts of Ganymede.** A NASA Hubble Space Telescope image of Ganymede's auroral belts (colored blue in the illustration) are overlaid on a Galileo orbiter image. The amount of rocking of the moon's magnetic field suggests that the moon has a subsurface saltwater ocean. *Source: NASA/ESA. https://www.nasa.gov/sites/default/files/thumb-nails/image/15-33i2.png.*

based on experiments in the laboratory that simulate salty seas, shows that the ocean and ice may be stacked up in multiple layers.

Ice comes in different forms depending on pressures. "Ice I," the least dense form of ice, which floats in the top. (See Section 1.4.7 and Fig. 1.52 for the phase diagram of pure water. Note that it is not those of the salty water.) As pressure increases, ice molecules become more tightly packed and thus denser. Because Ganymede's oceans are up to 800 km deep, they would experience more pressure than the Earth's oceans. The deepest and most dense form of ice in Ganymede may be "Ice VI."

The model predicts that a strange phenomenon might occur in the uppermost liquid layer, where ice floats upward. In this scenario, cold plumes cause Ice III to form. As the ice forms, salt precipitates out. The salt then sinks down while the ice "snows" upward. Eventually, this ice would melt, resulting in a slushy layer in Ganymede's structure (Clavin, 2014).

The fact that salty water stays at the bottom of the rocky seafloor, rather than ice, is favorable for the occurrence of Life. Researchers think that Life emerges through a series of chemical interactions at water—mineral interfaces, so a wet seafloor on Ganymede might be a key ingredient for Life there.

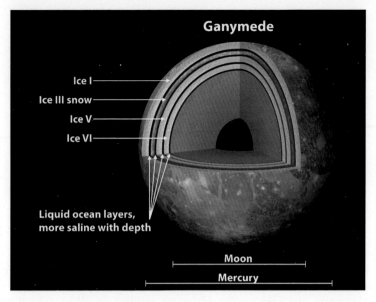

FIGURE 9.8 **A sandwich model of ice and oceans of Ganymede (artist's concept).** Ganymede is expected to have four layers of liquid ocean. *Source: NASA/JPL-Caltech. http:// photojournal.jpl.nasa.gov/figures/PIA18005_fig1.jpg.*

9.2.5 Callisto

Callisto orbits are farthest from Jupiter in the four Jovian moons. The radius of Callisto is 2400 km, slightly larger than that of the Moon (1740 km). Callisto's dark surface is pocked by numerous bright impact craters as shown in Fig. 9.9. The large Valhalla multi-ring structure (see Fig. 9.9) has a diameter of 3800 km, making it one of the largest impact features in the solar system. Although many crater rims exhibit bright icy "bedrock" material, a dark layer composed of hydrated minerals and organic components is seen inside many craters and in other low areas. Evidence of tectonic and volcanic activity, as seen on the other Galilean satellites, appears to be absent on Callisto (Dodd, 2015).

9.3 Moons of Saturn

9.3.1 Introduction

In this section, Saturn's moons—Enceladus, Titan, and Mimas—are picked up and explained. Especially, Enceladus emits water vapors, which means liquid water exists inside the moon. Although Saturn is farther from the Earth than Jupiter, and thus colder, it is amazing that

FIGURE 9.9   **The Valhalla crater on Callisto.** The image was taken by Voyager 1 in 1979. *Source: NASA/JPL-Caltech. https://upload.wikimedia.org/wikipedia/commons/d/d6/Valhalla_crater_on_Callisto.jpg.*

clear images of Enceladus are obtained, and that liquid water exists in Saturn's moons.

Furthermore, the ESA probe, Huygens landed on Titan, which has a dense yellow atmosphere. However, it is −180°C, and too cold for the Life to exist. Recently, the nearest moon of Saturn, Mimas was noticed to have unexplainable librational motion, and proposed to have the liquid internal ocean (Tajeddine et al., 2014).

9.3.2 Enceladus: a Saturn's moon of moving ice

Cassini launched in 1997 and entered orbit around Saturn in 2004. Enceladus was found to be an icy moon of Saturn (Fig. 9.10A). Cassini discovered that Enceladus has curtains of jet plumes (Fig. 9.10B). The jets were made of water vapor and organic molecules spraying from its south polar region. Cassini determined that the moon has a global ocean below between the core and the white ice layer, and likely hydrothermal activity, meaning it could support simple Life (Platt et al., 2014). Spacecrafts have flown to the surface of Enceladus before, but never observed such an active plume.

The flyby was not intended to detect Life, but it will provide powerful new insights about how habitable the ocean environment is within Enceladus. The scientists also expect to better understand the chemistry of the plume.

The flyby will help to solve the mystery of whether the plume is composed of column-like, individual jets, or icy curtain eruptions, or a combination of both. The answer would make clearer how the material is getting to the surface from the ocean below. Researchers are not sure

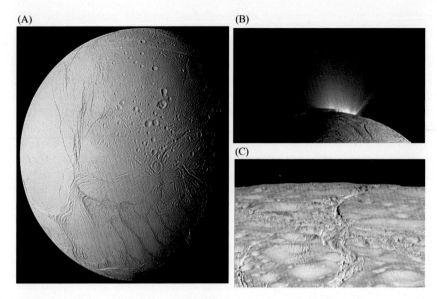

FIGURE 9.10 **Enceladus, jets of Enceladus, and the north pole of Enceladus.** (A) NASA's Cassini spacecraft took the image in 2005. Enceladus exhibits a bizarre mixture of softened craters and complex, fractured terrains. This is the anti-Saturn hemisphere on Enceladus. The gravity measurements suggest an ice outer shell and a low-density, rocky core with a regional water ocean sandwiched in between at high southern latitudes. Enceladus is 504 km in diameter. (B) This is the Cassini image of the geyser basin at the south pole of Enceladus. (C) The Cassini spacecraft captured this image of the moon's north pole. North on Enceladus is up. The view was acquired at a distance of ∼6000 km from Enceladus and a Sun-Enceladus spacecraft angle of 9 degrees. *Source: (A) NASA. https://upload.wikimedia. org/wikipedia/commons/f/fe/Titan_in_true_color_by_Kevin_M._Gill.jpg. (B) NASA/JPL-Caltech/ASI/USGS. http://www.jpl.nasa.gov/spaceimages/images/largesize/PIA17655_hires.jpg. (C) NASA/JPL-Caltech/ASI. https://photojournal.jpl.nasa.gov/jpegMod/PIA20710_modest.jpg. https://upload.wikimedia.org/wikipedia/commons/b/bc/Mimas*

how much icy material the plumes are actually spraying into space. The amount of activity has major implications for how long Enceladus might have been active.

Scientists expected the north polar region of Enceladus to be heavily cratered, based on low-resolution images from the Voyager mission, but high-resolution Cassini images show a landscape of strong contrasts (see Fig. 9.10C). Thin cracks cross over the pole. Before this Cassini flyby, scientists did not know if the fractures extended so far north on Enceladus.

Postberg et al. (2018) observed ice grains containing concentrated macromolecular organic materials with molecular masses above 200 atomic mass units. The data constrain the macromolecular structure of organics detected in the ice grains suggesting the presence of a thin organic-rich film on top of the oceanic water table. The bursting of bubbles allows the probing of Enceladus' organic materials in high concentrations.

Recently, Postberg et al. (2023) found sodium phosphates from Cassini's Cosmic Dust Analyzer mass spectra of ice grains emitted by Enceladus. Phosphorous (P) is necessary to Life for forming DNA, RNA, and membranes. Their observation and laboratory analog experiments showed that P is readily available in Enceladus's ocean in the form of orthophosphates, of which concentrations are at least 100 times higher than those in Earth's oceans. In addition, laboratory experiments showed that phosphorous would exist in oceans beyond the CO_2-rich fluid snowline with moderate temperatures. The higher solubility of calcium phosphate minerals than that of calcium carbonate in pH = ~ 10 would be the reason of high P.

9.3.3 Titan and atmospheres with organic molecules

Saturn's moon, Titan has a radius of about 2600 km (see Fig. 9.11A). Cassini has identified two forms of methane- and ethane-filled depressions that create distinctive features near Titan's poles. Apart from Earth, Titan is the only body in the solar system to possess surface lakes and seas, which have been observed by the Cassini spacecraft. There are numerous smaller, shallower lakes, with rounded edges and steep walls that are generally found in flat areas (see Fig. 9.11B). Cassini also has observed many empty depressions. The lakes are generally not associated with rivers, and thus are thought to fill up by rainfall and liquids feeding them from underground. Some of the lakes fill and dry out again during the 30-year seasonal cycle on Saturn and Titan. But exactly how the depressions hosting the lakes came about in the first place is poorly understood.

Titan has an atmosphere denser than that of the Earth, at 1.45 atm, which was revealed by the ESA's Titan lander, Huygens. It is made of nitrogen (97%), methane ($2.7\% \pm 0.1\%$), and hydrogen ($0.1\%-0.2\%$). Titan's surface temperatures are roughly $-180°C$, liquid methane and ethane, rather than water, dominate Titan's hydrocarbon equivalent of Earth's water. In 2005, the Huygens first detected benzene by mass spectrometry in the atmosphere of Titan. The detection of benzene suggested the existence of polycyclic aromatic hydrocarbons (PAHs) as potential nucleation sources for the growth of Titan's orange-brownish haze atmospheric layer.

It was found that Titan's lakes are reminiscent of karstic landforms on the Earth. These are terrestrial landscapes that result from erosion of dissolvable rocks, such as limestone and gypsum, in groundwater and rainfall percolating through rocks. Over time, this leads to features like

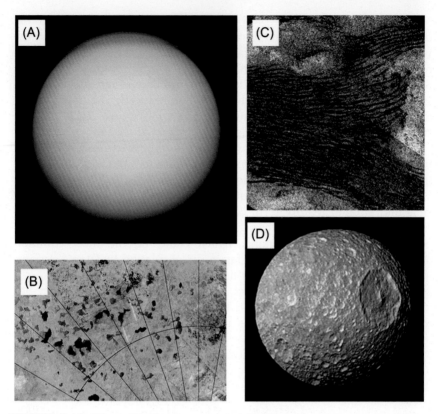

FIGURE 9.11 **Titan.** (A) Titan in true color. (B) The surface radar image of Titan. (C) Sand Sea on Titan. (D) Mimas and Herschel Crater. (A) NASA's Cassini spacecraft took a true color image in 2011. Titan has deep stmosphere so the outline is not sharp. (B) Radar images from Cassini revealed many lakes on Titan's surface, some filled with liquid, and some appearing as empty depressions. (C) Sands were carried from left to right. This image was taken from the Synthetic Aperture Radar on the Cassini spacecraft. Hundreds of sand dunes are visible as dark lines. (D) The Cassini spacecraft took Mimas images in 2010 from 9500 km, and this mosaic picture was made from the six photographs. Mimas is 396 km across, and Herschel Crater is 130 km wide. *Source: (A) NASA. https://upload.wikimedia.org/wikipedia/commons/f/fe/Titan_in_true_color_by_Kevin_M._Gill.jpg. (B) NASA/JPL-Caltech/ASI/USGS. http://www.jpl.nasa.gov/spaceimages/images/largesize/PIA17655_hires.jpg. (C) NASA/JPL-Caltech/ASI. https://photojournal.jpl.nasa.gov/jpegMod/PIA20710_modest.jpg. (D) NASA/JPL-Caltec/Space Science Institute. https://upload.wikimedia.org/wikipedia/commons/b/bc/Mimas_Cassini.jpg.*

sinkholes and caves in humid climates, and salt-pans where the climate is more arid. The rate of erosion creating such features depends on factors such as the chemistry of the rocks, the rainfall rate, and the surface temperature. While all these aspects clearly differ between Titan and the Earth, it is surprising that the underlying process may be surprisingly similar.

It was proposed that it would take around 50 Myrs to create a 100-m depression at Titan's relatively rainy polar regions, consistent with the youthful age of the moon's surface. The dissolution was a major cause of landscape evolution on Titan and could be the origin of its lakes. Scientists has calculated as how long it would take to form lake depressions at lower latitudes, where the rainfall is reduced. The much longer timescale of 380 My is consistent with the relative absence of depressions in these geographical locations (Baldwin et al., 2015).

Some of the organics are deposited in the liquid hydrocarbon lakes in the polar regions and form evaporite features when the lakes dry out in the Titan's methane/ethane cycle which resembles Earth's water cycle. Vu et al. (2014) demonstrated that some organic molecules (such as benzene and ethane) readily form co-crystals in Titan-relevant conditions. They showed Raman spectroscopic evidence for a new co-crystal between acetylene and butane, which could be the most common organic co-crystal discovered so far in direct relation to Titan's surface. Intermolecular interactions such as those in the acetylene—butane co-crystal could modify the kinetics and equilibria of various processes (dissolution, reprecipitation, etc.). Thus, the interaction may play a key role in the formation mechanisms and timescales of landscape evolution on Titan.

Cable et al. (2019) made a model that acetylene (C_2H_2) and butane (C_4H_{10}) would be the main components of such evaporite deposits by the following reactions:

$$C_2H_4 + h\nu \rightarrow C_2H_2 + 2\,H/H_2 \tag{9.1}$$

$$CH_3 + C_3H_7 \rightarrow C_4H_{10} + M \tag{9.2}$$

(M is a third neutral molecule probably N_2)

$$2C_2H_5 + M \rightarrow C_4H_{10} + M \tag{9.3}$$

Titan's equatorial dunes (see Fig. 9.11C) are the most monumental surface structures in our solar system. The chemical composition of the dark organics remains a fundamental, unsolved enigma. A key is that solid acetylene was detected near the dunes.

Abplanalp et al. (2019) found in laboratory simulation experiments that aromatics such as benzene, naphthalene, phenanthrene, phenylacetylene, and styrene (Fig. 9.12A—E), which were possible building blocks of the organic dune materials, can be efficiently synthesized via galactic cosmic ray exposure of low-temperature acetylene ices on Titan's surface. Thus, aromatic hydrocarbons are formed solely in Titan's atmosphere. These processes are also very important in revealing the origin and chemical composition of the dark surfaces of airless bodies in the outer solar system, where hydrocarbon precipitation from the

(A)   (B)   (C)   (D)   (E)

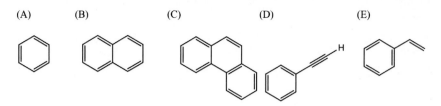

FIGURE 9.12   **Possible organics on Titan's equatorial dunes.** Chemical structures of (A) benzene; (B) naphthalene; (C) phenanthrene; (D) phenylacetylene and (E) styrene.

atmosphere cannot occur. This finding advanced our understanding of the distribution of carbon throughout our solar system such as on Kuiper belt objects like Makemake.

The fundamental mechanisms of the formation of PAHs in a low-temperature atmosphere of Titan have remained uncertain. Zhao et al. (2018) obtained persuasive evidence through laboratory experiments and computations that PAHs like anthracene and phenanthrene ($C_{14}H_{10}$) are synthesized via reactions involving 2-naphthyl radicals ($C_{10}H_7^{\bullet}$) with vinylacetylene ($CH_2{=}CH{-}C{\equiv}CH$) in low-temperature conditions (see Fig. 9.13). These basic reactions are fast without entrance barriers, and synthesize anthracene and phenanthrene through van der Waals complexes and submerged barriers. This easy reaction route to anthracene and phenanthrene are potential building blocks to complex PAHs and aerosols in Titan. This finding requires to change the concept that PAHs can only be formed under high-temperature conditions. The chemistry of Titan's atmosphere is based on simple elementary reactions on fundamental levels.

9.3.4 Mimas

Mimas is the nearest moon to Saturn, and resides in the ring separating the widest A and B Rings. Mimas has the second largest crater on the solar moons (Herschel Crater; see Fig. 9.11D), and the outlook resembles the "Death Star" which appeared in the 1977 film Star Wars. (Astonishingly, the Star Wars is three years earlier than Herschel Crater was found by NASA.) Mimas has a low density of 1.15 g cm^{-3}, possibly it is made of water ice with a small amount of rock. Mimas is ellipsoidal, and the longest axis is $\sim 10\%$ longer than the shortest axis. Herschel Crater wall is 5 km high, and its floor is 6 km deep (max. 10 km), and its center peak is 6 km high above the floor.

Since anomalous libration was found, it was proposed that the internal ocean should exist (Tajeddine et al., 2014). However, there is a counterargument that the libration should be caused by the mass anomaly

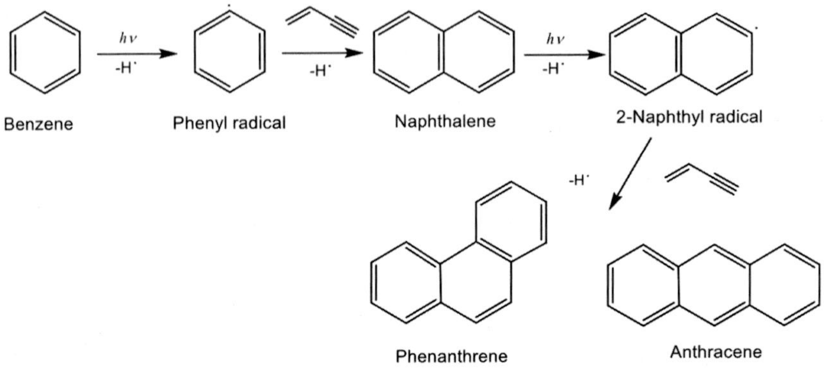

FIGURE 9.13   **Synthesis of PHAs on Titan.** Formation of polycyclic aromatic hydrocarbons (PHAs) by solar photons ($h\nu$) and vinylacetylene ($CH_2{=}CH{-}C{\equiv}CH$) at low temperature. *Source: Modified from Zhao, L., Kaiser, R.I., Xu, B., Ablikim, U., Ahmed, M., et al., 2018. Low-temperature formation of polycyclic aromatic hydrocarbons in Titan's atmosphere. Nat. Astron. 2, 973–979.*

by Herschel Crater, because there is no evidence of tidal stress which should cause surface cracking and other tectonic activities. Recently, Rhoden and Walker (2022) calculated that the tidal heating is large enough to keep the inner ocean, but small enough to keep the ice shell thickness of 24–31 km without the stress. They also proposed that the surface heat flux should be measured to prove the hypothesis[1].

References

Abplanalp, M.J., Frigge, R., Kaiser, R.I., 2019. Low-temperature synthesis of polycyclic aromatic hydrocarbons in Titan's surface ices and on airless bodies. Sci. Adv. 5, eaaw5841.

Baldwin, E., Landau, E., Dyches, P., 2015. The mysterious 'lakes' on Saturn's moon Titan. Cassini at Saturn. http://www.nasa.gov/feature/the-mysterious-lakes-on-saturns-moon-titan.

Cable, M.L., Vu, T.H., Malaska, M.J., Maynard-Casely, H.E., Choukroun, M., et al., 2019. A Co-crystal between acetylene and butane: a potentially ubiquitous molecular mineral on titan. ACS Earth Space Chem. 3, 2808–2815.

Castillo-Roges, J., 2015. Solar system exploration. http://solarsystem.nasa.gov/people/castillo-rogezj.

Clavin, W., 2014. Solar system exploration. http://solarsystem.nasa.gov/news/2014/05/01/ganymede-may-harbor-club-sandwich-of-oceans-and-ice.

Cook, J.-R., 2014. Solar system exploration. http://solarsystem.nasa.gov/news/2014/02/12/largest-solar-system-moon-detailed-in-geologic-map.

Dodd, S., 2015. Solar system exploration. http://solarsystem.nasa.gov/planets/callisto/indepth.

Lainey, V., Rambaux, N., Tobie, G., Cooper, N., Zhang, Q., Noyelles, B., Baillié, K., 2024. A recently formed ocean inside Saturn's moon Mimas. Nature 626, 280–282.

Lopes, L., 2015. Solar system exploration. http://solarsystem.nasa.gov/people/lopesr.

1 Lainey et al. (2024) reported the new evidence of the large ocean under the ice inside Mimas.

Platt, J., Brown, D., Bell, B., 2014. NASA space assets detect ocean inside Saturn moon. http://solarsystem.nasa.gov/news/2014/04/03/nasa-space-assets-detect-ocean-inside-saturn-moon.

Postberg, F., Khawaja, N., Abel, B., Choblet, G., Glein, C.R., et al., 2018. Macromolecular organic compounds from the depths of Enceladus. Nature 558, 564–568.

Postberg, F., Sekine, Y., Klenner, F., Glein, C.R., Zou, Z., Abel, B., et al., 2023. Detection of phosphates originating from Enceladus's ocean. Nature 618, 489–493.

Rhoden, A.R., Walker, M.E., 2022. The case for and ocean-bearing Mimas from tidal heating analysis. Icarus 376, 114872.

Tajeddine, R., Rambaux, N., Lainey, V., Chamoz, S., Richard, A., et al., 2014. Constraints on Mimas' interior from Cassini ISS libration measurements. Science 346, 322–324.

Vu, T.H., Cable, M.L., Choukroun, M., Hodyss, R., Beauchamp, P., 2014. Formation of a new benzene-ethane co-crystalline structure under cryogenic conditions. J. Phys. Chem. A 118, 4087–4094.

Zhao, L., Kaiser, R.I., Xu, B., Ablikim, U., Ahmed, M., et al., 2018. Low-temperature formation of polycyclic aromatic hydrocarbons in Titan's atmosphere. Nat. Astron. 2, 973–979.

Exosolar materials and planets

10.1 Life-related molecules observed in space and interstellar medium

10.1.1 Cyanonaphthalenes were found in the interstellar medium

Unidentified infrared emission is ubiquitous in many astronomical sources. The emission is widely attributed to collective emissions from polycyclic aromatic hydrocarbon (PAH) molecules; however, no such single species have been assigned in the interstellar medium (ISM).

Cordiner et al. (2021) detected two nitrile-groups—functionalized PAHs, 1- and 2-cyanonaphthalenes (1- and 2-CNNs; see Fig. 10.1) in the ISM from radio data of the Green Bank Telescope (GBT) by using spectral matched filtering. Both naphthalene derivative molecules were observed in the TMC-1 molecular cloud.

In situ gas-phase PAH formation pathways from smaller organic precursor molecules are discussed as follows: there are two major formation routes for the synthesis of naphthalene ($C_{10}H_8$);

$$C_6H_5 + CH_2CHC_2H \rightarrow C_{10}H_8 + H \tag{10.1}$$

$$C_6H_5 + CH_2CHCH_2 \rightarrow C_{10}H_{10} + H \tag{10.2}$$

$$\rightarrow C_{10}H_8 + 2H \tag{10.3}$$

Then, 1- and 2-CNNs are formed.

$$C_{10}H_8 + CN \rightarrow 1\text{-CNN} + H \tag{10.4}$$

$$C_{10}H_8 + CN \rightarrow 2\text{-CNN} + H \tag{10.5}$$

Introductory Astrochemistry
DOI: https://doi.org/10.1016/B978-0-443-23938-0.00010-8

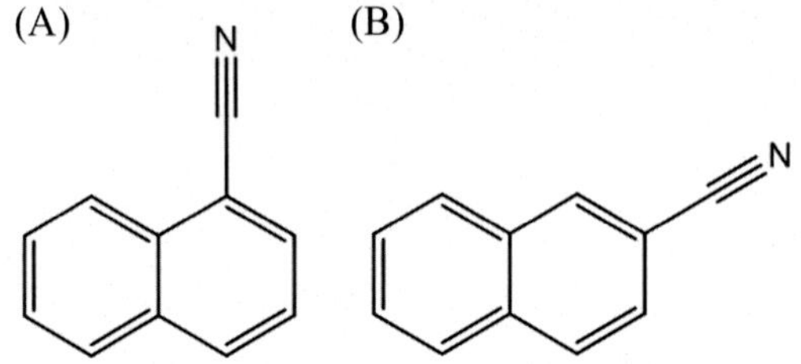

(A)   (B)

FIGURE 10.1 **Chemical structure of cyanonaphthalenes.** (A) 1-Cyanonaphthalene (1-CNN). (B) 2-Cyanonaphthalene (2-CNN).

10.1.2 Methanol signatures were found in the cold dark molecular disk and the star-forming regions

The composition of the material in protoplanetary disks is one of the essential parameters for determining whether habitable environments can be produced or not in the exoplanetary systems. When habitability is discussed, complex organic molecules, especially methanol (CH_3OH) is very important. Methanol primarily forms at low temperatures via the hydrogenation of CO ice on the surface of icy dust grains and is necessary for the formation of more complex molecules such as amino acids and proteins.

Booth et al. (2021) detected CH_3OH in a disk around a young, luminous A-type star, HD 100546. This disk is considered to be warm enough to melt an abundant reservoir of CO ice. Therefore, the CH_3OH does not form in situ, but it must have survived the earlier cold dark molecular cloud phase with complex-organic-molecule-rich ice. This means that at least some interstellar organic materials survive the disk-formation process and are incorporated into forming planets, moons, and comets. Therefore, prebiotic chemical evolution must occur in dark star-forming clouds.

Tan et al. (2020) also reported one of the highest sensitivity surveys for molecular lines in the range of 6.0—7.4 GHz. The observations were done with the 305-m Arecibo Telescope toward a sample of 12 intermediate-/high-mass star-forming regions. Many transitions of different molecules, including CH_3OH and OH were searched. The low rms noise of their data ($\sim$5 mJy for most sources and transitions; 1 Jy $= 10^{-26}$ W m^{-2}/Hz) allowed the detection of OH and 6.7 GHz CH_3OH absorption. A review of 6.7 GHz CH_3OH detections indicated a relation between absorption and radio continuum sources in high-mass star-forming regions. Absorption of excited OH transitions was also detected toward three sources.

10.1.3 One of the cell membrane components was found in space

Cell membranes are one of the most important parts of life because they keep the genes and metabolites in a bounded area (see Fig. 10.2A and B). All cell membranes are made of phospholipids (Fig. 10.2B), however, the nature of the first membranes and the origin of phospholipids are still unsolved.

Rivilla et al. (2021) found the presence of ethanolamine (EtA; $NH_2CH_2CH_2OH$; 2-aminoethanol) in the molecular clouds located in

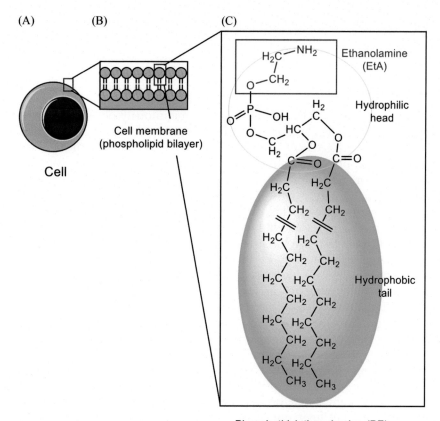

(A) (B) (C)

Cell

Cell membrane
(phospholipid bilayer)

Ethanolamine
(EtA)

Hydrophilic
head

Hydrophobic
tail

Phosphatidylethanolamine (PE)

FIGURE 10.2 Structure of cell membranes. (A) Schematic diagram of a cell. (B) Enlarged view of the cell membrane, which is made by a phospholipid bilayer. (C) Structure of the phospholipid, made by the hydrophilic head and the hydrophobic tail. The former is composed of ethanolamine (EtA), phosphoric acid, and glycerin and the latter is formed by two fatty acids. *Source: Modified from Rivilla, V.M., Jiménez-Serra, I., Martín-Pintado, J., Briones, C., Rodríguez-Almeida, L.F., et al., 2021. Discovery in space of ethanolamine, the simplest phospholipid head group. PNAS, 118, e2101314118.*

the Galactic Center, which is a component of the hydrophilic head of phospholipid in membranes (see Fig. 10.2C). Yebes (Guadalajara, Spain) 40-m telescope was used. The molecular column density of EtA in interstellar space was $N = (1.51 \pm 0.07) \times 10^{13} \, cm^{-2}$, implying a molecular abundance with respect to H_2 of $(0.9-1.4) \times 10^{-10}$.

EtA (see Fig. 10.2C) was also found in meteoritic materials, but it can be synthesized in the meteorite itself by the decomposition of amino acids. The proportion of the molecule with respect to water in the ISM was similar to the one found in the meteorite (10^{-6}). These results indicated that EtA forms efficiently in space and, if delivered onto early Earth, EtA could have contributed to the assembling and early evolution of primitive membranes.

10.1.4 Interstellar synthesis of phosphorus oxoacids

The phosphorus (V) oxidation state is ubiquitously observed in contemporary biomolecules such as phospholipids, adenosine diphosphate and triphosphate (ADP/ATP), and RNA/DNA. Phosphate diesters are the backbone of RNA and DNA, which are key molecules carrying genetic information. Monoesters of pyrophosphate ($P_2O_7^{4-}$) and triphosphate ($P_3O_{10}^{5-}$) play a critical role in cellular energy transfer as ADP and ATP, respectively. This causes an unresolved "phosphorus problem," of which the solubility of phosphate minerals such as apatite ($Ca_5(PO_4)_3(F,Cl,OH)$) are too low to supply bioavailable phosphorus for the first organisms on early Earth.

Cooper et al. (1992) found soluble phosphorus (III) molecules in the Murchison meteorite such as the alkylphosphonic acids ($RP(O)(OH)_2$); (R = methyl, ethyl, propyl, butyl), which can serve as soluble phosphorus compounds. Thus, prebiotic phosphorus chemistry could have been started by phosphorus (III) or the first organisms might have oxidized bioavailable phosphorus (III) to (V). This could overcome the "phosphorus problem," however, the synthetic routes to these phosphorus-bearing biomolecules from their precursors are still a long way. Thus, fundamental knowledge on the origins of phosphorus chemistry is critical to unravel how phosphorus biochemistry and life itself might have emerged on early Earth.

Turner et al. (2018) paid attention to a previously overlooked source of prebiotic phosphorus of interstellar phosphine (PH_3), which produces key phosphorus oxoacids, such as phosphoric acid (H_3PO_4), phosphonic acid (H_3PO_3), and pyrophosphoric acid ($H_4P_2O_7$), in interstellar analog ices exposed to ionizing radiation at temperatures as low as 5K. The molecular clouds with the ice enter circumstellar disks and are

(A) (B)

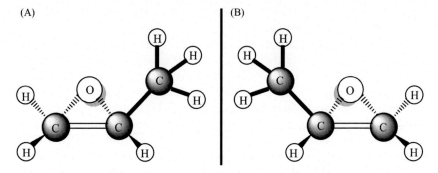

FIGURE 10.3 **Molecular structures of propylene oxide (CH_3CHCH_2O).** (A) and (B) are chiral molecules. As they are mirror images, they cannot be overlapped.

incorporated into planetesimals like proto-Earth. The synthesis of oxoacids is a key to solve the "phosphorus problem" and the origin of water-soluble prebiotic phosphorus compounds.

10.1.5 Discovery of chiral molecule in space

McGuire et al. (2016) discovered the interstellar chiral molecule propylene oxide (CH_3CHCH_2O), which is shown in Fig. 10.3. The chiral molecule was detected by radio astronomy towards the Sagittarius B2 North [SgrB2(N)] molecular cloud, the preeminent source for new complex-molecular detections in the ISM. The propylene oxide was observed using data from the publicity available Prebiotic Interstellar Molecular Survey project at the GBT, which gives high-resolution, high-sensitivity spectral survey data toward SgrB2(N) from 1 to 50 GHz. The chirality was separated by the radio telescope data of a rotational excitation temperature of $\sim 5K$ at a molecular density of $N_T = 1 \times 10^{13}\ cm^{-2}$ and a velocity of 64 km/s. However, by this method, it cannot be determined which molecule in Fig. 10.3A and B is enriched.

10.2 Exosolar planets in the habitable zone

10.2.1 Introduction to exosolar planets

Exoplanets have been found at a fast pace recently. There are mainly two methods to detect the exosolar planets. One is the Doppler spectroscopy, in which the radial velocity of the mother star changes by the gravity of the exoplanet. For example, the Sun changes 13 m/s by the movement of Jupiter, but only 9 cm/s by that of the Earth.

Modern spectrometers with telescopes can detect ~ 3 m/s. In this method, the mass of the exoplanet can be calculated.

The other method is transit photometry, where the change of the periodic brightness change of the mother star by crossing (transit) of the star is used. This method gives the relative size of the exoplanet to the mother star is obtained. In this method, there are many assumptions that both the star and the exoplanet are spherical, the orbit of the exoplanet is circular, the stellar disk is uniform, etc. The largest disadvantage is that the exoplanet's orbit plane needs to happen to face the observer perfectly, otherwise, the transit cannot be detected. This method is suitable for searching the planets around the white and brown dwarfs. The transit photometry can scan the large areas of the sky.

At the beginning of the search for exosolar planets, huge extrasolar planets orbiting near the mother star were primarily found, because they are easily found by the Doppler spectroscopy on ground-based search projects. Planets at Jovian mass within a few thousand light years were detected by this method. Many gas exoplanets orbiting near the mother star were found and called as "hot Jupiters." However, the efficiency for searching the extrasolar planets was not so high, because the Doppler shifts of stars were needed to be measured one by one.

After the special spacecraft for searching the exoplanets, Kepler was launched (see Section 1.2.8); the efficiency for searching the exoplanets drastically increased because the light intensities of ten thousand of stars were measured at once by the transit photometry, and the stars with periodic change of the light intensity were found automatically by the computer.

10.2.2 Exoplanet search projects

Many exoplanet search projects on the ground-based ones and space missions are running today. The projects that found more than 10 exoplanets by 2022 are shown in Table 10.1. In the ground-based projects, High Accuracy Radial Velocity Planet Searcher (HARPS), HAT, and WASP have found more than 100 exoplanets. After the Kepler space observatory (Section 1.2.8) began to work in 2009, the number of found exoplanets drastically increased. In space missions, Kepler and K2 have totally found more than 3000 exoplanets. Transiting Exoplanet Survey Satellite (TESS) also found more than 200 exoplanets.

In 2011, the Kepler team released a list of 2326 extrasolar planet candidates: 207 were similar in size to the Earth (the radius of the exoplanet is <1.25 R_E [R_E is the radius of the Earth]), 680 are super-Earth size (generally, the definition of the "super-Earth" is that the exoplanet mass is between 10

TABLE 10.1   List of exoplanet search projects.

	Abbreviation	Number of exoplanets found	Comment
(a) *Ground-based search projects*			
Anglo-Australian Planet Search	AAPS	∼30	
CARMENES		∼20	At Calar Alto Observatory
High Accuracy Radial Velocity	HARPS	>130	
Planet Searcher			
HARPS-N		∼20	
HATNet and HATSouth Projects	HAT	>100	
KELT		∼30	Finished
Magellan Planet Search Program		∼10	
Microlensing Follow-Up Network	MicroFUN	10	Merged with PLANET
Next-Generation Transit Survey	NGTS	20	At Paranal Observatory
Optical Gravitational Lensing Experiment	OGLE	∼20	
SuperWASP	WASP	∼200	
(b) *Space missions*			
SWEEPS	SWEEPS	16	2006, by HST
Convection, Rotation, and planetary Transits	CoRoT	34	2006−12, transit method
Kepler		∼3000	2009−13, transit method
Kepler's "Second Light"	K2	∼500	2013−18, by Kepler
Transiting Exoplanet Survey Satellite	TESS	>233	2018−

Note: (a) Ground-based search projects. The projects which found more than 10 exoplanets were chosen. (b) Space missions to search exoplanets. The author apologizes many projects are omitted because some data are not updated.

M_E and 1 or 1.9 or 5 M_E; however, in this report, the definition is the radius of the exoplanet is between 1.25 R_E and 2 R_E), 1181 are Neptune size (2−6 R_E), 203 are Jupiter size (6−15 R_E) and 55 are larger than Jupiter size (> 15 R_E). This means that only <10% is the size of the Earth.

In 2018, Kepler discovered 5011 exosolar planet candidates, and 2662 were confirmed. Of course, this result is biased, because the larger planets are more easily detected and confirmed by the doppler spectroscopy. On 1 May 2023, 5366 exosolar planets were confirmed.

10.2.3 Habitable zone

After finding many exoplanets, the interests of astronomers and astrobiologists have been changed into the existence of Life on exoplanets or "civilization" on the exoplanets. Thus, the circumstellar habitable zone (CHZ), or simply the habitable zone began to be discussed by astronomers and astrobiologists.

The concept of a CHZ was first introduced in 1913, by E. Maunder in his book "Are The Planets Inhabited?" (Lorenz, 2020). Then, in 1953, H. Strughold made the term "ecosphere," and discussed various "zones." H. Shapley also emphasized the importance of liquid water to life in the same year (Kasting, 2010). In 1960, Huang first used the term "habitable zone" as the planetary habitability (Planetary Habitability Laboratory) and extraterrestrial life (Huang, 1960).

The habitable zone is the range of orbits of an exoplanet around the mother star within which the planetary surface can keep liquid water with sufficient atmospheric pressure. The boundary of the CHZ is based on the Earth's position in its Solar System, and the amount of radiation energy compared to our Sun (Sol). The diagram showing the CHZ is shown in Fig. 10.4 with some exosolar planets. The horizontal axis is the starlight energy on the planet relative to that on Earth (%), and the vertical axis is the temperature of the mother star. The area between the red and blue lines is the optimistic habitable zone, and that between the yellow and blue lines is the conservative habitable zone. The habitable zone is the criterion for Life, and at least, the existence of liquid water is a prerequisite. The area between the red and yellow lines is considered to be hotter, resulting in the difficulty of the existence of liquid water. Venus, the Earth, and Mars are plotted for comparison. Details of each exosolar planets will be explained in Section 10.2.4.

The habitable zone is also called as the Goldilocks zone, which came from the children's fairy tale of "Goldilocks and the Three Bears." The term "Goldilocks zone" emerged in 1970s. In this story, a little girl chooses from sets of three items. She ignored too extreme ones and chose the one in the middle.

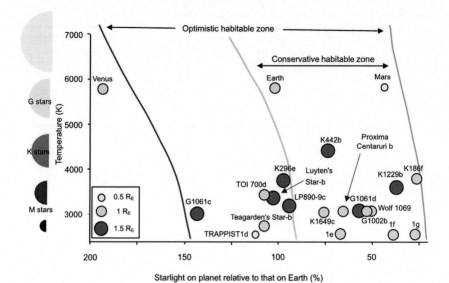

FIGURE 10.4 **Illustration of the habitable zone and exoplanets.** This figure shows the habitable zone of planets. The horizontal axis is the starlight on each planet relative to sunlight on the Earth shown in % (100% is our Sun). The vertical axis is the temperature (K) of each mother star of its planet. The semicircle indicates the color and relative size of the mother star. The circles indicate the planets in the habitable zone. The radii of the planets are divided into three sizes ($\sim$0.5 R_E, $\sim$1.0 R_E, and $\sim$1.5 R_E; R_E means the radius of the Earth), which are indicated by the size of the circle. The exoplanets found by the "Kepler" program start from "K" and the following number. Some of the habitable planets in Table 10.2 are plotted in the figure. *Source: Modified from the Figure by Chester Harman of https://upload.wikimedia.org/wikipedia/commons/thumb/f/f0/Diagram_of_different_habitable_zone_ regions_by_Chester_Harman.jpg/1024px-Diagram_of_different_habitable_zone_regions_by_ Chester_Harman.jpg.*

10.2.4 Earth Similarity Index

The Earth Similarity Index (ESI) is how similar the exoplanet is to the Earth. Schulze-Makuch et al. (2011) proposed ESI, which incorporates radius, density, escape velocity, and surface temperature of the planet. Unfortunately, ESI has little relation to the habitability of the exoplanet.

10.2.5 List of potentially habitable exoplanets

Table 10.2 is a list of potentially habitable exoplanets within the CHZ, smaller than 10 M_E (the mass of the Earth) and 2.5 R_E (the radius of the Earth), and have a chance of rocky. The list is based on the estimates of habitability by Habitable Exoplanet Catalog and data from NASA Exoplanet Archive. In this section, some exoplanets are chosen and explained. The exoplanet name starts from that of the mother star.

TABLE 10.2 List of potentially habitable exoplanets.

Star	Planet	Star type	Mass (M_E)	Radius (R_E)	Flux (F_E)	T_{eq} (K)	Period (d)	Distance (ly)
Sun	Earth	G2V	1.00	1.00	1.00	255	365.25	0
Teagarden's Star	b	M7V	>1.05	~1.02	1.15	264	4.91	12.5
TOI-700	d	M2V	~1.72	1.14	0.87	246	37.4	101
Kepler-1649	c	M5V	~1.20	1.06	0.75	237	19.5	301
TRAPPIST-1	d	M8V	0.39	0.78	1.12	258	4.05	41
Luyten's star	b	M3V	>2.89	~1.35	1.06	258	18.65	12.3
LP890-9	c	M6V		1.37	0.91	272	8.46	105
K-72	e	M?V	~2.21	1.29	1.30	261	24.2	217
Gliese 1061	d	M5V	>1.64	~1.16	0.69	218	13.0	12
Gliese 1002	b	M5V	>1.08	~1.03	0.67	231	10.3	15.8
Gliese 1061	c	M5V	>1.74	~1.18	1.45	275	6.7	12
Kepler-296	e	K7V	~2.96	1.52	1.00	276	34.1	737
Wolf 1069	b	M5V	>1.26	~1.08	0.65	250	15.6	31.2
TRAPPIST-1	e	M8V	0.69	0.92	0.65	230	6.1	41
Proxima Centauri	b	M5V	>1.27	~1.30	0.70	228	11.186	4.25
Kepler-442	b	K5V	~2.36	1.35	0.70	233	112.3	1193
Kepler-1229	b	M?V	~2.54	1.40	0.32	213	86.8	865
Kepler-186	f	M1V	~1.71	1.17	0.29	188	129.9	579

Note: This is a potentially habitable exoplanet according to NASA Exoplanet Archive. Exoplanets under 10 Earth masses and smaller than 2.5 Earth radii, and maybe rocky composition are chosen. "Planet" column shows the Planet number of each star. "~" means that it is estimated from the mass-radius relationship. "M?V" means the temperature of the star is undefined. Most data are from "The Habitable Exoplanets Catalog (HEC) (2022)" and NASA Exoplanet Archive.

The last small alphabet means the number of the planet: "a", "b", "c", ... indicate the first, the second and the third,... planets from the innermost one.

- Teegarden's star b: Teegarden's star is a red dwarf. The CARMENES project found two Earth-mass stars, b and c in the habitable zone, but c is the edge of the blue line of Fig. 10.4. Both exoplanets could have a dense atmosphere.

- TOI-700 d: TOI-700 is a red dwarf with low stellar activity. TOI-700 d is a near-Earth-sized exoplanet in the habitable zone, discovered by TESS. It seems to have an atmosphere and a magnetic field.
- Kepler-1649 c: Kepler-1649 is a red dwarf. Kepler-1649 d is an Earth-sized and possibly rocky exoplanet within the habitable zone. The director of K2 commented that "This planet is the most similar planet to Earth." However, the existence of the atmosphere is unclear.
- TRansiting Planets and Planetesimals Small Telescope (TRAPPIST)-1 b, c, d, e, f, and g: TRAPPIST-1 is an ultracool dwarf star (2566K), and has seven planets, b, c, d, e, f, g, and h. TRAPPIST-1 b, c, and d were Earth-sized planets found by the TRAPPIST project by the Belgium team. TRAPPIST-1 e, f, and g were found by the transit method by the Spitzer Space Telescope. TRAPPIST-1 d has a hydrogen-poor atmosphere, possibly habitable. It may be an "eyeball planet," which is a tidally locked planet always facing the same side to the star. (It is like our Moon which is facing the same side to the Earth.) TRAPPIST-1 e, f, and g may be rocky and habitable, and the surface temperatures will be $-7°C$, $-27.1°C$, and $0°C$, respectively. Green et al. (2023) reported that there is no atmosphere on TRAPPIST-1 b using James Webb Space Telescope (JWST).
- Luyten's star b (or Gliese 273 b): Luyten's star is a red dwarf. Luyten's star b is a super-Earth within the habitable zone, discovered by HARPS. It is one of the most Earth-like planets and may have an atmosphere.
- LP890-9 c: LP890-9 is a red dwarf. LP890-9 c was discovered by TESS, and a terrestrial planet at the edge of the conservative habitable zone. It is tidally locked and possibly has an Earth-like atmosphere and climate.
- K2-72 e: K2-72 is a red dwarf with a temperature of 3360K. K2-72 e was found by the K2 mission. It may be rocky, and tidally locked.
- Gliese 1061 c and d: Gliese 1061 is a red dwarf star. Its mass is 12.5% of that of the Sun, and only 0.2% luminous. Both Gliese c and d are potentially habitable. If they have similar composition of the atmosphere, the equilibrated temperatures of their surfaces are $2°C$ and $-23°C$, respectively.
- Gliese 1002 b: Gliese 1002 is a red dwarf star at 15.8 ly away from the Sun. It is a quiet red dwarf and, thus, does not release flares, which could harm the atmosphere. It has two planets, and Gliese 1002 b will be potentially habitable.
- Kepler-296 e and f: Kepler-296 is a binary star system composed of a K-type main sequence star and a red dwarf star. Kepler-296 e and f are within the habitable zone.
- Wolf 1069 b: Wolf 1069 is a red dwarf star. Wolf 1069 b was discovered by the radial velocity measurement, within the habitable zone and the Earth-mass planet.

- Proxima Centauri b: Proxima Centauri forms the larger triple star system, Alpha Centauri, and is the nearest star of 4.2 ly away from the Sun. Proxima Centauri is a red dwarf and a flare star, which could strip away the atmosphere of the planet. Proxima Centauri b is the nearest planet within the habitable zone found by the Doppler spectroscopy, and may be tidally locked.

10.2.6 Characters of host stars of planets in habitable zones

As seen in Table 10.2 and Section 10.2.4, most habitable planet hosting stars are red dwarf stars. It may be partly the result of the detection method, because the transit method can detect more easily the Earth-sized planets in the dark small star system. The brightness of the Sun drops only 0.008% when the Earth crosses in front of the Sun. Therefore, most habitable planets are detected in the dark star systems.

There are two definitions of the red dwarf. One definition means M-type main sequence star (the maximum temperature of 3900K and $0.6 \, M_\odot$ [the mass of the Sun]), and the other includes both M-type and K-type main sequence stars (the maximum temperature of 5200K and $0.8 \, M_\odot$). The red dwarf is shining by the nuclear fusion of hydrogen. The temperature and mass criterion of the red dwarf include the brown dwarf, however, it is not shining by the nuclear fusion, therefore, the brown dwarf is not included in the red dwarf category.

As the whole red dwarf is convective, ^4He formed by nuclear fusion does not concentrate around the core. Therefore, only the hydrogen burning occurs inside the star, and the star keeps a constant luminosity for trillions of years, until all the fuel is used.

It is also observed that all red dwarfs have high metallicity, meaning that the red dwarfs contain heavier elements than hydrogen and helium. As the lifespan of the red dwarf is longer than the life of the universe, the low metallicity of red dwarfs such as the first generation stars could be included. However, low metallic red dwarfs are rare.

10.2.7 Magnetic field of a planet and Life

When a planet has a liquid metallic core, the magnetic field is generated by the so-called "Dynamo effect." When a planet is solidified perfectly, there is no magnetic field around the planet. In other words, the inside of the planet can be imaged when the planet has a magnetic field. For example, Mars has lost its magnetic fields, because all the core is thought to be solidified.

When a planet has a magnetic field, it is very advantageous for Life to flourish. Because the magnetic field decelerates the fast solar wind which is made of fast-moving energetic particles and is harmful to living things. Such energetic particles are forced to move around and along the magnetic field lines, which go from north to south poles. Please assume a vertically stood magnet field at the north and south poles, which corresponds to a planet. When charged particles come, remember Fleming's left-hand rule. The charged particles go around the magnetic fields and are trapped, and finally go to either the north or south poles.

10.3 Atmospheric molecules in exoplanets

10.3.1 Aerosol composition of hot giant exoplanets

Atmospheres of exoplanets in wide ranges of temperatures, masses, and ages commonly contain aerosols. These aerosols affect transmitted, reflected, and emitted light from exoplanets, obscuring the understanding of exoplanet thermal structure and composition. If the dominant aerosol composition is known, interpretation of observations of exoplanets would be easy, and theoretical understanding of their atmospheres would proceed. A variety of aerosol compositions have been proposed, such as metal oxides and sulfides, iron, chromium, sulfur, and hydrocarbons. However, the relative contributions of these species to exoplanet aerosol opacity are unknown.

Gao et al. (2020) found that the aerosol composition of giant exoplanets observed in transmission was dominated by silicates and hydrocarbons. By constraining an aerosol microphysics model with trends in giant exoplanet transmission spectra, they concluded that silicates dominated aerosol opacity above planetary equilibrium temperatures of 950K due to low nucleation energy barriers and high elemental abundances. In contrast, hydrocarbon aerosols dominated below 950K due to an increase in methane abundance. Their results seem robust to variations in planet gravity and atmospheric metallicity within the range of most giant transiting exoplanets. They predicted that spectral signatures of condensed silicates in the mid-infrared should be most prominent for hot ($>1600K$), and low-gravity ($<10\,m/s^2$) objects.

10.3.2 Flares affect the compositions of exoplanets' atmospheres

Low-mass stars show vigorous magnetic activity in the form of large flares and coronal mass ejections. Such space weather events may affect the habitability and observational fingerprints of exoplanetary atmospheres.

Chen et al. (2021) performed a three-dimensional coupled chemistry-climate model simulation to study the effects of time-dependent stellar activity on rocky planet atmospheres orbiting G, K, and M dwarf stars. They used observed data from the MUSCLES (Measurements of the Ultraviolet Spectral Characteristics of Low-mass Exoplanetary Systems) campaign and the TESS (the Transiting Exoplanet Survey Satellite), and tested a range of rotation period, magnetic field strength, and flare frequency assumptions. They found that recurring flares make the atmospheres of planets around K and M dwarfs into chemical equilibria that are different from their pre-flare regimes, though the atmospheres of G dwarf planets quickly return to their baseline states. Interestingly, simulated O_2-poor and O_2-rich atmospheres experiencing flares produce similar mesospheric nitric oxide abundances, suggesting that stellar flares highlight undetectable chemical species. Applying a radiative transfer model to their chemistry-climate model, they concluded that chemical features, such as NO_2, N_2O, and HNO_3, would be detected by future transmission instruments.

10.3.3 Methane, ammonia, and acetylene were found on a hot Jupiter

Hot Jupiters (see Section 2.3.7), which are hot gaseous giant exoplanets because of orbiting too close to their parent stars, were found ~ 20 years ago. The chemical and physical properties of planetary atmospheres under extreme irradiation conditions have been estimated. Previous observations of hot Jupiters as they transit in front of their host stars showed that water vapor and carbon monoxide existed in their atmospheres, in which chemical equilibrium with the solar composition was assumed. Both molecules and hydrogen cyanide were found in the atmosphere of HD 209458b, a well-studied hot Jupiter (with an equilibrium temperature around 1500K), but no ammonia was found.

Giacobbe et al. (2021) observed HD 209458b and concluded that not only water (H_2O), carbon monoxide (CO), and hydrogen cyanide (HCN), but also methane (CH_4), ammonia (NH_3), and acetylene (C_2H_2) should exist with a statistical significance of 5.3–9.9 standard deviations for each molecule. Atmospheric models to explain the detected species indicated a carbon-rich chemistry with a C/O ratio of ≥ 1, which is higher than the solar value of 0.55. Existing models of atmospheric chemistry of the planet formation and migration scenarios suggested that HD 209458b should formed far from its present place and migrated inwards later. Other hot Jupiters could also show richer chemistry than previously found. Furthermore, the frequently made assumption that they have solar-like and oxygen-rich compositions has become doubtful.

10.3.4 Water vapor was found in the exoplanet

The presence of water is the most important indicator of habitable conditions of warm terrestrial planets. Observations from space and the ground have shown that water, which is the main carrier of oxygen, is the most abundant molecular species after hydrogen in the atmospheres of gaseous extrasolar planets. Water is also a tracer of the origin and the evolution of planets. Atmospheres of small and relatively cold planets are most interesting to observe, however, no atmospheric spectral signatures have so far been detected.

Super-Earth planets lighter than 10 Earth masses around later-type stars may provide the first opportunity to study spectroscopically the characteristics of such planets, as they are best suited for transit observations. Tsiaras et al. (2019) detected a spectroscopic signature of water in the atmosphere of K2-18b, which is a planet of eight Earth masses in the habitable zone of an M dwarf, with high statistical confidence (Atmospheric Detectability Index = 5.0, $\sim 3.6\sigma$). In addition, the derived mean molecular weight suggests an atmosphere still containing some hydrogen. The data were observed by the Hubble Space Telescope/ Wide Field Camera 3 and analyzed with their publicly available algorithms. Although the suitability of M dwarfs to host habitable worlds is still under discussion, K2-18 b offers a unique opportunity to gain insight into the composition and climate of habitable zone planets.

10.3.5 Metal-rich atmosphere for GJ 1214b

The planets between the Earth and Neptune size are found in extrasolar systems. The larger size of this group (called as "sub-Neptunes") is proposed to be distinguished by having a hydrogen-dominated atmosphere. GJ 1214b is such a sub-Neptune which has been observed extensively using transmission spectroscopy to test this hypothesis. However, the measured spectra are featureless, and thus inconclusive. It is partly because of the presence of high-altitude aerosols in the planet's atmosphere.

Kempton et al. (2023) used its spectroscopic thermal phase curve obtained with the JWST in the mid-infrared. Both the dayside and nightside spectra (average brightness temperatures of 553K $\pm$ 9K and 437K $\pm$ 19K, respectively) showed more than 3σ evidence of absorption features, with H_2O as the most likely cause. The measured global thermal emission indicated that Bond albedo of GJ 1214b was 0.51 $\pm$ 0.06. Comparison between the spectroscopic phase curve data and three-dimensional models of GJ 1214b showed that the planet has a high metallicity atmosphere blanketed by a thick and highly reflective layer of clouds.

References

Booth, A.S., Walsh, C., van Scheltinga, J.T., van Dishoeck, E.F., Ilee, J.D., et al., 2021. An inherited complex organic molecule reservoir in a warm planet-hosting disk. Nat. Astron. 5, 684—690.

Chen, H., Zhan, Z., Youngblood, A., Wolf, E.T., Feinstein, A.D., Horton, D.E., 2021. Persistence of flare-driven atmospheric chemistry on rocky habitable zone worlds. Nat. Astron. 5, 298—310.

Cooper, G.W., Onwo, W., Cronin, J.R., 1992. Alkyl phosphonic acids and sulfonic acids in the Murchison meteorite. Geochim. Cosmochim. Acta 56, 4109—4115.

Cordiner, M.A., Kalenskii, E.H., Siebert, M.A., Willis, E.R., Xue, C., Remijan, A.J., et al., 2021. Detection of two interstellar polycyclic aromatic hydrocarbons via spectral matched filtering. Science 371, 1265—1269.

Gao, P., Thorngren, D.P., Lee, G.K.H., Fortney, J.J., Morley, C.V., et al., 2020. Aerosol composition of hot giant exoplanets dominated by silicates and hydrocarbon hazes. Nat. Astron. e4, 951—995.

Giacobbe, P., Brogi, M., Gandhi, S., Cubillos, P.E., Bonomo, A.S., et al., 2021. Five carbon- and nitrogen-bearing species in a hot giant planet's atmosphere. Nature 592, 205—208.

Green, T.P., Bell, T.J., Ducrot, E., Dyrek, A., Lagage, P.-O., Fortney, J.J., 2023. Thermal emission from the Earth-sized exoplanet TRAPPIST-1 b using JWST. Nature 618, 39—42.

Huang, S.-S., 1960. Life-supporting regions in the vicinity of binary systems. Astronom. Soc. Pacific 72, 106—114.

Kasting, J.F., 2010. How to Find a Habitable Planet. Princeton University Press.

Kempton, E.M.R., Zhang, M., Bean, J.L., Steinrueck, M.E., Piette, A.A.A., et al., 2023. A reflective, metal-rich atmosphere for GJ 1214b from its JWST phase curve. Nature 620, 67—71.

Lorenz, R., 2020. Maunder's work on planetary habitability in 1913: early use of the term "habitable zone" and a "Drake equation" calculation. Res. Notes AAS 4, 79.

McGuire, B.A., Carroll, P.B., Loomis, R.A., et al., 2016. Discovery of the interstellar chiral molecule propylene oxide (CH_3CHCH_2O). Science 352, 1449—1452.

NASA Exoplanet Science Institute. NASA Exoplanet Archive. Available from: https://exoplanetarchive.ipac.caltech.edu/. (Accessed 11 July 2023).

Planetary Habitability Laboratory, University of Puerto Rico at Arecibo, 2023. The Habitable Exoplanets Catalog. Available from: https://phl.upr.edu/projects/habitable-exoplanets-catalog. (2023) (Accessed 5 January).

Rivilla, V.M., Jiménez-Serra, I., Martín-Pintado, J., Briones, C., Rodríguez-Almeida, L.F., et al., 2021. Discovery in space of ethanolamine, the simplest phospholipid head group. PNAS 118, e2101314118.

Schulze-Makuch, D., Méndez, A., Fairén, A.G., von Paris, P., Turse, C., Boyer, G., et al., 2011. A two-tiered approach to assess the habitability of exoplanets. Astrobiology 11 (10), 1041—1052.

Strughold, H., 1953. The Green and Red Planet: A Physiological Study of the Possibility of Life on Mars. University of New Mexico Press.

Tan, W.S., Araya, E.D., Lee, L.E., Hofner, P., Kurtz, S., et al., 2020. High-sensitivity observations of molecular lines with the Arecibo Telescope. Mon. Not. R. Astron. Soc. 497, 1348—1364.

Tsiaras, A., Waldmann, I.P., Tinetti, G., Tennyson, J., Yurchenko, S.N., 2019. Water vapour in the atmosphere of the habitable-zone eight-Earth-mass planet K2-18 b. Nat. Astron. 3, 1086—1091.

Turner, A.M., Bergantini, A., Abplanalp, M.J., Zhu, C., Góbi, S., et al., 2018. An interstellar synthesis of phosphorus oxoacids. Nat. Commun. 9, 3851.

Index

Note: Page numbers followed by "*f*" and "*t*" refer to figures and tables, respectively.

A

A0106, 317
AA. *See* Alpha Aecer (AA)
Abiotic graphite, 241−242
Absorption of lights, 39
Accretion-ejection model, 138
Acetone (CH_3COCH_3), 315−316
Acetonitrile (CH_3CN), 315−316
Acetylene (C_2H_2), 352, 370
  on hot Jupiter, 370
Achondrites, 83−84
Acid(s), 186
  acid-hydrolyzed hot water extract, 316
  residues, 85−86
Active volcanic plume of Io, 342*f*
Adenosine diphosphate (ADP), 360
Adenosine triphosphate (ATP), 360
ADP. *See* Adenosine diphosphate (ADP)
Aerodynamic drag, 147
Aerosol composition of hot giant
  exoplanets, 369
AFM. *See* Atomic force microscopy (AFM)
AGB stars. *See* Asymptotic giant branch
  stars (AGB stars)
Ala. *See* Alanine (Ala)
Alanine (Ala), 255−256
Alcohols, electrochemistry of brines and
  metal-rich minerals on Mars reduced
  CO_2 to, 283−284
Aldoses, 106−107, 251
ALH84001. *See* Allan Hills 84001 (ALH84001)
Aliphatic carboxylic acids, 254
Alkyl chain, 256
Alkylcarbazoles, 310
Alkylphosphonic acids, 360
Alkylpiperidines, 316
Alkylpyridines, 310
Alkylquinolines, 310
Allan Hills 84001 (ALH84001), 270−272,
  279, 281
  discovery of Life in Martian meteorite,
    272−273

fossils of microorganisms found in
  Martian meteorite, 270−282
ALMA. *See* Atacama Large Millimeter
  Array (ALMA)
Alpha Aecer (AA), 200
α-alanine, 253
α-amino-n-butyric acid (α-ABA), 255−256,
  312
α-decay, 35−37
AM. *See* Angular momentum (AM)
Amino acids, 88−91, 90*f*, 91*t*, 253−254, 358
  comparison between amino acids in A-
    and C-series samples, and
    implications, 312−313
  precursors, 305
Ammonia (NH_3), 29, 254−255, 305, 319, 370
  on hot Jupiter, 370
Amorphous graphite, 235
Angular momentum (AM), 204−205
Anion exchange chromatography, 159
Anthracene, 353
Apache Point Observatory (APO), 26
Apache Point Observatory Galactic
  Evolution Experiment (APOGEE), 27
  APOGEE-2, 27
Apatite [$Ca_3(PO_4)_2$], 237, 272, 360
Apex chert
  evidence for Earth's oldest fossils in, 235
    carbonaceous microfossils in thin
      section, 236*f*
  oldest microfossils in, 234−235
APO. *See* Apache Point Observatory (APO)
APOGEE. *See* Apache Point Observatory
  Galactic Evolution Experiment
  (APOGEE)
Apollo group asteroids, 22
Apollo program, 151
APPI. *See* Atmospheric photoionization
  (APPI); Atmospheric pressure
  photoionization (APPI)
Aqueous alteration minerals, 224
Aragonite, 75

Archaea, 106
  road from protocell to, 112−113
Archaean Earth
  perspective of atmospheric evolution
    from Hadean to, 227−228
  transportation of materials on Archaean
    Earth by late heavy bombardment, 228
Argon (Ar), 42
Aromatics, 352−353
Artificial intelligence, 96−97
Asp. *See* Aspartic acid (Asp)
Aspartic acid (Asp), 253, 255−256
Asteroid(s), 86, 227, 251−252, 294−298
  basic knowledge on, 84−88
    calcium−aluminum rich inclusions,
      84−85
    chondrules globular shape, 85
    CI chondrites and trace elemental
      abundance of Earth, 87
    evidence of early formation and
      differentiation of protoplanets, 85
    Palermo scale and Torino scale, 87−88
    presolar grains, 85−86
    Yarkovsky effect, 86
    Yarkovsky−O'Keefe−Radzievskii−
      Paddack effect, 86−87
  belts, 24t
  classification of, 22−26
    asteroid number and naming of, 23−26
    classification of Tholen, 22−23
    motion of asteroid, 24f
    orbital classification, 22
    SMASS classification, 23
  definition of, 22
  distribution of, 23f
  explorations, 289−298
    of asteroid exploration missions,
      289−291
    Dawn spacecraft, 293
    Deep Impact spacecraft, 292
    Galileo spacecraft, 291
    Lucy spacecraft, 294
    missions, 291−294
    NEAR-Shoemaker spacecraft, 291−292
    Psyche spacecraft, 293−294
  formation of, 142−143
    evolution of solar system, 143f
  life-related materials in asteroids and
    comets by laboratory experiments,
    318−327
  organic materials and life-related
    materials in, 305−306

photographs of asteroids targeted by
    sample-return missions, 301f
  D/H ratio on comets and origin of
    water on Earth, 294−295
  identification of water emission from
    main-belt comet, 295
  mystery of bright spots on Ceres, 296
  origin of volatiles on Earth, 295
  sample-return missions from, 298−305
Asteroidal collision with Earth
  DART project, 297f
  defense from, 296−297
Astrochemical inorganic materials,
    68−69
Astrochemistry, 29−31, 43, 45, 57
  radioactive isotopes and half-lives used
    in, 38t
Astronomical unit (AU), 3
Astronomy, 3, 12−29, 39
  classification of asteroids, 22−26
  definition of asteroids and comets, 22
  Doppler effect and redshift, 18−21
  electromagnetic spectrum and
    observatories, 13f
  gamma-ray, 13−15
  Herschel space observatory, 28
  infrared, 17
  Kepler space observatory, 28
  Lyman-α lines, 29
  metallicity of stars, 21
  optical, 17
  planetary system, 29
  radio, 18
  SDSS, 26−27
  spacecraft for, 14f
  Stellar classification, 18
  ultraviolet, 15−16
  X-ray, 15
Astrophysical simulations of giant impact,
    168−169
  model calculation results for accretion
    efficiency, 168f
Astrophysics, 1−12
  classical mechanics, 3−8
  high-temperature and high-pressure
    experiments, 8−12
  neutron cross-section, 12
  Roche limit, 12
  SI units, 1−3
  spacecraft for, 14f
  three-body problem and Lagrange
    points, 4−5

Astroscience
  astronomy, 12−29
  astrophysics, 1−12
  inorganic astrochemistry, 29−68
  mineralogy and petrology, 68−88
  organic astrochemistry, 88−114
  statistics, 114−118
Asymptotic giant branch stars (AGB stars), 18
Atacama Desert, 17
Atacama Large Millimeter Array (ALMA), 17
Atacama Large submillimeter Array, 17
Ataxite, 83
Atmosphere of Jupiter, 338
Atmospheric molecules
  in exoplanets, 369−371
    aerosol composition of hot giant exoplanets, 369
    flares affect compositions of exoplanets' atmospheres, 369−370
    metal-rich atmosphere for GJ 1214b, 371
    methane, ammonia, and acetylene found on hot Jupiter, 370
    water vapor in exoplanet, 371
Atmospheric photoionization (APPI), 283
Atmospheric pressure photoionization (APPI), 313−314
Atomic force microscopy (AFM), 280
Atomic number, 31−34
ATP. See Adenosine triphosphate (ATP)
AU. See Astronomical unit (AU)
Aufbau principle, 33−34, 34f
Aug. See Augite (Aug)
Augite (Aug), 71−73

B
Bacteria, 106
  road from protocell to, 112−113
Bacterial microfossils, 235
Bacterial ribosomes, 101
Banded iron formation (BIF), 239−240
Barium carbonate, 280
Baryon Oscillation Spectroscopic Survey (BOSS), 27
Basaltic breccias, 269
Basalts of Moon (BM), 203
BB. See Big Bang (BB)
Below ice surface, 339, 343
Benzene, 352−353
β-alanine (β-Ala), 253, 255−256, 313

β-aminoisobutyric acid (β-AIBA), 255−256
BIF. See Banded iron formation (BIF)
Big Bang (BB), 121
  elemental synthesis in BB and first stars, 127−128
  nucleosynthesis, 125
  theory, 121−126
    artist illustrations of spacecraft for observation of CMBR, 122f
    nine-year Wilkinson microwave anisotropy probe, 122f
    timeline of universe, 124t
Binary system, 135
Biogenic detritus, 240
Biological macromolecules, 279
Biomarkers, magnetite crystals not, 277−278
Biotite porphyroblasts, 240
Black Hole Mapper, the, 27
BM. See Basalts of Moon (BM)
Bosons, 6
BOSS. See Baryon Oscillation Spectroscopic Survey (BOSS)
Brines on Mars reduced CO_2 to alcohols, electrochemistry of, 283−284
BSE. See Bulk silicate Earth (BSE)
BSM. See Bulk silicate Moon (BSM)
Bulk analysis, 43−44, 44f
  methods, 45−46
    schematic diagram of MC-ICP-MS, 48f
    schematic diagram of Q-pole type ICP-MS, 46f
    schematic diagram of TIMS, 47f
Bulk silicate Earth (BSE), 159, 178, 183, 185−186, 194, 204−207, 220−221
Bulk silicate Moon (BSM), 198−199
Butane (C_4H_{10}), 352

C
^{13}C-depleted carbon microparticles in >3700 Ma sea-floor sedimentary rocks, 240
C-type asteroids. See Carbonaceous meteoritic type asteroids (C-type asteroids)
C0107, 317
Ca aluminate hibonite ($CaAl_{12}O_{19}$), 171
CAIs. See Calcium−aluminum rich inclusions (CAIs)
Calcium carbonates ($CaCO_3$), 75
Calcium−aluminum rich inclusions (CAIs), 84−85, 171

Callisto, 339–341, 347
  Valhalla crater on, 348*f*
Canonical nucleobase syntheses from
  DAMN by laser-induced dielectric
  breakdown, 325
Carbon, hydrogen, oxygen, nitrogen, and
  sulfur (CHONS), 313
Carbon (C), 205–207, 235–236, 251–253,
  306
  abundances and speciation analyses of C
    and S by pyrolysis and combustion,
    307
  carbon-bearing fluids, 239
  elemental synthesis of light elements up
    to, 128–130
    photograph of oldest star, 128*f*
  isotope
    fractionation, 236–237
    ratios, 234
  isotopic fractionation, 105, 235
  molecules, 253
  reservoir, 255
Carbon dioxide (CO_2), 254
  assimilation, 263
  electrochemistry of brines and metal-rich
    minerals on Mars reduced CO_2 to
    alcohols, 283–284
Carbon monoxide (CO), 370
Carbon tetrachloride (CCl_4), 114
Carbonaceous chondrites, 79, 84, 175–176,
  180, 217, 251–252, 315
  meteorites, 233
  organic materials and life-related
    materials in, 305–306
Carbonaceous meteoritic type asteroids (C-
  type asteroids), 22–23
Carbonates, 75, 296
  carbonate-silicate cycle, 227
  globules, 274
  reduction reactions, 256–258
Carbonation on early Mars, organic
  synthesis associated with
  serpentinization and, 281–282
Carbonyl compounds, 305
Carcinogens, 97
Cassini, 336*t*, 348
  cosmic dust analyzer, 350
  spacecraft, 343
Cassini–Huygens, 335, 337*f*, 338
CDF. *See* Cumulative distribution function
  (CDF)
Celestial body, 12

Cell membrane, 109–110, 112
  components found in space, 359–360,
    359*f*
Ceres
  craters, 296
  mystery of bright spots on, 296
CGRO. *See* Compton Gamma Ray
  Observatory (CGRO)
Characteristic X-ray, 39
Chassignites, 269
Chemistry of organic materials,
  338–339
Chemo-fossil, 238
China National Space Administration
  (CNSA), 266
Chiral molecule in space, discovery of, 361
Chondrites, 83–84, 194–196
Chondritic asteroids, 176
Chondritic materials, 215
Chondritic meteorites, 187–188
CHondritic Uniform Reservoir (CHUR
  reservoir), 58–59
Chondrules, 85
  globular shape, 85
CHONS. *See* Carbon, hydrogen, oxygen,
  nitrogen, and sulfur (CHONS)
Chromite ($FeCr_2O_4$), 272
Chromium, 193, 369
  isotope ratios, 193–194
  isotopic composition of lunar rocks,
    193–194
Chromosomes, 110
CHUR reservoir. *See* CHondritic Uniform
  Reservoir (CHUR reservoir)
CHZ. *See* Circumstellar habitable zone
  (CHZ)
CI chondrites and trace elemental
  abundance of Earth, 87
Circularly polarized synchrotron radiation
  (CPSR), 319–320
Circumstellar habitable zone (CHZ), 364
Classic experiments, 253
  Miller experiment, 253, 254*f*
  primordial soup theory, 253
Classical mechanics, 3–8
  elementary particles in standard
    models, 6*f*
  fine structure of ^4He atom, 7*f*
  validity of, 5*f*
Classification of Tholen, 22–23, 25*t*
CLM. *See* Continental lithospheric mantle
  (CLM)

Clustered Regularly Interspaced Short Palindromic Repeats-Crispr Associated protein 9 (CRISPR-Cas9), 103
  gene editing, 103, 104f
CMBR. *See* Cosmic microwave background radiation (CMBR)
$C_nH_{2n-7}N$ homologs, 310
CNSA. *See* China National Space Administration (CNSA)
Coesite, 74
Cold dark molecular disk, methanol signatures in, 358
Combustion, abundances and speciation analyses of C and S by, 307
Comets, 251–252, 294–298
  D/H ratio on comets and origin of water on Earth, 294–295
  defense from asteroidal collision with Earth, 296–297
  definition of, 22
  explorations, 289–298
    Dawn spacecraft, 293
    Deep Impact spacecraft, 292
    Galileo spacecraft, 291
    Lucy spacecraft, 294
    missions, 291–294
    NEAR-Shoemaker spacecraft, 291–292
    overview of asteroid exploration missions, 289–291
    Psyche spacecraft, 293–294
  identification of water emission from main-belt comet, 295
  MMX mission, 297–298
  mystery of bright spots on Ceres, 296
  organic materials and life-related materials in, 305–306
  origin of life-related materials in asteroids and comets by laboratory experiments, 318–327
  origin of volatiles on Earth, 295
  sample-return missions from, 298–305
Compton Gamma Ray Observatory (CGRO), 13–15
Computer simulations
  origin of life-related molecules by, 259–260
  computer simulations of six generations, 260f
Concordia diagram, 64, 65f
Condensation process, 171
  of elements after giant impact, 170–171
Confocal Raman spectroscopy, 105, 238
Constant force, 3

Continental lithospheric mantle (CLM), 183
Conventional carbon–hydrogen–oxygen–nitrogen–sulfur isotopic astrochemistry, 29–31
Copper, 189–191
  isotopic composition of lunar rocks, 189–191
Core formation age from Hf–W systematics, 225–226
CoRoL. *See* Corotation limit (CoRoL)
Corotation limit (CoRoL), 204–205
Cosmic gas, 138–139
Cosmic microwave background radiation (CMBR), 121
Cosmic rays, 219
Cosmochemists, 56
CPSR. *See* Circularly polarized synchrotron radiation (CPSR)
Crust–mantle system, 229
Crystal fractionation, 231
Crystallization, 161–162, 229
  products, 219
Cumulative distribution function (CDF), 116–117
Cyanobacterial microfossils, 235
Cyanobacterium-like microorganisms, 235
Cyanonaphthalenes found in interstellar medium, 357
  chemical structure of, 358f
2-Cyanonaphthalenes (2-CNNs), 357
1-Cyanonaphthalenes (1-CNNs), 357
Cytoplasm, 101, 112

D

D,L-alanine ($C_3H_7NO_2$), 316
D,L-valine ($C_5H_{11}NO_2$), 316
D,L-α-amino-n-butyric acid ($C_4H_9NO_2$), 316
D,L-β-amino-n-butyric acid ($C_4H_9NO_2$), 316
D/H ratio. *See* Deuterium to hydrogen ratio (D/H ratio)
Dark ages, 125
DART. *See* Double Asteroid Redirection Test (DART)
Dawn spacecraft, 293
DCM. *See* Dichloromethane (DCM)
Deep Impact mission, 292
Deep Impact spacecraft, 292
Deep sea volcanism, basic geological knowledge on, 78–79
Deoxyribo-nucleic acid (DNA), 93–96, 360
  synthesis of proteins from, 98–99
    genetic codes, 100t

Deoxyribose, 94–95
DESI-OT-MS. *See* Desorption Electrospray
     Ionization-Orbitrap-Mass Spectrometry
     (DESI-OT-MS)
Desorption Electrospray Ionization-
     Orbitrap-Mass Spectrometry (DESI-OT-
     MS), 309
     organic material study using, 309–310
Deuterated methanol (CD$_3$OD), 114
Deuterated water (D$_2$O), 114
Deuterium to hydrogen ratio (D/H ratio),
     294–295
     on comets and origin of water on Earth,
          294–295
Di. *See* Diopside (Di)
2,3-diaminofumaronitrile (DAFN), 325
2,3-diaminomaleonitrile (DAMN), 325
     canonical nucleobase syntheses from, 325
Diamond anvil cell, 11–12, 11*f*
Dichloromethane (DCM), 313–314
     extracts, 315, 315*f*
Dichotomy, 199
Diethyl ether (C$_2$H$_5$OC$_2$H$_5$), 315–316
Differential scanning calorimetry (DSC),
     282
Digital image analysis, 235
Digital terrain models (DTMs), 304
Diopside (Di), 71–73
Disk formation process, 358
DNA. *See* Deoxyribo-nucleic acid (DNA)
Doppler effect
     of light, 20
     and redshift, 18–21, 20*f*
Doppler spectroscopy, 361–362
Double Asteroid Redirection Test (DART),
     296–297
Double helix, 97
     DNA, 98*f*
Double spike method, 193, 196
DSC. *See* Differential scanning calorimetry
     (DSC)
DTMs. *See* Digital terrain models (DTMs)
Dynamo effect, 368

E

Early Mars, organic synthesis associated
     with serpentinization and carbonation
     on, 281–282
Early sample-return missions from space,
     298–300
     genesis spacecraft, 300
     Stardust spacecraft, 299

Earth
     CI chondrites and trace elemental
          abundance of, 87
     D/H ratio on comets and origin of water
          on, 294–295
     defense from asteroidal collision with,
          296–297
     Earth/Moon impact flux ratio, 223
     icy planet fell on Earth's ocean, 253–259
     late veneer, 215–224
     oldest evidence of Life on Earth,
          234–245
          ^{13}C-depleted carbon microparticles in
               > 3700 Ma sea-floor sedimentary
               rocks, 240
          back to Isua Supracrustal Belt, Western
               Greenland, 241–242
          carbon isotopic fractionation, 235
          counterarguments to Isua Supracrustal
               Belt structures as stromatolites,
               243–244
          evidence for Earth's oldest fossils in
               Apex cherts, 235
          evidences of Life older than 3800 Ma
               at Isua supracrustal belt and Akilia
               island, 237–238
          geological evidence of recycling of
               altered crust in Hadean, 240–241
          objection to earliest Life of Akilia
               Island, 238–240
          oldest microfossils in Apex chert,
               234–235
          origin of life back to 4.1 billion-year-
               old, 242
          stromatolites, 242–243
          stromatolites in Isua Supracrustal Belt
               oldest Life in 3700 Ma, 243
     oldest geological records on Earth,
          224–234
          age estimates for early Earth's
               differentiation events, 225*t*
          application of [142]Nd isotope
               systematics to oldest crust on Earth,
               229–231
          core formation age from Hf–W
               systematics, 225–226
          enstatite chondrites source materials of
               Earth, 233–234
          first water on early Earth from oxygen
               isotopic data of zircon by HR-SIMS,
               232–233
          oldest zircon on Earth, 231–232

perspective of atmospheric evolution from Hadean to Archaean Earth, 227–228

transportation of materials on Archaean Earth by Late Heavy Bombardment, 228

U–Pb age of Earth, 228–229

origin of volatiles on, 295

oxygen in hematite on Moon could come from, 160–161

similarity between chemical compositions of Moon and Earth's mantle, 154–158

condensation temperature *vs.* depletion of elemental concentration, 155f

metal/silicate partition coefficients, 157f

synthetic experiments of life-related molecules on, 253–260

classic experiments, 253

icy planet fell on Earth's ocean, 253–259

origin of life-related molecules by computer simulations, 259–260

transportation of life-related molecules of extraterrestrial origin on, 251–253

transportation of life-related molecules of space origin onto, 251–252

Earth Similarity Index (ESI), 365

Earth-Moon system, 166, 205, 223

giant impact model for formation of, 166–170

astrophysical simulations of giant impact, 168–169

history of giant impact model, 166–168

latest model proving similarity between proto-Earth and Theia, 170

peculiarities of moon, 166

standard model, 169–170

eBOSS. *See* Extended BOSS (eBOSS)

EBSD. *See* Electron backscatter diffraction (EBSD)

EC. *See* Electron capture (EC)

ECs. *See* Enstatite chondrites (ECs)

Edgeworth-Kuiper belt, 29

EDS. *See* Energy dispersive spectrometry (EDS)

Ejecta of giant impact, 224

Electrochemistry of brines and metal-rich minerals on Mars reduced CO_2 to alcohols, 283–284

Electromagnetic field, 125

Electromagnetic radiation, 113

Electromagnetic spectrum and observatories, 13f

Electromagnetism, 123

Electron

beam, 43–44

degeneration, 134–135

microscopy, 235

shielding, 113

Electron backscatter diffraction (EBSD), 244

Electron capture (EC), 34–37

EC-like asteroids, 234

Electron probe micro analyzer (EPMA), 43–44, 47–48, 50f, 70, 306

Electrospray ionization (ESI), 283, 313–314

ESI-APPI-FT-ICR/MS, 316

Electrostatic attraction, 96–97

Electroweak epoch, 123–125

Elemental analytical methods, 43–52

bulk analysis methods, 45–46

bulk analytical methods, 44f

comparison of, 43f

detailed classification of spot analyses, 45f

spot analysis methods, 47–52

Elemental astrochemistry, 166

Elemental fractionation, 61

Elementary particles, 6

evaporation and condensation of elements after giant impact, 170–171

in standard models, 6f

Elements, 7, 31–34

classic classification of, 53, 54f

elemental synthesis

in Big Bang and first stars, 127–128

heavier than Fe and supernova, 132–134, 134f

of heavy elements up to Fe, 130–132

of light elements up to carbon, 128–130

formation of proto-Earth in solar nebular, 137–149

neutron star merger, 135–137, 137f

origin, 126–137

cosmic microwave background radiation, 121

and evolution of universe, 121–126, 123f, 126f

universe, 121–126

Elements (*Continued*)
  symbol, 31–34
  synthesis of Li, Be, and B, 134
  three forces are sustaining star in fine
    balance, 127, 127f
  type Ia supernova, 134–135, 136f
    aged binary system, 135f
Embryos, 146, 170
Emission of lights, 39
Enantiomers of life-related materials,
  selection of, 327
Enceladus, 347–350
Endoplasmic reticulum, 110
Energy dispersive spectrometry (EDS), 47
Energy spectra, absorption and emission of
  lights and, 40f
Enstatite (MgSiO₃), 71–73, 171
Enstatite chondrites (ECs), 175–176,
  179–180, 194–196, 233
  data, 179–180
  source materials of Earth, 233–234
Enzyme, 99
EPB. *See* Eucrite parent body (EPB)
EPMA. *See* Electron probe micro analyzer
  (EPMA)
EPOXI. *See* Extrasolar Planet Observation
  and Deep Impact Extended Investigation
  (EPOXI)
Error magnification, 69f
Escherichia coli, 102
ESI. *See* Earth Similarity Index (ESI);
  Electrospray ionization (ESI)
ESI high-resolution mass spectrometry
  (ESI-HRMS), 317
ESI-HRMS. *See* ESI high-resolution mass
  spectrometry (ESI-HRMS)
Ethanol (C₂H₅OH), 315–316
Ethanolamine (EtA), 359–360
Ethylamine (C₂H₅NH₂), 316
Eucaryotic ribosomes, 101
Eucrite parent body (EPB), 154
Eukaryote cell, schematic structure of,
  109–110, 109f
Eukaryotes, road from protocell to,
  112–113
Europa, 339–340, 342–343
  Jupiter's icy moon, 344f
  support life, 339, 343
Europa Clipper mission, 339, 343
Evaporation of elements after giant impact,
  170–171
Evolution model, 141

Exoplanet
  atmospheric molecules in, 369–371
  flares affect compositions of exoplanets'
    atmospheres, 369–370
  illustration of habitable zone and, 365f
  search projects, 362–364, 363t
  water vapor found in, 371
Exosolar
  inconsistency between solar system and
    exosolar systems found by Kepler
    program, 147–149
  planets, 361–362
    characters of host stars of planets in
      habitable zones, 368
    ESI, 365
    exoplanet search projects,
      362–364
    in habitable zone, 361–369, 365f
    list of potentially habitable exoplanets,
      365–368
    magnetic field of planet and life,
      368–369
Exploration, 294
  of Jupiter and Saturn, 335–339
    explorers of Jupiter and Saturn, 337f
    list of Jupiter and Saturn explorer
      missions, 336t
  of mars, 265–267
    Mars sample return mission, 267
    searching for life-related materials and
      Life on Mars, 265–267
Exponential law, 66
Extended BOSS (eBOSS), 27
Extinct nuclides, 59–61, 60t
Extraction resins, 174
Extrasolar Planet Observation and Deep
  Impact Extended Investigation (EPOXI),
  292

F

FAN. *See* Ferroan anorthosite (FAN)
FD. *See* Fluorescence detector (FD)
Feldspar group, 73
Felsic igneous rocks, 231
Felsic volcanism, 242
Femtosecond laser, 52
Fermions, 6
Ferric iron–bearing minerals, 160
Ferroan anorthosite (FAN), 151–153, 161,
  163
  age of, 163
  samples, 186

Ferrosilite (Fs), 71–73
Filamentous microbial fossils, 235
Fischer projection, 103–104, 104f
Fischer-Tropsch-type synthesis, 235
Fischer–Tropsch reaction, 105
Fission model, 156, 166
Fluorescence detector (FD), 314–315
Fluorophile elements, 159
Flyby, 348–349
Formamide-based synthesis, 324
Forsterite (Mg_2SiO_4), 171, 255
Fossils
    isochron, 59–61
        plot for $^{182}Hf-^{182}W$ system, 62f
        plot for $^{26}Al-^{26}Mg$ system, 60f
    of microorganisms found in Martian
        meteorite, ALH 84001, 270–282
        carbonate globules and magnetite
            crystals, 274
        discovery of Life in Martian meteorite,
            ALH84001, 272–273
        ineffectiveness of morphological
            similarities, 279–280
        magnetite crystals not biomarkers,
            277–278
        morphology, 280–281
        nanobacteria not Life, 278–279
        organic synthesis associated with
            serpentinization and carbonation on
            early Mars, 281–282
        PAH, 273–274
        schematic diagrams showing
            morphologies of crystals, 277f
        structures resembling terrestrial life,
            274–277
Fourier transform-ion cyclotron resonance
    mass spectrometry (FT-ICR/MS), 283,
    313–314
Fourier-transform infrared spectroscopy
    (FTIR spectroscopy), 315
Fractional crystallization, 200–201, 203
Fs. See Ferrosilite (Fs)
FT-ICR/MS. See Fourier transform-ion
    cyclotron resonance mass spectrometry
    (FT-ICR/MS)
FTIR spectroscopy. See Fourier-transform
    infrared spectroscopy (FTIR
    spectroscopy)
Full width at half maximum (FWHM), 312
Fusion crusts, 80–83
FWHM. See Full width at half maximum
    (FWHM)

G
Gabbronorites, 154
Galactic Center, 359–360
Galactic cosmic ray (GCR), 200
    spallation process, 200
Galilean moons, 338, 343–345
Galilean satellites
    cutaway views of possible internal
        structures of, 341f
    of Jupiter, 340f
Galileo Galilei, 291, 336t, 337f, 338
Gallium (Ga), 185–186
    isotopic composition of lunar rocks,
        185–186, 187f
Gamma-ray astronomy, 13–15
Ganymede, 339, 343–346, 345f
    Auroral belts of, 346f
    sandwich model of ice and oceans of,
        347f
Garnet group, 74
Garnet porphyroblasts, 240
Gas chromatography, 42
Gas chromatography-isotope ratio mass
    spectrometry (GC-IRMS), 270
Gas chromatography-mass spectrometry
    (GC-MS), 294–295
Gas giant planets, formation of, 142–143
Gas-free scattered planetesimals, 227
Gaseous disk, 147
Gaseous nitrogen, 255
Gaseous planets, 147
GBT. See Green Bank Telescope (GBT)
GC-IRMS. See Gas chromatography-isotope
    ratio mass spectrometry (GC-IRMS)
GC-MS. See Gas chromatography-mass
    spectrometry (GC-MS)
GCR. See Galactic cosmic ray (GCR)
GCWM. See General Conference on
    Weights and Measures (GCWM)
General Conference on Weights and
    Measures (GCWM), 1
Genesis spacecraft, 300
Genetic information transfer system, 110
Geochemistry, 29–31
Giant impact model, 198
    history of, 166–168
GJ 1214b, metal-rich atmosphere for, 371
Gliese 1002, 367
Gliese 1061, 367
Glutamic acid (Glu), 255–256
Glycine ($C_2H_5NO_2$), 253–256, 299, 305, 313,
    316

Gold (Au), 70
Goldilocks zone. *See* Habitable zone
Grand Tack models, 137, 144–145, 147
  formation of inner planets in Grand Tack
    model, 145–146
Grand unification epoch, 123
Graphite
  globules, 238
  particles, 242
Green Bank Telescope (GBT), 357
Greenhouse gases, 227
Greigite (Fe_3S_4), 277–278
Gypsum, 350–351

H
Habitable zone, 364
  characters of host stars of planets in, 368
  illustration of, 365f
Hadean
  geological evidence of recycling of
    altered crust in, 240–241
  perspective of atmospheric evolution from
    Hadean to Archaean Earth, 227–228
Hadron epoch, 125
Hafnium, 61, 163
HARPS. *See* High Accuracy Radial Velocity
  Planet Searcher (HARPS)
Hartley-2, 292
Harvard spectral classification, 18, 19t
Hayabusa spacecraft, 300
  photographs of asteroids targeted by
    sample-return missions, 301f
Hayabusa2, 306
  spacecraft, 301–303
    illustrations of sample-return
      spacecraft, 302f
    photographs of returned samples from
      C-type asteroid, 302f
Heating model, 224
Heavy element, 184–185
  elemental synthesis of heavy elements up
    to Fe, 130–132
    onion shell structure in red giant star,
      131f
Heavy impactor flux, 228
Heavy isotopes, 193–194
  of potassium, 189
Heavy nuclei, 35–37, 166
Hedenbergite (Hd), 71–73
Helium (He), 42, 126–127
Helix, 97
  DNA, 98f

Hematite (Fe_2O_3), 160
  oxygen in hematite on Moon, 160–161
Herschel Space Observatory, 295
Hertzsprung–Russell diagram (H–R
  diagram), 18, 19f
Heteroatom, 318–319
Heterogeneous accretion, 219
Hexahedrite, 83
Hexahistidine-tag (6xHis-tag), 102
  synthesis of proteins with, 102
Hexane extracts, 315, 315f
HF. *See* Hydrofluoric acid (HF)
$^{182}Hf–^{182}W$ isotope system, 61–62
  fossil isochron plot for $^{182}Hf–^{182}W$
    system, 62f
HFSE. *See* High field strength elements
  (HFSE)
Hf–W isotopic systems, 225
Hf–W systematic, core formation age from,
  225–226
Higgs boson, 6–7
High Accuracy Radial Velocity Planet
  Searcher (HARPS), 362
High field strength elements (HFSE), 159
  constraints from, 159
High pressure–temperature experiments, 207
High resolution (HR), 51
High-performance liquid chromatography
  (HPLC), 42, 314–315
High-performance liquid chromatography
  coupled with ESI high-resolution mass
  spectrometry (HPLC/ESI-HRMS), 317
High-pressure experiments, 8–12
  diamond anvil cell, 11–12
  Kawai-type high-pressure apparatus,
    8–11
  piston cylinder apparatus, 8
High-resolution mass spectrometry
  (HRMS), 313–314
High-resolution secondary ion mass
  spectrometry (HR-SIMS), 51, 231, 233,
  236–237, 306
  first water on early Earth from oxygen
    isotopic data of zircon by, 232–233
  schematic diagram of, 52f
High-temperature experiments, 8–12
  diamond anvil cell, 11–12
  Kawai-type high-pressure apparatus,
    8–11
  piston cylinder apparatus, 8
Highly siderophile element (HSE), 154, 215,
  219–220, 222–223, 225–226

Homochirality, 327
Hot giant exoplanets, aerosol composition of, 369
Hot Jupiters, 362
  methane, ammonia, and acetylene found on, 370
Hot water extract, 316
HPLC. *See* High-performance liquid chromatography (HPLC)
HPLC/ESI-HRMS. *See* High-performance liquid chromatography coupled with ESI high-resolution mass spectrometry (HPLC/ESI-HRMS)
HR. *See* High resolution (HR)
H−R diagram. *See* Hertzsprung−Russell diagram (H−R diagram)
HR-SIMS. *See* High-resolution secondary ion mass spectrometry (HR-SIMS)
HRMS. *See* High-resolution mass spectrometry (HRMS)
HSE. *See* Highly siderophile element (HSE)
HST. *See* Hubble space telescope (HST)
Hubble constant, 121
Hubble space telescope (HST), 17, 126
Hydrated materials, 233
Hydrated Na carbonates, 296
Hydration, 159−160
Hydrocarbons, 369
Hydrodynamic simulations, 146
Hydrofluoric acid (HF), 159
Hydrogen (H), 126−127, 205−207, 251, 281
  bond, 96−97, 97f
Hydrogen chloride in Mars's atmosphere, 268−269
Hydrogen cyanide (HCN), 370
Hydrothermal systems, 235, 242
Hydroxyl alcohols, 106−109
Hydroxyl aldehydes, 106−109
Hydroxyl ketones, 106−109
Hydroxyl-bearing minerals (OH-bearing minerals), 303

I

IACT. *See* Imaging Atmospheric Cherenkov Telescopes (IACT)
IAU. *See* International Astronomical Union (IAU)
ICP-MS. *See* Inductively coupled plasma mass spectrometer (ICP-MS)
ICP-QMS, 45
Icy comet hits icy comet in space, synthesis of life-related materials, 321−323
Icy planet
  fell on Earth's ocean, 253−259
    implication for life-related molecules in early Earth and origin of life, 259
    implication for prebiotic Earth, 256−259
    run products of impact experiments, 255−256, 257t
    shock-recovery experiments, 255
  and icy/silicate planet collisions, 323−327
    from birth of sun to LHB, and LRM, 323−324
    experiments of laser-induced dielectric breakdown, 324−325
    four canonical nucleobase syntheses from DAMN by LIDB, 325
    LIDB experiments, 325−327
ID method. *See* Isotope dilution method (ID method)
ID-MC-ICP-MS. *See* Isotope dilution-multicollector ICP mass spectrometry (ID-MC-ICP-MS)
IDPs. *See* Interplanetary dust particles (IDPs)
Igneous rocks, 77
Ilmenite (FeTiO$_3$), 75
IMAC. *See* Immobilized metal affinity chromatography (IMAC)
Imaging Atmospheric Cherenkov Telescopes (IACT), 13−15
Immobilized metal affinity chromatography (IMAC), 102
  using Ni-NTA resin, purification of protein by, 102
Impact−shock experiments for synthesis of life-related materials, 318−319
In situ gas-phase PAH formation pathways, 357
Incompatible elements, 164
Inductively coupled plasma mass spectrometer (ICP-MS), 41
Inflationary epoch, 123
Infrared (IR), 17
  astronomy, 17
    telescopes on Earth, 16f
  lights, 12
  telescopes, 141
Inorganic analysis, 307
  details of mineralogy and, 307−309
  of Ryugu samples, 307−309

Inorganic astrochemistry, 29–68
^{182}Hf–^{182}W isotope system, 61–62
   absorption and emission of lights, 39
      absorption and emission of lights and
         energy spectra, 40f
   age dating by radioactive isotopes, 56–57
      model age, 58–59
   atomic number, element symbol, and
      element, 31–34, 32t
      Madelung rule or Aufbau principle,
         34f
      periodic table of elements, 33f
      stable isotopic abundances of each
         element, 37f
   classic classification of elements, 53
   decays of radioactive elements, 34–39
      change of neutron and atomic
         numbers by radioactive decay, 38f
      radioactive isotopes and half-lives
         used in astrochemistry, 38t
   elemental analytical methods, 43–52
   extinct nuclides and fossil isochron,
      59–61
      fossil isochron plot for ^{26}Al–^{26}Mg
         system, 60f
   isotope dilution method, 66–68
   mass discrimination correction laws, 64–66
   mass spectrometry, 40–42
   rare Earth element pattern, 53–55
   single zircon dating by Pb–Pb method,
      63–64
   trace element geochemistry, 55–56
Inorganic carbon, 236–237
Inorganic chemistry of samples recovered
   from asteroid Ryugu, 306–309
Inorganic compounds, 253
Inorganic ions, 279
Inorganic mass spectrometry, 41–42
Inorganic minerals, 280
Institute for Planetary Materials (IPM), 8
Institute for Study of Earth's Interior (ISEI), 8
Interior seas, 338–339
Internal standardization technique,
   171–174
International Astronomical Union (IAU),
   23–26
International system of units (SI units), 1
Interplanetary dust particles (IDPs),
   251–252
Interstellar ice, syntheses of life-related
   materials by ultra-violet irradiation on,
   319–321

Interstellar medium (ISM), 313, 357
   cyanonaphthalenes were found in, 357
   life-related molecules observed in,
      357–361
Interstellar phosphine (PH$_3$), 360–361
Interstellar space, 359–360
Interstellar synthesis of phosphorus
   oxoacids, 360–361
Io moon, 341–342
   active volcanic plume of, 342f
   rock, 339
Ion
   beam, 43–44
   gradients, 112
   source, 40–41
Ionized hydrogen, 140
IPM. See Institute for Planetary Materials
   (IPM)
IR. See Infrared (IR)
Iron, 196, 369
   cores, 169–170
   meteorite material, 221
   meteorites, 75, 85
      constraints from, 85
Iron (Fe)
   elemental synthesis of heavy elements up
      to, 130–132
   isotopic composition of lunar rocks,
      196–198, 197f
Iron sulfide catalysts, 258
ISB. See Isua Supracrustal Belt (ISB)
ISEI. See Institute for Study of Earth's
   Interior (ISEI)
ISM. See Interstellar medium (ISM)
Isochron, 57, 231
   measurements, 163
Isoleucine (Ile), 255–256
Isopropylamine [(CH$_3$)$_2$CHNH$_2$], 316
Isotope, 34
   fractionation, 171
Isotope dilution method (ID method),
   66–68, 67f
   basics of, 66–68
   error magnification, 69f
   error propagation in, 68
Isotope dilution-multicollector
   ICP mass spectrometry
   (ID-MC-ICP-MS), 159
Isotopic compositions from volatile to
   refractory elements, 203–204
Isotopic system, 163, 219
Isua Supracrustal Belt (ISB), 230, 234, 237

counterarguments to ISB structures as
stromatolites, 243–244
evidences of Life older than 3800 Ma at
ISB and Akilia island, 237–238
stromatolites in ISB oldest Life in 3700
Ma, 243
Western Greenland, back to, 241–242
Italian Space Agency (ASI), 296–297

J
James Webb Space Telescope (JWST), 17,
126, 295, 367
JAXA, 305
Jovian atmosphere, 338
Jovian satellites, 338–339
Juice, 335, 336t
Juno, 335, 336t, 337f, 338
microwave radiometer, 338
JunoCam, 338
Jupiter
composition, 338
contrasting moons of, 339–347
exploration of, 335–339
explorers of, 337f
Galilean Satellites of, 340f
gravity, 338
icy moon Europa, 344f
icy moons explorer, 337f, 338–339
JWST. *See* James Webb Space Telescope
(JWST)

K
K2–72, 367
Kawai-type apparatus, 157–158
Kawai-type high-pressure apparatus, 8–11
conceptual image of guide-block, 10f
in IPM, 9f
sample assembly, 10f
KBOs. *See* Kuiper belt objects (KBOs)
Kepler program, 147
inconsistency between our solar
system and exosolar systems by,
147–149
Kepler space observatory, 28
illustration of, 28f
Kepler-1649, 367
Kepler-296, 367
Ketoses, 106–107, 251
Kinetic energy, 3–4, 141
KREEP rocks, age of, 164
Kuiper belt objects (KBOs), 29

L
LA. *See* Laser ablation (LA)
LA-ICP-MS, 52, 53f
Laboratory experiments, origin of life-
related materials in asteroids and comets
by, 318–327
Lagrange points, 4–5, 4f
Lambda-Cold Dark Matter model, 121
Lanthanum chromite (LaCrO$_3$), 8–11
Large energy, 251
Largest spacecraft NASA made for
planetary mission, 343
Las Campagnas Observatory (LCO), 26
Laser ablation (LA), 52
Laser beam, 43–44
Laser-Induced Dielectric Breakdown
(LIDB), 324
canonical nucleobase syntheses from
DAMN by, 325
experiments, 324–327
plasma experiments, 325
Last universal common ancestor (LUCA),
112–113
Late heavy bombardment (LHB), 137,
143–144, 228, 252, 323–324
in Nice model, 143–144
transportation of materials on Archaean
Earth by, 228
Late veneer, 166, 215–224
constraints
on late veneer from oxygen isotope
ratios, 223–224
for late veneer from W isotopes,
219–223
from platinum group elements,
215–219
ejecta of giant impact, 224
impactor, 224
planetesimals, 223
Late-accreted materials, 222
LC-FD/HRMS. *See* Liquid chromatography
with fluorescence and HRMS (LC-FD/
HRMS)
LC-FD/TOFMS. *See* Liquid
chromatography with fluorescence
detection and TOFMS (LC-FD/TOFMS)
LCO. *See* Las Campagnas Observatory
(LCO)
Lepton epoch, 125
Leucine (Leu), 255–256
LHB. *See* Late heavy bombardment
(LHB)

LICIACube. *See* Light Italian CubeSat for Imaging of Asteroid (LICIACube)

LIDB. *See* Laser-Induced Dielectric Breakdown (LIDB)

Life, magnetic field of, 368–369

Life-related materials (LRMs), 265, 270, 305, 323–324
  in asteroids and comets, 305–306
  in atmosphere on Mars, 267–269
    high concentrations of O_2 in Martian brines, 269
    hydrogen chloride found in Mars's atmosphere, 268–269
    liquid water beneath Mars ice cap, 268
    methane on Mars, 268
  organic materials and LRM in recovered samples from asteroid Ryugu, 309–318
  origin of LRMs in asteroids and comets by laboratory experiments, 318–327
    impact-shock experiments for synthesis of LRMs, 318–319
    syntheses of LRMs by ultra-violet irradiation on interstellar ice, 319–321
    synthesis of LRMs icy comet hits another icy comet in space, 321–323
  searching for life-related materials and Life on Mars, 265–267
  selection of enantiomers of, 327

Life–related molecules (LRMs), 251
  implication for life–related molecules in early Earth and origin of life, 259
  observed in space and interstellar medium, 357–361
    cell membrane components found in space, 359–360
    cyanonaphthalenes found in interstellar medium, 357
    discovery of chiral molecule in space, 361
    interstellar synthesis of phosphorus oxoacids, 360–361
    methanol signatures found in cold dark molecular disk and star-forming regions, 358
  origin of life-related molecules by computer simulations, 259–260
  synthetic experiments of life-related molecules on Earth, 253–260
  transportation of life-related molecules of extraterrestrial origin on Earth, 251–253
    transportation of life-related molecules of space origin onto Earth, 251–252
  on Venus atmosphere, 263–264
    phosphine on Venus atmosphere, 263

Light carbon, 237–238

Light elements up to carbon, elemental synthesis of, 128–130

Light Italian CubeSat for Imaging of Asteroid (LICIACube), 296–297

Limestone, 350–351

Linear law, 65

Linear α-amino acids, 322

Liquid chromatography, 42

Liquid chromatography with fluorescence and HRMS (LC-FD/HRMS), 314–315

Liquid chromatography with fluorescence detection and TOFMS (LC-FD/TOFMS), 314–315

Liquid metal, 156

Liquid water, 241
  beneath icy crust, 342–343
  beneath Mars ice cap, 268
  exploration of Jupiter and Saturn, 335–339
  four contrasting moons of Jupiter, 339–347
  moons of Saturn, 347–354
  ocean, 339–340

Lithium, 192
  isotopic composition of lunar rocks, 192–193
  synthesis of, 134

Lithophile trace elements, 55

LMO. *See* Lunar magma ocean (LMO)

Local Volume Mapper, the, 27

Low-mass stars, 369

LP890–9, 367

LRMs. *See* Life-related materials (LRMs); Life–related molecules (LRMs)

^{176}Lu–^{176}Hf systematics to West Greenland samples, application of, 230

LUCA. *See* Last universal common ancestor (LUCA)

Lucy spacecraft, 294
  illustrations of, 294*f*

Lu–Hf isotope systems, 230

Lunar
  disk models, 204
  materials, 159
  oxygen isotopic ratios, 177–178
  rocks, 183
    Cr isotopic composition of, 193–194

Cu isotopic composition of, 189–191
Fe isotopic composition of, 196–198
Ga isotopic composition of, 185–186
K isotopic composition of, 186–189, 189f
Li isotopic composition of, 192–193
Mg isotopic composition of, 198–200
Rb isotopic composition of, 184–185
Si isotopic composition of, 194–196
Ti isotopic composition of, 200–202
V isotopic composition of, 200
Zn isotopic composition of, 183–184
volcanic rocks, 217
W isotopic ratios and percentage of late veneer, 219–221
Lunar magma ocean (LMO), 199–200, 202, 204
Luyten's star, 367
Lyman-α lines, 29

M
Macromolecular carbon (MMC), 284
Macromolecular organic matter in Ryugu samples, 317–318
Madelung rule, 33–34, 34f
Magma ocean
  crystallization, 165
  model, 161
Magmatic oxygen isotopic ratios, 233
Magmatic process, 240–241
Magnesium (Mg), 198, 279
  age of Mg-suite rocks, 165
  isotopic composition of lunar rocks, 198–200, 199f
  Mg-containing organic compounds, 316
  Mg-suite cumulate rocks, 162
  Mg–Ca carbonates, 296
Magnesium carbonates (MgCO$_3$), 75
Magnesium perovskite, 230
Magnesium silicate ((Mg, Fe)SiO$_3$), 75
Magnetic field, 338–339
  of planet and life, 368–369
Magnetic resonance imaging (MRI), 113–114
Magnetite (Fe$_3$O$_4$), 255, 277–278
  crystals, 274
  not biomarkers, 277–278
  particles, 278
Magnetofossils, 274
Main-belt comet, identification of water emission from, 295

MaNGA. *See* Mapping Nearby Galaxies at APO (MaNGA)
Mapping Nearby Galaxies at APO (MaNGA), 27
Marine carbonate production, 243
Mars, 42
  diversity of organic geochemistry on mars glimpsed from Tissint meteorite, 282–283
  electrochemistry of brines and metal-rich minerals on mars reduced CO$_2$ to alcohols, 283–284
  exploration of mars, 265–267
  express, 267
  fossils of microorganisms found in Martian meteorite, ALH 84001, 270–282
  global surveyor, 267
  hydrogen chloride found in Mars's atmosphere, 268–269
  LRMs in atmosphere on, 267–269
  methane on, 268
  organic compounds in Martian meteorites, 269–270
  sample return mission, 267
  searching for life-related materials and Life on, 265–267
  Tissint fell, 282–284
Mars Advanced Radar for Subsurface and Ionosphere Sounding (MARSIS), 268
MARSIS. *See* Mars Advanced Radar for Subsurface and Ionosphere Sounding (MARSIS)
Martian brines, high concentrations of O$_2$ in, 269
Martian canals, 265
Martian meteorites, organic compounds in, 269–270
Martian moons, 298f
Martian Moons eXploration (MMX), 297–298
  mission, 297–298, 298f
MARVELS. *See* Multi-object APO Radial Velocity Exoplanet Largearea Survey (MARVELS)
Maskelynite (NaAlSi$_3$O$_8$), 272
Mass discrimination correction laws, 64–66
Mass fraction laws, 64–65
Mass fractionation process, 181
Mass number, 34
Mass spectrometric evolved gas analysis (MSEGA), 282

Mass spectrometry (MS), 40−42, 255
  inorganic, 41−42
  organic, 42
Mass-dependent fractionation, 177
MAVEN, 267
MC-ICP-MS. *See* Multicollector Inductively
  Coupled Plasma Mass Spectrometry
  (MC-ICP-MS)
Mean motion resonance (MMR), 22, 144
Metal oxides, 369
Metal reequilibration, 229
Metal-rich atmosphere for GJ 1214b, 371
Metal-rich minerals on Mars reduced CO_2
  to alcohols, electrochemistry of, 283−284
Metallic cores, 197−198
Metallic iron, 156, 255−256
  alloy, 171
  asteroids, 22−23
Metallic melts, 217−219
Metallicity of stars, 21
Metalliferous hydrothermal vein, 235
Metalloprotein, 93
Metal−silicate fractionation, 197−198
Metamorphism, basic geological
  knowledge on, 78−79
Meteorite(s), 69, 80−83
  basic knowledge on, 84−88
    calcium−aluminum rich inclusions,
      84−85
    chondrules globular shape, 85
    CI chondrites and trace elemental
      abundance of Earth, 87
    evidence of early formation and
      differentiation of protoplanets, 85
    Palermo scale and Torino scale, 87−88
    presolar grains, 85−86
    Yarkovsky effect, 86
    Yarkovsky−O'Keefe−Radzievskii−
      Paddack effect, 86−87
  classification of, 80−84, 81*f*
    photographs of sliced meteorites, 82*f*
Meteoritic materials, 253−254
Methane (CH_4), 29, 370
  on hot Jupiter, 370
  on Mars, 268
Methanol (CH_3OH), 254, 305, 313−316, 319,
  321, 358
  extracts, 316
  formaldehyde formic acid, 322
  signatures in cold dark molecular disk
    and star-forming regions, 358
Methyl formate ($HCOOCH_3$), 315−316

Methyl α-amino acids, 322
Methylamine (CH_3NH_2), 316
Methylammonium lead triiodide
  ($CH_3NH_3PbI_3$), 75
Micro particles, 217−219
Micro-FTIR spectroscopy, 317−318
Micro-Raman spectroscopy, 235
Micro-X-ray fluorescence, 243−244
Microfossils, 234, 236−237
Microprobe two-step laser mass
  spectrometer (mL^2Ms), 273
Mid-oceanic ridge basalts (MORBs), 78, 178
Milky Way, 21, 128
Milky Way Mapper, the, 27
Miller experiment, 252−253, 254*f*
Miller−Urey experiment, 253
Mimas, 347−348, 353−354
Mineralogy, 68−88
  basic geological knowledge on
    sedimentary rocks, metamorphism,
    and deep sea volcanism, 78−79
  basic knowledge on meteorites and
    asteroids, 84−88
  carbonates, 75
  characterization of volcanic rocks and
    ultramafic rocks, 77−78
  classification of meteorite, 80−84, 81*f*
  details of mineralogy and inorganic
    analytical results, 307−309
  determination of major elements, 79
    X-ray fluorescence spectrometry, 79,
      80*f*
  and inorganic chemistry of samples
    recovered from asteroid Ryugu,
    306−309
  nickel−iron alloy, 75
  oxide minerals, 74−75
  partition coefficients, 79−80
  phase diagram of water, 75−77, 76*f*
  rock-forming minerals, 70−74
  solid solution and phase diagram, 70
    phase diagram of not perfect solid
      solution made of materials, 71*f*
  thin section, 69−70
Minerals, 70−71, 170−171
Minor Planet Center (MPC), 23−26
Mitochondria, 110
MMC. *See* Macromolecular carbon (MMC)
MMR. *See* Mean motion resonance (MMR)
MMX. *See* Martian Moons eXploration
  (MMX)
Model age, 58−59

Moderately volatile elements (MVEs), 204
Molecular oxygen, 28
Molecular water on moon, 159–160
Molten iron, 157
Monazites, 230–231
Monochromatic light, 105
Monte-Carlo method, 304
  simulations, 207
Moon, 194–196, 217
  age of moon, 161–165
    age estimates for early lunar
      differentiation events, 162t
    age of FANs, 163
    age of KREEP rocks, 164
    age of Mg-suite rocks, 165
    age of moon mantle differentiation by
      ^{146}Sm-^{142}Nd method, 165
    zircons ages, 163–164
  characteristics of materials of moon,
    151–161
    constraints from high field strength
      element, 159
    rocks collected from Moon, 151–154,
      153f
    sample-return programs of moon
      rocks, 151
    similarity between chemical
      compositions of Moon and Earth's
      mantle, 154–158
  comparison of primitive upper mantle
    with, 217
  constraints from stable isotope
    astrochemistry for giant impact,
    170–202
  constraints from three oxygen isotopes
    for origin of, 174–183
  formation age of, 221–223
  four contrasting moons of Jupiter,
    339–347, 340f
    Callisto, 347
    Europa, 342–343
    Ganymede, 343–346
    Io, 341–342
  giant impact model for formation of
    Earth-Moon system, 166–170
  peculiarities of, 166
  recent findings on Moon's surface,
    159–161
    molecular water on moon, 159–160
    oxygen in hematite on Moon could
      come from Earth, 160–161

of Saturn, 347–354
  Enceladus, 348–350
  Mimas, 353–354
  possible organics on Titan's equatorial
    dunes, 353f
  synthesis of PHAs on Titan, 354f
  titan and atmospheres with organic
    molecules, 350–353
MORBs. *See* Mid-oceanic ridge basalts
  (MORBs)
MPC. *See* Minor Planet Center (MPC)
MRI. *See* Magnetic resonance imaging
  (MRI)
MS. *See* Mass spectrometry (MS)
MSEGA. *See* Mass spectrometric evolved
  gas analysis (MSEGA)
Mudstone, 238
Multi-anvil apparatus. *See* Kawai-type
  high-pressure apparatus
Multi-object APO Radial Velocity
  Exoplanet Largearea Survey
  (MARVELS), 27
Multi-reservoir model, 181
Multicollector Inductively Coupled Plasma
  Mass Spectrometry (MC-ICP-MS), 42, 46,
  48f, 61
Murchison meteorite, 360
MVEs. *See* Moderately volatile elements
  (MVEs)

N

N,N-dimethylglycine, 312
n-propylamine ($C_3H_7NH_2$), 316
N-TIMS. *See* Negative thermal ionization
  mass spectrometry (N-TIMS)
Nakhla (Martian meteorites), 284
Nakhlites, 269
Nano-liquid chromatography/high-
  resolution mass spectrometry
  (nanoLC/HRMS), 313–314
Nanobacteria
  hypothesis, 279–280
  not Life, 278–279
nanoLC/HRMS. *See* Nano-liquid
  chromatography/high-resolution mass
  spectrometry (nanoLC/HRMS)
Naphthalene ($C_{10}H_8$), 352–353, 357
2-naphthyl radicals, 353
NASA, 265, 298–299, 305, 339, 343
  NASA-ESA Mars sample return, 267
  spacecraft, 289

Natrite (Na_2CO_3), 296
^{142}Nd isotope systematics to oldest crust on Earth, application of, 229−231
  application of ^{146,147}Sm−^{142,143}Nd and ^{176}Lu−^{176}Hf systematics to West Greenland samples, 230
  application of ^{146,147}Sm−^{142,143}Nd systematics to Ujaraaluk unit in NGB, Canada, 230−231
  application of ^{146,147}Sm−^{142,143}Nd systematics to West Greenland samples, 229
Near-Earth asteroids (NEAs), 22
Near-Earth objects (NEOs), 87, 296−297
NEAR-Shoemaker spacecraft, 291−292
  photographs of asteroids and core of comet, 290f
NEAs. See Near-Earth asteroids (NEAs)
Nebula, 141
Negative thermal ionization mass spectrometry (N-TIMS), 219−220
Neon (Ne), 42
NEOs. See Near-Earth objects (NEOs)
Neutral element, 31
Neutrino decoupling, 125
Neutron
  cosmic rays, 163
  cross-section, 12
  neutron-rich isotopes, 132−133
  star merger, 135−137, 137f
New Horizons, 336t, 337f
Newton's second law, 3
NGB. See Nuvvuagittuq Greenstone Belt (NGB)
Ni-NTA. See Nitrilotriacetic acid (Ni-NTA)
Nice model, 137, 142−143, 228
  late heavy bombardment in Nice model, 143−144
Nickel, 256
  nickel−iron alloy, 75
Nitrile-groups—functionalized PAHs, 357
Nitrilotriacetic acid (Ni-NTA), 102
  purification of protein by immobilized metal affinity chromatography using, 102
Nitrogen (N), 205−207
  oxides, 258
193nm ArF excimer laser, 52
Noble gases, 42
Nonproteinogenic amino acids of β-alanine ($C_3H_7NO_2$), 316
Norites, 154

Nuclear magnetic resonance, 113−114
  proton nuclear magnetic resonance, 114
Nuclear reaction, 129
Nuclei, 224
Nucleic acids, 99, 318−319
  found in Ryugu samples, A0106 and C0107, 317
Nucleobases, 253−254, 258
  synthesis, 325−327
Nucleotides, 97
Nuvvuagittuq Greenstone Belt (NGB), 230
  Canada, application of ^{146,147}Sm−^{142,143}Nd systematics to Ujaraaluk unit in, 230−231
  isochrons for Ujaraaluk samples, 232f
NWA 1950 (Martian meteorites), 284

O

Observational astronomy, 13
Ocean beneath icy crust, 343
Octahedrite, 83
Oddo−Harkins rule, 87
Olivine [(Mg, Fe)SiO_4], 71, 184, 272
  grains, 299
  group, 71
  olivine-orthopyroxene-clinopyroxene, 78
Oparin-Haldane hypothesis, 253
Optical astronomy, 17
Optical imaging technique, 105
Optical microscope, 83, 235
Optical spectrum, 128
Orbital classification, 22
  asteroid belts, 24t
  distribution of asteroid, 23f
Organic astrochemistry, 29−31, 88−114
  amino acids, 88−91, 90f
  characterization of present life, 106
  confocal Raman spectroscopy, 105
  CRISPR-Cas9 gene editing, 103, 104f
  Fischer projection, 103−104, 104f
  Fischer−Tropsch reaction and carbon isotopic fractionation, 105
  helix and double helix, 97
  hydrogen bond, 96−97, 97f
  hydroxyl alcohols, hydroxyl aldehydes, and hydroxyl ketones, 106−109
  nuclear magnetic resonance, 113−114
  protein data bank, 93
  proteins, 91−92
  purification of protein by immobilized metal affinity chromatography using Ni-NTA resin, 102

purines, pyrimidines, DNA, and RNA, 93–96
nucleotides, 96f
schematic structure of eukaryote cell, 109–110, 109f
short story for origin of life, 110–113
peptide + nucleic acid world, 111
peptide world, 110
protocell, 112
RNA world, 111–112
road from protocell to archaea, bacteria, and eukaryotes, 112–113
stromatolites, 106, 107f
synthesis
from RNA to protein, 101
of proteins from DNA, 98–99, 99f
of proteins with hexahistidine-tag, 102
tholin, 104–105
transcription, 99–100
Organic carbon, 227, 236–237, 270
Organic compounds, 104–105
in Martian meteorites, 269–270
Organic detritus, 240
Organic geochemistry on Mars glimpsed from Tissint meteorite, diversity of, 282–283
Organic mass spectrometry, 42
Organic materials, 42, 228
and life-related materials in asteroids and comets, 305–306
organic materials and life-related materials in carbonaceous chondrites, 305–306
and life-related materials in carbonaceous chondrites, 305–306
and life-related materials in recovered samples from asteroid Ryugu, 309–318
comparison between amino acids in A- and C-series samples, and implications, 312–313
hexane and dichloromethane extracts, 315
hot water extract, 316
icy planet and icy/silicate planet collisions, 323–327
macromolecular organic matter in Ryugu samples, 317–318
methanol extracts, 316
nucleic acid, uracil found in Ryugu samples, A0106 and C0107, 317
organic material study using DESI-OT-MS, 309–310
organic material study using UHPLC-OT-MS, 310–312
overview, 313–315
selection of enantiomers of life-related materials, 327
systematic analyses of organic materials for asteroid Ryugu, 309–313
systematic study of organic materials, 313–317
study using UHPLC-OT-MS, 310–312
organic molecules found in water-soluble fraction of Ryugu sample, 311f
Organic matter, 238–239, 252
Organic molecules, Titan and atmospheres with, 350–353
Organic pyrolysis, 252
Organic synthesis associated with serpentinization and carbonation on early Mars, 281–282
Opx serpentine magnetite, 281
Serpentine talc magnesite, 281–282
Organics, 352
Origin of moon
C, N, and S delivered giant impact, 205–207
constraints from three oxygen isotopes for, 174–183
Cr isotopic composition of lunar rocks, 193–194, 193f
Cu isotopic composition of lunar rocks, 189–191
$\Delta^{17}O$—new oxygen isotope ratio presentation, 177
development of accuracy of oxygen isotopic ratios, 178–180
Fe isotopic composition of lunar rocks, 196–198, 197f
Ga isotopic composition of lunar rocks, 185–186, 187f
isotopic compositions from volatile to refractory elements, 203–204
K isotopic composition of lunar rocks, 186–189, 189f
Li isotopic composition of lunar rocks, 192–193
lunar oxygen isotopic ratios, 177–178
Mg isotopic composition of lunar rocks, 198–200, 199f
oxygen isotopic evidences, 178–180

Origin of moon (*Continued*)
  Rb isotopic composition of lunar rocks, 184–185, 185*f*
  recent oxygen isotopic studies, 181–183
  Si isotopic composition of lunar rocks, 194–196
  synestia model, 204–205
  three oxygen isotopes support giant impact, 174–177
  Ti isotopic composition of lunar rocks, 200–202, 202*f*
  V isotopic composition of lunar rocks, 200, 201*f*
  Zn isotopic composition of lunar rocks, 183–184, 184*f*
Orthogneisses, 229
Orthopyroxene [(Mg, Fe)SiO$_3$], 184, 272, 281
  serpentine magnetite, 281
Orthopyroxenite, 269
OSIRIS–REx sample-return mission, 303–305
Oxidative reactions, 263
Oxide minerals, 74–75
  perovskites, 75
  silica minerals, 74
  spinels, 74
  titanium-related oxides, 75
Oxidizing processes, 160
Oxygen (O), 174, 251, 306
  fugacity, 158, 215
  in hematite on Moon could come from Earth, 160–161
  high concentrations of O$_2$ in Martian brines, 269
  isotope, 170
    constraints from three oxygen isotopes for origin of moon, 174–183
    constraints on late veneer from, 223–224
    measurements, 161
    ratios, 83, 174, 223
  isotopic data, 233
    first water on early Earth from oxygen isotopic data of zircon by HR-SIMS, 232–233
  isotopic evidences, 178–180
  isotopic ratios, 174
    development of accuracy of, 178–180
  isotopic studies, 181–183
  O-burning process, 131
  O$_2$-producing photoautotrophs, 235
Ozone-related reaction, 175

P

PAHs. *See* Polycyclic aromatic hydrocarbons (PAHs)
Palermo scale, 87–88
Palladium, 217
PDB. *See* Pee Dee Belemnite (PDB); Protein Data Bank (PDB)
PDF. *See* Probability density function (PDF)
Pee Dee Belemnite (PDB), 236
Pelagic sediments, 78–79
Pentoses, 107–109
Peptide + nucleic acid world, 111
Peptide Nucleic Acids (PNA), 111
Peptide sequence, 110
Peptide world, 110
Periodic table, 31
  of elements, 33*f*
Perovskites, 75
Petrology, 68–88
  basic geological knowledge on sedimentary rocks, metamorphism, and deep sea volcanism, 78–79
  basic knowledge on meteorites and asteroids, 84–88
  carbonates, 75
  characterization of volcanic rocks and ultramafic rocks, 77–78
  classification of meteorite, 80–84, 81*f*
  determination of major elements, 79
    X-ray fluorescence spectrometry, 79, 80*f*
  nickel–iron alloy, 75
  oxide minerals, 74–75
  partition coefficients, 79–80
  phase diagram of water, 75–77, 76*f*
  rock-forming minerals, 70–74
  solid solution and phase diagram, 70, 71*f*
  thin section, 69–70
PGE. *See* Platinum group elements (PGE)
Phase diagram, 70
Phenanthrene (C$_{14}$H$_{10}$), 352–353
Phenylacetylene, 352–353
Philae, 292
Phobos, 298
Phosphate minerals, 360
Phosphine (PH$_3$), 263–264
  on Venus atmosphere, 263
Phospholipids, 279, 360
Phosphonic acid (H$_3$PO$_3$), 360–361
Phosphorous (P), 350
Phosphorus (V) oxidation state, 360
Phosphorus oxoacids, 360–361
  interstellar synthesis of, 360–361

Photo-spectrometry, 39
Photo-spectroscopy, 39
Photodisintegration process, 133–134
Photon epoch, 125
Photosynthesis, 227, 235, 263
Pigeonite (Pig), 71–73
Pioneer 10, 336*t*, 337*f*, 338
Pioneer 11, 336*t*, 337*f*, 338–339
Piston cylinder apparatus, 8, 9*f*
PIXL. *See* Planetary instrument for X-ray
    lithochemistry (PIXL)
Plagioclase, 153–154, 161
Planck constant, 113
Planck epoch, 123
Planck spacecraft, 121
Planet
    characters of host stars of planets in
        habitable zones, 368
    magnetic field of, 368–369
    planet–impactor, 170
Planetary accretion model, 181, 227
Planetary instrument for X-ray
    lithochemistry (PIXL), 243–244
Planetary isotope, 197–198
Planetary migration, 145
Planetary system, 29, 30*t*, 141, 170
Planetesimals, 141, 170
Planktonic organisms, 240
Platinum group elements (PGE), 171, 215
    comparison of primitive upper mantle
        with moon, 217
    constraints from, 215–219
    platinum group element depletion of
        primitive upper mantle, 217–219
    primitive upper mantle, 216–217
Plume activity, 343
Plutonium-238 (^{238}Pu), 335
PNA. *See* Peptide Nucleic Acids (PNA)
Polycyclic aromatic hydrocarbons (PAHs),
    270–274, 313–314, 350, 357
    chemical structures of, 275*f*
    synthesis on Titan, 354*f*
Polyhistidine tag, 102
Polypeptide
    chains, 101
    polymer molecule, 92
Polyubiquitin chain, 101
Possibility of life, 340, 345
Potassium, 161–162, 186
    isotopic composition of lunar rocks,
        186–189, 189*f*
    isotopic fractionation, 186–187

Potentially habitable exoplanets, list of,
    365–368, 366*t*
Power law, 65
p–p chain reaction. *See* Proton–proton
    chain reaction (p–p chain reaction)
Pre-solar disk, 141
Pre-solar nebula to solar system,
    141–142
Prebiotic Earth
    implication for, 256–259
    non-proteinogenic amino acids, 258*f*
Prebiotic nucleobases, 259
Presolar grains, 85–86
Primary ion beam, 49–51
Primitive upper mantle (PUM), 216–217
    elemental ratios of platinum group
        elements, 217*t*
    HSE concentrations of, 216*t*
    platinum group element depletion of,
        217–219
Primordial soup theory, 253
Probability density function (PDF),
    116–117
Procellarum, 154
Progressive metamorphism, 79
Proline (Pro), 255–256
Prometheus, 341–342
Promethium (Pm), 31
Propylene oxide (CH_3CHCH_2O), 361
Protein Data Bank (PDB), 93, 94*f*
Protein(s), 91–92, 279, 318–319, 358
    combination of three amino acids by
        peptide bonds, 92*f*
    folding process, 101
    purification by immobilized metal
        affinity chromatography using Ni-
        NTA resin, 102
    reagents for protein recombination, 103*f*
    synthesis
        from DNA, 98–99, 99*f*
        with hexahistidine-tag, 102
        from RNA to, 101
Proto-asteroidal materials, 303
Protocell, 112
    bacteria, and eukaryotes, road from
        protocell to archaea, 112–113
    membranes, 112
Proto–Earth
    formation of proto-Earth in solar nebular,
        137–149
    heavier, 200
    system spinning, 189

Protolunar
  disk, 169
  isotopic system, 169
Proton
  nuclear magnetic resonance, 114
  proton-detected NMR, 283
  proton-rich-isotope synthesis process,
    132
Proton—proton chain reaction (p—p chain
  reaction), 128
Protoplanets, evidence of early formation
  and differentiation of, 85
Protosolar nebular (PSN), 313
Proxima Centauri, 368
PSN. *See* Protosolar nebular (PSN)
Psyche spacecraft, 293—294
  illustrations of, 294*f*
PUM. *See* Primitive upper mantle (PUM)
Purines, 93—96
  and pyrimidines, 95*f*
Pyrimidines, 93—96, 259
  purines and, 95*f*
Pyrite (FeS$_2$), 242—243, 272
Pyrolysis, abundances and speciation
  analyses of C and S by, 307
Pyrophosphate (P$_2$O$_7$$^{4-}$), 360
Pyrophosphoric acid (H$_4$P$_2$O$_7$), 360—361
Pyroxene, 182—183, 193—194
  group, 71—73

Q
Q-pole type ICP-MS, schematic diagram of,
  46*f*
Quantum mechanics, 31
Quantum theory of light, 5—6
Quark epoch, 125
Quartz-pyroxene-magnetite, 238

R
Radio astronomy, 18
Radio waves, 18
Radioactive decay, 57
Radioactive elements, decays of, 34—39
Radioactive isotopes, 34—35
  age dating by, 56—57
  astrochemistry, 31
  and half-lives used in astrochemistry, 38*t*
  isochron plot for ^{147}Sm—143Nd system, 58*f*
  systems, 225—226
Radioisotope Thermoelectric Generators
  (RTGs), 335

Radioisotopes (RI), 34
Raman scattering, 105
Raman spectroscopy, 241—242, 312
Rare earth element (REE), 53—55, 161
  pattern, 53—55
  of meteorites, 54*f*
Rayleigh distillation, 189
Rayleigh fractionation law, 66
Recombinant DNA, 102
Red dwarf, 368
Redshift, 18—21, 20*f*
Reductive reactions, 263
REE. *See* Rare earth element (REE)
Reference line (RL), 177
Reflectance spectra of Ryugu's surface, 303
Refractory elements, 171
  isotopic compositions from volatile to,
    203—204
Reionization, 125
Relative standard deviation, 115
Remote-sensing, 303
  data, 303
Repulsion, 127
Retrogressive metamorphism, 79
RI. *See* Radioisotopes (RI)
Ribonucleic acid (RNA), 93—96, 360
  synthesis from RNA to protein, 101
  world, 111—112
Ribosome, 101
Ring system, 338
RL. *See* Reference line (RL)
RNA. *See* Ribonucleic acid (RNA)
Robotic missions, 151
Roche limit, 12
Rock(s)
  collected from Moon, 151—154, 153*f*
    Apollo programs and collected sample
      amounts, 153*t*
    Apollo spacecrafts for sample-return
      mission of moon, 152*f*
  rock-forming minerals, 70—74
    eight oxygens, 73, 73*f*
    four oxygens, 71
    six oxygens, 71—73
    twelve oxygens, 74
Rosetta, 292, 294—295
RTGs. *See* Radioisotope Thermoelectric
  Generators (RTGs)
Rubidium (Rb), 184—185
  isotopic composition of lunar rocks,
    184—185, 185*f*
  Rb—Sr isotope system, 59

Run products of impact experiments, 255–256, 257t
Ryugu
  mineralogy and inorganic chemistry of samples recovered from asteroid, 306–309
  organic materials and life-related materials in recovered samples from asteroid, 309–318
  samples, 309
    abundances and speciation analyses of C and S by pyrolysis and combustion, 307
    details of mineralogy and inorganic analytical results, 307–309
    inorganic analysis of, 307–309
    macromolecular organic matter in, 317–318
    organic molecules found in water-soluble fraction of, 311f
    systematic analysis of, 306–307
    systematic analyses of organic materials for asteroid, 309–313

S

Sagittarius B2 North [SgrB₂(N)], 361
SAM. See Sample Analysis at Mars (SAM)
Sample Analysis at Mars (SAM), 282
Sample-return spacecraft, illustrations of, 302f
Sample–return missions
  from asteroids, 300–305
    Hayabusa spacecraft, 300
    Hayabusa2 spacecraft, 301–303
    OSIRIS-REx sample-return mission, 303–305
  from comets and asteroids, 298–305
    early sample-return missions from space, 298–300
    sample-return missions from asteroids, 300–305
San Carlos olivine (SC olivine), 180
Sandstone, 238
Sarcosine (Sar), 255–256
Saturn
  exploration of, 335–339
  explorers of, 337f
  missions, 339
  moons of, 347–354
    of moving ice, 348–350
SB1. See Spectroscopic binary (SB1)
SC olivine. See San Carlos olivine (SC olivine)

Scalar boson, 6–7
Scanning analysis, 47–48
Scanning electron microscope (SEM), 47, 69
  SEM-EDX, 70, 306
  SEM-Energy dispersive spectrometry, 47
  schematic diagram of, 49f
SDSS. See Sloan Digital Sky Survey (SDSS)
SE. See Standard error (SE)
Second-generation planets, 147
Secondary ion mass spectrometry (SIMS), 43–44, 49, 306
  schematic diagram of, 51f
Sedimentary rocks, 232–233
  basic geological knowledge on, 78–79
Sediments, 238
SEGUE. See Sloan Extension for Galactic Understanding and Exploration (SEGUE)
SEM. See Scanning electron microscope (SEM)
Separation method, 78
Serine (Ser), 255–256, 313
Serpentine talc magnesite, 281–282
Serpentinization
  organic synthesis associated with serpentinization and carbonation on early Mars, 281–282
  reaction, 105
SF. See Spontaneous fission (SF)
Shergottites, 269
Shock-recovery experiments, 255, 256t
Shorter wavelength laser, 52
SI units, 1–3
  base units, 2t
  defining constants, 2t
  prefixes, 2t
Siderophile elements, 156
Silica, 280
  minerals, 74
  silica-saturated rocks, 231
Silicate (S), 169–170
  delivered giant impact, 205–207
  gas disk, 167
  moon, 221–222
  planet collisions, 323–327
  powders, 79
  reequilibration, 229
  S-type asteroids, 22–23
  shell, 339
  silicate-metal-melt, 204
Silicon (Si), 194
  isotopic composition of lunar rocks, 194–196
  Si-burning process, 132

Silver (Ag), 70
SIMS. *See* Secondary ion mass spectrometry (SIMS)
Single zircon dating by Pb—Pb method, 63—64
  concordia diagram, 65*f*
  Zircon crystal, 63*f*
Sloan Digital Sky Survey (SDSS), 26—27
  map showing observable Universe in rainbow colors, 27*f*
  SDSS-III, 27
  SDSS-IV, 27
  SDSS-V, 27
Sloan Extension for Galactic Understanding and Exploration (SEGUE), 26—27
  SEGUE-2, 27
^{146}Sm-^{142}Nd method, 165
  age of moon mantle differentiation by, 165
^{146,147}Sm—^{142,143}Nd systematics to Ujaraaluk unit in NGB, Canada, application of, 230—231
^{146,147}Sm—^{142,143}Nd systematics to West Greenland samples, application of, 229
^{146,147}Sm—^{142,143}Nd systematics to West Greenland samples, application of, 230
Small Main-belt Asteroid Spectroscopic Survey classification (SMASS classification), 22—23, 26*t*
Small solar system bodies (SSSB), 22
SMASS classification. *See* Small Main-belt Asteroid Spectroscopic Survey classification (SMASS classification)
SMOW. *See* Standard mean ocean water (SMOW)
Soai autocatalytic reaction system, 327
Sodium phosphates, 350
SOFIA. *See* Stratospheric Observatory for Infrared Astronomy (SOFIA)
Solar nebular
  evolution of molecular clouds to solar nebula, 138—141, 139*f*
  formation of gas giant planets and asteroids, 142—143
  formation of inner planets in Grand Tack model, 145—146
  formation of proto-Earth in, 137—149
  grand tack model, 144—145
  inconsistency between our solar system and exosolar systems found by Kepler program, 147—149
  late heavy bombardment in Nice model, 143—144

from pre-solar nebula to solar system, 141—142
Solar panels, 338
Solar polar orbiter, 338
Solar system, 84—85, 130, 137, 165—166, 219, 233
  formation, 221—222
  pre-solar nebula to, 141—142
Solid solution, 70
Solidification, 228
Soluble organic materials (SOM), 306, 313
SOM. *See* Soluble organic materials (SOM)
Space
  cell membrane components found in, 359—360
  debris collides, 141
  early sample-return missions from, 298—300
  life-related molecules observed in, 357—361
  space-based telescopes, 227
  synthesis of life-related materials icy comet hits another icy comet in, 321—323
Spacecraft for astronomy and astrophysics, 14*f*
Special relativity theory, 5—6
Spectrometry method, 39
Spectroscopic binary (SB1), 128
Spectroscopic investigations, 227
Spectroscopic thermal phase curve, 371
Spinels, 74, 193—194
  triangular prism for, 74*f*
Spin—spin coupling, 114
Spitzer Space Telescope (SST), 17, 126
Spontaneous fission (SF), 34—35
Spot analysis methods, 43—44, 47—52, 306
  detailed classification of, 45*f*
  schematic diagram
    of EPMA, 50*f*
    of HR-SIMS, 52*f*
    of LA-ICP-MS, 53*f*
    of SIMS, 51*f*
    of sputtering phenomenon, 50*f*
Spot elemental analysis, 43—44
Sputtering, 49
  schematic diagram of sputtering phenomenon, 50*f*
SSB method. *See* Standard-sample bracketing method (SSB method)
SSRL. *See* Stanford Synchrotron Radiation Lightsource (SSRL)
SSSB. *See* Small solar system bodies (SSSB)

SST. *See* Spitzer Space Telescope (SST)
Stable isotope astrochemistry, 31
  constraints for giant impact, 170−202
  constraints from three oxygen isotopes
    for origin of moon, 174−183
  evaporation and condensation of
    elements after giant impact, 170−171
  oxygen isotopic ratios, 174
  popular, 171−174, 173*f*
Stable isotopic abundances of each element,
  37*f*
Standard error (SE), 117
Standard mean ocean water (SMOW), 180
Standard model, 169−170
Standard-sample bracketing method (SSB
  method), 171−174
Stanford Synchrotron Radiation
  Lightsource (SSRL), 242
Star(s), 125
  metallicity of, 21
  star-forming regions, methanol
    signatures found in, 358
Stardust mission, 298−299
Stardust spacecraft, 299
  photographs of particle collectors from
    comet and the solar wind, 299*f*
Statistics in astroscience, 114−118
  average and standard deviation, 115
  normal distribution, 116
  standard error, 117
  student's t-distribution, 117−118
Stellar classification, 18
  Harvard spectral classification, 18
  H−R diagram, 18
Stishovite, 74
Stony-iron meteorites, 83
Stratospheric Observatory for Infrared
  Astronomy (SOFIA), 160
Strecker-cyanohydrin synthesis, 322
Stromatolites, 106, 107*f*, 242−243
  counterarguments to Isua Supracrustal
    Belt structures as, 243−244
  in Isua Supracrustal Belt oldest Life in
    3700 Ma, 243
Styrene, 352−353
Sub-Neptunes, 371
Sugars, 318−319
  molecules, 320−321
Sulfides, 369
Sulfur (S), 205−207, 306, 369
  abundances and speciation analyses of C
    and S by pyrolysis and combustion, 307
  aerosol, 263

Sulfur monoxide dimmers, 263
Sulfuric acid, 263
Sun, 196
Sunlight, 253
Super-Earth planets, 362−364, 371
Supernova models, 136
  elemental synthesis heavier than Fe and,
    132−134
Synestia model, 204−205
Synthetic reactions, 263−264

T
Tandem mass spectrometry, 311
Technetium (Tc), 31
Tectonic activity, 338
Teegarden's star, 366
Telescopes on Earth, 16*f*
TEM. *See* Transmission electron microscope
  (TEM)
Ternary diagram, 71−73
Terrestrial fractionation (TF), 307−308
Terrestrial fractionation line (TFL), 175,
  177−178
Terrestrial magnetite particles, 274
Terrestrial nitrogen species, 255
Terrestrial-planet-forming region, 146
TESS. *See* Transiting Exoplanet Survey
  Satellite (TESS)
TF. *See* Terrestrial fractionation (TF)
TFL. *See* Terrestrial fractionation line (TFL)
TG-DSC-MSEGA.
  See Thermogravimetry−DSC−MSEGA
  (TG-DSC-MSEGA)
Theia, latest model proving similarity
  between proto-Earth and, 170
Thermal ionization mass spectrometer
  (TIMS), 41, 45, 192
  schematic diagram of, 47*f*
Thermal neutron, 12
Thermodynamic consideration, 239
Thermogravimetry−DSC−MSEGA (TG-
  DSC-MSEGA), 282
Tholin, 104−105
Three oxygen isotopes support giant
  impact, 174−177
Three-body problem, 4−5
Three-dimension (3D)
  coupled chemistry climate model
    simulation, 370
  HPLC with high-sensitivity FD, 314−315
  molecular structural images, 238
Time-of-flight mass spectrometry (TOFMS),
  313−314

TIMS. *See* Thermal ionization mass
  spectrometer (TIMS)
Tissint fell (least contaminated Martian
  meteorite), 282–284
  diversity of organic geochemistry on
    Mars glimpsed from Tissint meteorite,
    282–283
  electrochemistry of brines and metal-rich
    minerals on Mars reduced CO_2 to
    alcohols, 283–284
Titan, 347–348
  equatorial dunes, 352
    possible organics on, 353*f*
  with organic molecules, 350–353
  synthesis of PHAs on, 354*f*
Titanium (Ti), 200–201
  isotopic composition of lunar rocks,
    200–202, 202*f*
  titanium-related oxides, 75
TOFMS. *See* Time-of-flight mass
  spectrometry (TOFMS)
TOI-700, 367
Torino scale, 87–88
Trace element
  astrochemistry, 31
  geochemistry, 55–56, 56*f*
  pattern, 56
Transcription, 99–100, 100*t*
Transiting Exoplanet Survey Satellite
  (TESS), 362
TRansiting Planets and Planetesimals Small
  Telescope-1 (TRAPPIST-1), 367
Transmission electron microscope (TEM),
  241–242
Transmission X-ray microscopy, 242
Transuranium elements, 31
TRAPPIST-1. *See* TRansiting Planets and
  Planetesimals Small Telescope-1
  (TRAPPIST-1)
Triphosphate ($P_3O_{10}^{5-}$), 360
Triple oxygen isotope equilibrium, 182
Triple-alpha process, 130
Tungsten carbide (WC), 8–11
Tungsten isotopic materials, 223
Turbiditic sediments, 78–79
2D gas chromatography with time-of-flight
  mass spectrometry (GCxGC-TOFMS),
  313–314

U
UHPLC. *See* Ultra-high performance liquid
  chromatography (UHPLC)

Ultra-high performance liquid
  chromatography (UHPLC), 255
  coupled with tandem mass spectrometry,
    255
  UHPLC-orbitrap-mass spectrometry,
    309–310
  organic material study using, 310–312
Ultramafic rock, 78
  characterization of, 77–78
  ternary diagram of, 78*f*
Ultraviolet (UV), 17
  astronomy, 15–16
  light, 252
  spectra, 263
  syntheses of life-related materials by UV
    irradiation on interstellar ice, 319–321
Ulysses, 336*t*, 338
Unidentified infrared emission, 357
Uniform reservoir (UR), 58–59
Unsaturated alkylpyridines, 310
UR. *See* Uniform reservoir (UR)
Uracil found in Ryugu samples, 317
Uranium (U), 31
  isotopes, 64
  U-rich minerals, 230
  U–Pb
    age of Earth, 228–229
    isotopic systems, 225
    method, 228
UV. *See* Ultraviolet (UV)

V
Valhalla crater on Callisto, 348*f*
Valine (Val), 255–256
Vanadium (V), 200
  isotopic composition of lunar rocks, 200,
    201*f*
Vaporization, 197–198
Venus, 263
  life-related molecules on Venus
    atmosphere, 263–264
  phosphine on Venus atmosphere, 263
Very high-K (VHK), 154
Very Large Array (VLA), 18
Very low-Ti (VLT), 154
4Vesta, 293
VHK. *See* Very high-K (VHK)
Vigorous magma ocean, 166
Viking spacecrafts, 265–266
Visible light astronomy, 17
VLA. *See* Very Large Array (VLA)
VLT. *See* Very low-Ti (VLT)

Volatile(s), 178
  compounds, 316
  delivery, 205–207
  elements, 156, 171–174, 185, 188, 205
  origin of volatiles on Earth, 295
  to refractory elements, isotopic
    compositions from, 203–204
  volatile-element-depleted bodies, 186
  volatile-poor asteroids, 145–146
Volcanic rocks, 77, 234–235
  characterization of, 77–78
Voyager 1, 336*t*, 337*f*, 338, 348*f*
Voyager 2, 336*t*, 337*f*, 338

W
W isotope
  constraints for late veneer from, 219–223
    discussion on formation age of moon,
      221–223
    lunar W isotopic ratios and percentage
      of late veneer, 219–221
    problems in early works, and
      solutions, 219
  ratios, 62
Water (H_2O), 29, 41–42, 159–160, 186,
  234–235, 319, 370–371
  D/H ratio on comets and origin of water
    on Earth, 294–295
  identification of water emission from
    main-belt comet, 295
  materials, 228
  phase diagram of, 75–77, 76*f*
  rich planetesimals, 227
  vapor found in exoplanet, 371
Water vapor, 343, 347–348
Wavelength dispersive spectrometry
  (WDS), 47–48
WDS. *See* Wavelength dispersive
  spectrometry (WDS)
Weak forces, 8
West Greenland samples
  application of [146, 147]Sm–[142, 143]Nd and
    [176]Lu–[176]Hf systematics to, 230
  application of [146, 147]Sm–[142, 143]Nd
    systematics to, 229
White dwarf, 129

Wide-field Infrared Survey Explorer
  (WISE), 17
Wilkinson Microwave Anisotropy Probe
  (WMAP), 121, 122*f*
WISE. *See* Wide-field Infrared Survey
  Explorer (WISE)
WMAP. *See* Wilkinson Microwave
  Anisotropy Probe (WMAP)
Wolf 1069, 367
Wollastonite (Wo), 71–73

X
X-ray
  astronomy, 14*f*, 15
  telescopes, 141
X-ray fluorescence spectrometry (XRF), 79,
  80*f*
X-type asteroids, 22–23
Xenon (Xe), 42
XRF. *See* X-ray fluorescence spectrometry
  (XRF)
Xylitol, 103–104

Y
Yarkovsky effect, 86
Yarkovsky–O'Keefe–
  Radzievskii–
  Paddack effect, 86–87
YBCO. *See* Yttrium barium copper oxide
  (YBCO)
Yellowish terrain, 342–343
Yttrium barium copper oxide (YBCO), 75

Z
Zircon(s)
  ages, 163–164
  crystal, 63*f*
  crystallizes, 163
  by HR-SIMS, first water on early Earth
    from oxygen isotopic data of, 232–233
  relative frequency of zircons on moon,
    164*f*
Zirconium, 159
Zn isotopic composition of lunar rocks,
  183–184, 184*f*

Printed in the United States
by Baker & Taylor Publisher Services